W9-BWX-535
CONTEMPORARY PSYCHOLOGY
DEPARTMENT OF PSYCHOLOGY
UNIVERSITY OF TEXAS
AUSTIN, TX 78712
FEB 21 1984

Benchmark Papers in Behavior

Volume

A BENCHMARK® Book Series

LEARNING IN ANIMALS

Edited by
ROBERT W. HENDERSEN
University of Illinois at Urbana-Champaign

Hutchinson Ross Publishing Company
Stroudsburg, Pennsylvania

Benchmark Papers in Behavior, Volume 14
Library of Congress Catalog Card Number: 81-2247
ISBN: 0-87933-348-0

84 83 82 1 2 3 4 5
Manufactured in the United States of America.

LIBRARY OF CONGRESS CATALOGING IN PUBLICATION DATA
Main entry under title:
Learning in animals.
(Benchmark papers in behavior; 14)
Includes indexes.
1. Learning in animals. I. Hendersen, Robert W. II. Series.
QL785.L47 156′.35 81-2247
ISBN 0-87933-348-0 AACR2

Distributed world wide by Academic Press,
a subsidiary of Harcourt Brace Jovanovich,
Publishers.

CONTENTS

PART II: SYSTEMATIC EXPERIMENTAL ANALYSES OF LEARNING PROCESSES IN ANIMALS

PART III: COMPARATIVE STUDY OF LEARNING AND INTELLIGENCE

SERIES EDITORS' FOREWORD

It was not too many years ago that virtually all research publications dealing with animal behavior could be housed within the covers of a very few hard-bound volumes that were easily accessible to the few workers in the field. Times have changed! The present-day students of behavior have all they can do to keep abreast of developments within their own area of special interest, let alone in the field as a whole; and of course we have long since given up attempts to maintain more than a superficial awareness of what is happening "in biology," "in psychology," "in sociology," or in any of the broad fields touching upon or encompassing the behavioral sciences.

It was even fewer years ago that those who taught animal behavior courses could easily choose a suitable textbook from among the very few that were available; all "covered" the field, according to the bias of the author. Students working on a special project used *the* text and *the* journal as reference sources, and for the most part successfully covered their assigned topics. Times have changed! The present-day teacher of animal behavior is confronted with a bewildering array of books to choose among, some purported to be all-encompassing, others confessing to strictly delimited coverage, and still others being simply collections of recent and profound writings.

In response to the problem of the steadily increasing and overwhelming volume of information in the area, the Benchmark Papers in Behavior was launched as a series of single-topic volumes designed to be some things to some people. Each volume contains a collection of what an expert considers to be *the* significant research papers in a given topic area. Each volume, then, serves several purposes. To teachers, a Benchmark volume serves as a supplement to other written materials assigned to students; it permits in-depth consideration of a particular topic while at the same time confronting students (often for the first time) with original research papers of outstanding quality. To researchers, a Benchmark volume serves to save countless hours digging through the various journals to find *the* basic articles in their area of interest; often the journals are not easily available. To students, a Benchmark volume provides a readily accessible set of original papers on the topic, a set that forms the core of the more extensive bibliography that they are likely to compile; it also permits them to see at first hand what an "expert" thinks is important in the area, and to react accordingly. Finally, to librarians, a Benchmark volume represents a collection of important papers from many diverse sources, thus making readily available materials that might otherwise not be economically possible to obtain or physically possible to keep in stock.

The choice of topics to be covered in this series is no small matter. Each of us could come up with a long list of possible topics and then search for potential volume editors. Alternatively, we could draw up long lists of recognized and prominent scholars and try to persuade them to do a volume on a topic of their choice. For the most part, we have followed a mix of both approaches: match a distinguished researcher with a desired topic, and the results should be outstanding. So it is with the present volume.

Dr. Hendersen is a scholar and critical thinker with a unique expertise in the area of animal learning. He is an exciting and enthusiastic teacher who reaches deeply into the historical roots of psychology. This volume provides the reader with Dr. Hendersen's valuable insights regarding the origins of contemporary learning theory. The history of animal learning, as it unfolds in this volume, furnishes evidence of the ontogeny of methodological and theoretical concepts prevalent in the study of behavior. Today we find many of these concepts integrated into areas concerned with the control of human behavior; areas as diverse as behavioral medicine and advertising. Few are more knowledgeable or have a more systematic view of the extensive literature dealing with the history of animal learning than Dr. Hendersen. We are pleased that he has compiled these relevant historical papers and consider the volume a valuable addition to the Benchmark Papers in Behavior series.

MARTIN W. SCHEIN
STEPHEN W. PORGES

PREFACE

The papers reprinted in this volume are a small sample of a century's progress in the study of how nonhuman animals change with experience. The topics of the papers vary widely, reflecting both the complexity of learning processes and the diverse motivations that have led people to study them. This diversity is especially apparent in a collection of "benchmark" papers, many of which have introduced new phenomena and new techniques. Despite discontinuities, there are common threads running through these papers, and I have tried to help the reader keep track of these in my editorial comments accompanying the papers.

It is easy to explain why each of the papers reprinted here was selected for inclusion. Each has proved to be very influential. Most of these papers have inspired major lines of research activity, and most have been widely cited.

It is vastly more difficult to explain why scores of other obvious candidates for "benchmark" status are not to be found here. There are too many "classic" papers in the field of animal learning for them all to be included in one book (or, for that matter, ten books). I have tried to choose representative papers that have had powerful influence. I have preferred empirical contributions to theoretical papers, and I have largely sidestepped, for want of space to do it justice, the strong thread of epistemological rhetoric that runs through much of the literature of animal learning.

Some of the earliest papers reprinted here are virtually unknown today. They were very influential in their time, and their inclusion is justified by the impact they left on the developing field.

Many people helped me select papers for this volume, and I thank them for their suggestions. Discussions with J. Stephen Hazel, Richard Albin, Jacqueline Dawley, and Susan Mineka were especially helpful. Students in my "Theories of Learning" class at the University of Illinois clarified many issues for me in a class discussion of what makes a paper a "benchmark" contribution.

Knowing something about how animals learn has helped me appreciate both the degree to which humans share characteristics with other animals and the degree to which humans are unique. Studying how animals learn helps us understand the creatures with whom we share our world, but it also helps provide the perspective necessary for understanding ourselves.

ROBERT W. HENDERSEN

CONTENTS BY AUTHOR

LEARNING IN ANIMALS

INTRODUCTION

"If no organic being excepting man had possessed any mental power, or if his powers had been of a wholly different nature from those of the lower animals," wrote Charles Darwin in *The Descent of Man* (1871), "then we should never have been able to convince ourselves that our high faculties had been gradually developed. But it can be shown that there is no fundamental difference of this kind." The lines of research that arose from early attempts to determine the validity of Darwin's faith in the "mental power" of "lower animals" form the veins from which the papers reproduced in this book were mined. Central to this enterprise have been the questions of how, what, and why animals learn.

As a scientific endeavor the study of animal learning has proceeded for roughly a century, with the bulk of activity occupying the last fifty years. In that relatively short span of time approaches have changed dramatically, so that the questions asked and the methodologies employed by contemporary students of animal learning are often strikingly different from those of their predecessors. The study of animal learning is a field in transition, now as much as ever before. In order to understand contemporary ideas fully, it is necessary to consult the past, to re-examine from whence our ideas have come. This book is intended to aid in the achievement of this goal.

In the late nineteenth century most interest in animal learning was based on a desire to understand, and to establish the plausibility of, "mental evolution." In the twentieth century the traditions initiated by Pavlov and Thorndike have taken the field in other, very different directions. The structure imposed on thinking about animal learning by these early workers is a key to understanding much of the contemporary status of the field.

LEARNING AND MENTAL EVOLUTION

The reasons for a century's sustained interest in animal learning have shifted as theories and methodologies have grown, diverged, and developed. Much of the early interest in the ways that experience affects animal behavior derived from a belief that learning could serve as a crucial indicant of mind. If one accepts, as did many early students of animal behavior, that comparative psychology is the study of mental evolution, then it is critical to have a criterion by which to assess the existence of mental processes in animals. Arguing that mental adjustment must be distinguished from reflex action, Romanes (1883) proposed as his criterion of mind the following: "Does the organism learn to make new adjustments, or to modify old ones, in accordance with the results of its own individual experience?" Romanes insisted that evidence of learning through individual experience is "the best available evidence of conscious memory leading to intentional adaptation."

As the nineteenth century waned, reasoned opposition to mentalistic characterizations of animal behavior began to grow. Nevertheless, the validity of demonstrations of learning as indicants of mentation in animals was widely accepted, even among those who urged conservatism in the imputing of mental processes to animals (for example, Morgan, 1894). As observers became increasingly cautious in ascribing mind to animals, theoretical attention came to focus more and more strongly on demonstrations of learning. The modifiability of animal behavior posed a serious problem for those who sought to characterize it solely in mechanistic terms.

For example, in 1900 Jacques Loeb, while insisting that animal psychology must be as analytic an enterprise as brain physiology, suggested that some animals possess consciousness. "Consciousness," he carefully argued, "is only a metaphysical term for phenomena which are determined by associative memory." He further asserted, "If an animal can be trained, if it can learn, it possesses associative memory." Learning provided Loeb with a criterion of consciousness that avoided the subjectivist flavor pervading many of his contemporaries' interpretations of animal behavior (for example, Hobhouse, 1901; Washburn, 1908; Smith, 1915). When later he was able to suggest that a forced movement theory of conduct could account even for those aspects of behavior commonly interpreted as "free will," reference to "consciousness" could be avoided entirely (as in Loeb, 1918). The development that allowed learning phenomena to be discussed plausibly in mechanistic terms was the discovery by Ivan Petrovich Pavlov of the conditional reflex.

PAVLOV'S CONTRIBUTION

A profound impact was produced by Pavlov's reports of an extensive series of experiments that detailed how dogs drool. Why did these scores of slobbering dogs induce scientists to flock to Pavlov, George Bernard Shaw to satirize him, churchmen to denounce him, visionary novelists to exploit him, and the general public to fear his work?

Salivation in anticipation of food was not a new discovery. Not even the suggestion that psychological phenomena could be accounted for in terms of reflexology was new: I. M. Sechenov had tried to extend a reflexive analysis to voluntary behavior in *Reflexes of the Brain* (1863). Sechenov's account, which had a clear influence on Pavlov, was largely conjectural. Pavlov's contribution was to support this conjecture with empirical observations. He showed that "psychic" secretion could be studied successfully with the same methodology used to study reflexive salivation.

As it turned out, Pavlov's "conditional reflexes" were often very different from the spinal reflexes studied by vivisectionist physiologists. A new set of laws applied to the development and disappearance of conditional reflexes. Nevertheless, these laws could be determined empirically, in an intact animal, by treating the behavior as though it were reflexive. The conditional reflex, then, was not so much the discovery of a new kind of behavior as it was a new application of a serviceable methodology to the analysis of this behavior.

Pavlov's methodological influence has proved pervasive, extending to virtually all aspects of the study of learning in animals. Perhaps the most important feature of this methodology is the assumption that learning processes can be studied in an isolated, uncontaminated form by removing systematically as many extraneous influences on an animal's behavior as possible. Well before the development of the electronic jungles that have become standard for the control of experiments in many learning laboratories, the Pavlovians designed clever systems for delivering stimuli to animals in such a way that the immediate influence of the experimenter could be eliminated from the experimental situation. (For example, they used as a stimulus the sound of running water, which could be controlled remotely by turning a valve.) Such a strategy follows logically from Sechenov's belief that animals (including humans) are purely reactive creatures, always behaving in response to stimulation. By removing all but the particular features that are objects of study, the argument goes, one can isolate and purify learning processes for detailed examination.

Some of the most brilliant contributions to our understanding of animal learning processes have resulted from ingenious application of Pavlov's strategy. Learning processes have been analyzed, teased apart from one another, and systematically studied. One must be careful not to let the obvious fruitfulness of this general strategy obscure totally its limitations, however. For example, a strategy based so strongly on the belief that behavior is always in reaction to a stimulus imposes constraints on the theoretical analysis of learning processes. Such constraints are in many cases no doubt justified, but one would like them to be imposed by a careful consideration of the behavior, rather than by the strategy used to study the behavior. Furthermore, the phenomena that have been studied successfully through the strategy of simplification by environmental constraint may be strongly influenced by the techniques used to analyze them. A laboratory setting need not mimic nature in order to elucidate learning processes, but the restraints inherent in strict laboratory control of behavior may impose a structure on experimental results that limits their generality. Once such limitations are recognized, Pavlov's strategy can be used appropriately and productively. The power of the strategy is displayed most clearly in the progress that has been made under its influence.

THORNDIKE'S CONTINUITY HYPOTHESIS

Further structure was placed on the study of animal learning by the work of E. L. Thorndike. Even before Pavlov began studying conditional salivation in dogs, Thorndike had begun to study experimentally the learning processes of chicks and cats. His work convinced him that associative processes could successfully be examined in animals, and it led him to a view of the evolution of mind that has had enormous influence. In 1901, writing in the *Popular Science Monthly*, Thorndike proposed:

> If the facts did eventually corroborate it, we should have an eminently simple genesis of human faculty. . . . We should say:
> "The function of intellect is to provide a means of modifying our reactions to the circumstances of life, so that we may secure pleasure, the symptom of welfare. . . . The intellectual evolution of the race consists in an increase in the number, delicacy, complexity, permanence, and speed of formation of. . . associations. In man this increase reaches such a point that an

apparently new type of mind results, which conceals the real continuity of the process."

Thorndike's hypothesis about the evolution of mind, and particularly about the relationship between associative processes in man and those in other species, has important implications for the study of learning. The general acceptance (more often implicit than explicit) that the hypothesis enjoyed among many comparative psychologists for nearly fifty years has profoundly affected the kinds of questions asked by people studying learning in animals. If learning capacities are comparable along continous dimensions of the sort Throndike proposed (for example, speed of formation, permanence of associations), then the choice of species to study is not a particularly important one. Differences among species in learning capacities, according to this hypothesis, are of degree rather than of kind. Animals' minds are like human minds, only simpler.

Given this orientation it is not surprising that much of the work examining animal learning processes has been motivated by the desire to understand *human* functioning. If animals are simpler psychological beings, rather than qualitatively different creatures, then understanding how they change with experience should illuminate how humans learn as well. Thus, several theoretical analyses of animal learning processes have borrowed heavily from accounts of human functioning. For example, parts of theories of aversive learning in animals derive loosely from theories of human anxiety, and theories of human anxiety, in turn, have been revised because of data from experiments with rats, dogs, pigeons, and monkeys. Similarly, techniques developed initially to train animals have been modified and applied in human educational and therapeutic settings.

The Thorndikean view of the relationships among learning processes in different species seems, today, naive and misleading. Among the flaws for which it has been roundly criticized are these: It encourages focusing attention on a much too limited number of species (Beach, 1950; Lockard, 1968); it ignores adaptive specializations in learning processes (Rozin and Kalat, 1971; Shettleworth, 1972); it assumes more generality to learning processes than is justified (Seligman, 1970); and it leads to inappropriate comparisons among unrelated species (Hodos and Campbell, 1969).

Although the several flaws in Thorndike's hypothesis make it untenable in the form in which it was stated, the optimism it helped generate for a potentially comprehensive understanding of learning processes across a range of species remains as a valuable hertiage. People continue to search for learning processes common to a range of species, to study animal learning in order to provide insight into human functioning, and to seek to understand how simple learning processes are related to more complicated ones. Today's vastly more detailed understanding of brain physiology, evolutionary biology, and, most importantly, behavior itself, gives the modern investigator remarkable advantages that were not available to the early students of animal learning. We appreciate the crucial contributions made by those who built the foundation for the study of learning in animals all the more because the advances were made with but few of the technical, methodological, theoretical, and empirical resources available now. The seminal contributions of Pavlov and Thorndike provided the foundation on which a science of animal learning could build. Most of the papers reprinted in this volume reflect, in multifarious ways, this heritage.

REFERENCES

Beach, F. A., 1950, The snark was a boojum, *Am. Psychologist* **5**:115–124,

Darwin, C., 1871, *the Descent of Man*, Appleton, New York.

Hobhouse, L. T., 1901, *Mind in Evolution*, Macmillan, New York.

Hodos, W., and C. B. G. Campbell, 1969, *Scala naturae:* Why there is no theory in comparative psychology, *Psychol. Rev.* **76**:337–350.

Lockard, R. B., 1968, The albino rat: A defensible choice or a bad habit?, *Am. Psychologist* **23**:734–742.

Loeb, J., 1900, *Comparative Physiology of the Brain and Comparative Psychology*, Putnam, New York.

Loeb, J., 1918, *Forced Movements, Tropisms, and Animal Conduct*, Lippincott, Philadelphia.

Morgan, C. L., 1894, *An Introduction to Comparative Psychology*, W. Scott Ltd., London.

Romanes, G. J., 1883, *Mental Evolution in Animals*, Kegan, Paul, Trench, & Co., London.

Rozin, P., and J. W. Kalat, 1971, Specific hungers and poison avoidance as adaptive specializations of learning, *Psychol. Rev.* **78**:458–486.

Sechenov, I. M., 1965, *Reflexes of the Brain*, S. Belsky, trans., MIT Press, Cambridge, Mass. Original edition, 1863.

Seligman, M. E. P., 1970, On the generality of the laws of learning, *Psychol. Rev.* **77**:406–418.

Shettleworth, S. J., 1972, Constraints on learning, in *Advances in the Study of Behavior*, vol. 4, D. S. Lehrman, R. A. Hindes and E. Shaw, eds. Academic Press, New York.

Smith, E. M., 1915, *The Investigation of Mind in Animals*, Cambridge University Press, Cambridge, England.

Thorndike, E. L., 1901, The evolution of the human intellect, *Pop. Sci. Monthly*.

Washburn, M. F., 1908, *The Animal Mind: A Textbook of Comparative Psychology*, Macmillan, New York.

Part I

EARLY OBSERVATIONS OF LEARNING

Editor's Comments on Papers 1 and 2

1 MÖBIUS
Excerpts from *The Movements of Animals and Their Psychic Horizon*

2 PECKHAM and PECKHAM
Excerpts from *Some Observations on the Mental Powers of Spiders*

These papers are creative first steps toward the experimental study of learning in animals. They are distinguished from other early observations of animal learning processes because they are experiments, rather than "animal stories," with manipulations performed to assess how behavior changes with experience.

Möbius (Paper 1) reports an experiment actually performed by Herr Oekonomierath Amtsberg, using a form of punishment training. A glass partition separated a pike from its prey. After months of bumping against the glass, the pike ceased to attack the minnows. Remarkably, even when the partition was removed, the pike continued to leave the minnows unmolested. This demonstration received a lot of attention in its day, although it is rarely cited today. It was described by Darwin (1874) as an example of how an animal may show reasoning different from that of a man. The experiment was successfully repeated with perch by Triplett (1901). Hobhouse (1901) regarded the experiment as evidence of the stupidity of the pike, while Washburn (1908) cited it as evidence that punishment can inhibit instinctive acts.

The Peckhams' experiments with spiders (Paper 2) were also enormously influential, and were cited in most of the early texts that dealt with animal behavior. The process they studied is known today as "habituation." Their experiment is conducted with care. As researchers in the twentieth century came to focus almost exclusively on associative learning processes, the Peckhams' contribution was ignored for a time. As increasing interest comes to bear on nonassociative learning processes, including habituation, the importance of this early experiment is apparent.

REFERENCES

Darwin, C., 1874, *The Descent of Man*, 2nd. ed., Murray, London.
Hobhouse, L. T., 1901, *Mind in Evolution*, Macmillan, London.
Triplett, N. B., 1901, The educability of the perch, *Am. J. Psychol.* **12**: 354–360.
Washburn, M. F., 1908, *The Animal Mind: A Textbook of Comparative Psychology*, Macmillan, New York.

1

THE MOVEMENTS OF ANIMALS AND THEIR PSYCHIC HORIZON

Karl Möbius

This article was translated expressly for this Benchmark volume by Russell W. Snyder, University of Illinois at Urbana-Champaign, from pp. 113, 121–122 of Die Bewegungen der Thiere und ihr psychischer Horizont. Naturw. Ver. Schleswig-Holstein Schr. **1**:*113–129 (1873)*

When one knows the form and the origin of all the individual parts of a *machine*, and when one has seen how these parts mesh together in their working, then everything essential that one can know about a machine has been exhausted.

If one has followed the development of an *animal* from egg to development of all its organs; if one has been further able to determine which physical and chemical laws these organs operate according to, that the legs move the body according to the laws of mechanics, that the heart operates like a pumping apparatus, that the constituents of food are changed in the blood according to the laws of chemistry, that in the transparent parts of the eye light is refracted just as in curved lenses; nevertheless one is far from explaining what it is that goes on inside of an animal. For an animal is *alive*; it is a machine that perceives its own motion by its own force according to external conditions.

How is it, however, that we can maintain of animals that they feel, i.e., that they experience something within themselves which any one of us can perceive directly only within himself?

It is the *movements* of animals on which we base this assumption. It is from the type and manner of movements, how they occur and follow one another, that we draw this conclusion as to the different psychic acts in the higher and lower animals.

Since I begin with the highest animals and proceed to the lowest, I wish to attempt to make this clear by means of a series of facts.

[*Editor's Note:* Material has been omitted at this point.]

For *training* the higher organization of vertebrates seems to be necessary.

In the case of fishes, the lowest of the vertebrates, we have proof that cannot be doubted that they are capable of adapting their activities to previous perceptions.

A pike, which consumes all manner of small fish, was placed in an aquarium, to which he grew accustomed and was then separated from all of the other fish in it by means of a glass plate. Each time he swam into the glass he smashed his jaw into it, and indeed sometimes with such force that he came to rest on his back as if dead. He came to again, however, and repeated his foraging attacks, though more and more seldomly; after three months he had ceased these actions entirely. After he had been separated in this manner for six months, the glass plate was removed, and the pike was once again free to roam around the other fish in the aquarium. He immediately chased after these but did not attack any, but rather seemed to always stop approximately one inch from them and contented himself by feeding with the other fish on the meat thrown into the aquarium. He was thus trained to spare those inhabitants of the aquarium with which he was *acquainted.* If a new, strange fish was placed in the aquarium, the pike did not respect him but rather swallowed him immediately. After he had repeated this more than forty times in conjunction with the advanced forbearance of his aquarium-mates, he had to be removed from the aquarium due to his size.*

The training of the pike was not based on the fact that the pike reasoned; it was only the drilling of a definite direction of the will as a result of uniformly repetitive sensual perceptions. Indulgence of those fishes known to him shows especially that the pike behaved without deliberation. The sight of them evoked in him the natural instinct to devour them, but it also awakened in him at the same time the conception of pain that he had often experienced because of them as well as the conception that it is impossible to attain these, the desired ones. These conceptions attained a greater magnitude in him than the instinct to prey on others, and thus repressed this instinct momentarily. The same sensual impression, stemming from the individual fish was, in his soul, again and again the onset of the same series of psychic acts. He had to submit to this series running its own course, like a machine, again and again; naturally, however, as an animated machine that has the advantage over a mechanical machine such that it can adapt its work itself to cases previously unencountered, which is unattainable for mechanical machines. In the pike organism, this glass plate was certainly such an instance that had never been encountered before.

Our path from the rhizopoda to the fishes was one from the more simple to the more complex, i.e., to those animals that are closer and closer to the human organization.

With every step of progress in the corporal organization of animals, especially in the development of motor and sensory organs and of the central parts of the nervous system, the psychic life is moved to a higher and higher level. But inside of each organizational level the same thing is developed in a very different way, just as we must conclude from the often quite different actions of animals that belong to a systematic group.

*This interesting experiment with the pike was undertaken by Mr. Amtsberg, Minister of Economics in Stralsand, Germany. I am indebted to him for a written description of this experiment, as well as his permission to publish it.

2

Reprinted from pp. 383–385, 390–396 of *J. Morphol.* 1:383–419 (1887)

SOME OBSERVATIONS ON THE MENTAL POWERS OF SPIDERS.

GEORGE W. AND ELIZABETH G. PECKHAM.

INTRODUCTION.

THE differences of structure between a man and a spider are so numerous and profound that he who infers the mental state of a spider from a given action should not be in haste to make positive statements and broad generalizations. A critical study of many of the current anecdotes concerning animal intelligence would prevent their use as data for comparative psychology, at least until after their confirmation by competent observers. Up to that time they have, as Romanes says, only the value of suggestions. How far, for example, are "personal preconceptions" responsible for both facts and inferences in Dr. Brookes' assertion that *Epiblemum scenicum* "has been sometimes seen in the act of instructing its young ones how to hunt"? and also that "whenever an old one missed its leap, it would run from the place and hide itself in some crevice, as if ashamed of its mismanagement"?[1] After having observed spider after spider building a new web on the eve of a storm, how shall we explain the statement, that "when a storm threatens, the spider, which is very economical with its valuable spinning material, spins no web, for it knows that the storm will tear it in pieces, and waste its pains, and it also does not mend a web which has been torn; if it is seen spinning or mending, on the other hand, fine weather may be generally reckoned on"?[2] This would be, no doubt, the wisest way for spiders to act under the circumstances, and Dr. Büchner is in very illustrious company when he — unconsciously, of course — orders the actions of such simple creatures in full accord with the higher reason.

Lange has well said that the core of all the numerous cau-

[1] Bingley's *Animal Biography*, Vol. III., p. 455.
[2] Romanes' *Animal Intelligence*, p. 211.

tionary measures of the scientific method lies just in the neutralizing of the influence of the observer's subjectivity. The subjective element cannot, of course, be eliminated; but the observer should keep facts and inferences separate, and should, in addition, state the particular action, among the many, which is the external sign of the mental state which he believes to be proved by the experiment. Lange's words on the subject are worthy of immortal memory: —

"Where external observation shows us primarily only movements, gestures, and actions, the interpretation of which is liable to error, we may, nevertheless, carry out a comparatively very exact procedure, since we can easily subject the animal to experiments, and put it into positions which admit of the most accurate observation of each fresh emotion, and the repetition or suspension, as we will, of each stimulus to a psychical activity. Thus is secured that fundamental condition of all exactness; not, indeed, that error is absolutely avoided, but certainly that it can be rendered harmless by method. An exactly described procedure with an exactly described animal can always be repeated, by which means our interpretation, if it is due to variable bye-conditions, is at once corrected, and at all events thoroughly cleared from the influence of personal preconceptions, which have so great a share in so-called self-observation." [1]

We have felt that it might properly be demanded of us that we give the generic and specific names of every spider experimented upon, and also that we so describe our methods that the experiments can be repeated by any one who desires to test the validity of our conclusions.

Our rule has been not only to repeat an experiment many times, but to repeat it under as many different conditions as possible. The histologist often finds it necessary to adopt complicated and tiresome methods in order to demonstrate a single fact. So, also, we have found that to learn anything of the mental processes of spiders the way is long and beset with difficulties. To use the words of Ribot: "Many of these investigations, we shall see, pertain to very modest questions, and it is probable that the partisans of the old psychology will find the work too great for results so small. But those who give allegiance to the methods of the positive sciences will not com-

[1] *History of Materialism*, Vol. III., p. 178.

plain of this. They know how much effort the smallest questions require; how the solution of small questions leads on to the solution of great ones, and how barren of results it is to discuss great problems before the small ones have been solved." [1]

[*Editor's Note:* Material has been omitted at this point.]

HEARING.

Our first experiments in this direction consisted in shouting, clapping our hands, and whistling close to spiders which were at rest in their webs. They gave no sign of hearing anything. We felt, however, that this was not enough to warrant us in concluding that they were deaf, since there is nothing in the habits of these spiders that would lead them to make any active response to loud noises, even supposing they did hear them. *A. vittata*, when standing on a finger, jumped to one side when "*bang*" was shouted in a loud voice, with the head turned away; and when we whistled, it stood on the tip of its abdomen with its head held high. With this exception we failed to discover, by these means, anything about the hearing of spiders.

Fortunately a better method was suggested to us by the experiment of Mr. C. V. Boys with a tuning-fork on the garden-spider.[1]

We began a new series of experiments by sounding three tuning-forks near a large female of *E. strix* as she stood in the centre of her web. Two of the forks, A and C, were small, while B was large. The spider did not notice the two small forks, but when the large one was sounded she raised her first legs almost vertically, holding them as though ready to ward off an attack, and looking much like a boxer in an attitude of defence. The B fork was again sounded, and again the legs were raised. As a control experiment the fork, when not in vibration, was brought into the same relation to the spider. No notice was taken of it. The fork was again sounded, and held behind and above her cephalothorax. She extended her legs as before. The experiment was repeated with the fork still. She paid no attention to it. The fork was sounded and

[1] *German Psychology of To-day*, p. 14.

[1] *Nature*, XXIII., pp. 149–150.

brought to one side of her, when she not only moved the first legs, but also the leg of the second pair on the side toward the fork. It would tire the reader unnecessarily were we to describe the check experiments that were made after each observation, but we felt their importance, and never failed to make them; in fact, our check experiments were more numerous than our direct ones.

Later on tests similar to those given above were made on a smaller spider, a female of *E. labyrinthea*, as she stood in her web. In this instance she responded to all the forks, A, B, and C.

Second and third large individuals of *E. strix* acted as the first had done, responding to the large fork, but not to the small ones. On the other hand, five small individuals of *strix* were much excited by the small forks. Subsequent observation left no room for doubt that the large spiders, with few exceptions, only attended to the sound produced by the large forks.

To show the results of our experiments and also the way in which we worked we quote from our notes.

July 14. — Held the big fork, in vibration, over a large male of *E. insularis*, an inch and a half away. He threw up his first legs, making frantic efforts to reach it. When the fork was removed he settled down quietly in his web. This was repeated ten times, always with the same result. A female of this species acted as the male had done, but seemed less excited by the vibrations. Unless the fork was sounding neither spider paid any attention to it.

July 18. — Held the large fork, in vibration, near a female of *E. infumata* standing quietly on a wire screen. She did not move. Repeated the test with the fork, at first vibrating and then still, ten times without result. She was then placed in the web of another spider, and the B fork was brought near her as she stood there. She appeared frightened, and at once threw up the first and second pairs of legs. The fork was next held behind and to one side, so that she could not see it; but she seemed to hear it, since she turned toward the fork and almost fell backward in her efforts to reach it. The fork was now held in front of her again, when she moved her legs as before. This experiment was repeated many times with like results. To hear the fork when she could not see it evidently excited her more than to both hear and see it. The presence of the fork, when not in vibration, brought no response, nor did rapidly moving it to and fro in front of her attract her attention.

August 13. — The large fork, in vibration, was held near a female of *A. riparia*. She at once gave the usual sign that she heard it. It was next held behind her, and entirely out of her sight, when she quickly turned in the direction of the sound.

August 14. — Tried a new species, a young *Phillyra mammeata* Hentz. When the vibrating C fork approached she lifted first one, and then the other, of the anterior legs.

Were it necessary, we could cite a great many similar experiments which had like results, to show that certain spiders indicate that they hear a vibrating tuning-fork by characteristic movements of the legs. Another set of spiders, however, manifested their perception of the sound in a different way. With these the approach of a vibrating fork seemed to cause greater alarm, making them drop from the web and keep out of sight for a longer or shorter time. However, after one of these spiders had been subjected to the experiment several times, it would, instead of dropping, raise its legs in the manner described above.

For example, when the vibrating C fork approached a female of *E. labyrinthea* as she stood in her web, she fell. This was repeated eleven times, the spider falling each time, but at the twelfth she merely raised her first legs.

A few days after this experiment we found a more excitable spider of the same species. Not until she had fallen out of the web twenty-two times, at the approach of the fork, could she restrain the impulse to drop. It was apparent, however, after the seventh or eighth time, that she was less startled by the sound than at first, since the distance that she fell and the period of time that elapsed before she returned to the web grew shorter and shorter in the later experiments. At first she fell fifteen or eighteen inches, and remained at the end of her line for several minutes, while toward the last she fell only an inch or two, and immediately ran back to the web. After the twenty-second trial she only held up her legs as the fork approached. Finally, completely worn out and disgusted, she retreated to a neighboring branch, drew in her legs, and remained sullenly unresponsive to all further attempts.

We shall now give a series of notes which describe an attempt to teach a very interesting and docile little female spider of the species *Cyclosa conica* Menge to listen composedly to the vibration of the tuning-fork. We first saw her on July 18, when we marked her with a spot of scarlet paint, that there might be no question of mistaken identity; and by the time that we lost her, a month later, we had come to have a very friendly interest in all that concerned her. Her web was about five feet from the ground, in the branches of a cedar-tree. Across it was stretched a line of bits of rubbish, dead insects, and cocoons, and in the middle of this stood the little spider, bearing so close a resemblance, in color and shape, to the other parts of the line that she was almost indistinguishable. So perfect was the mimicry, that even after we had visited her day after day for weeks, we frequently thought, at the first glance, that our spider was gone.

Her record stands as follows: —

July 18. — *C. conica* fell from the web three times when the vibrating C fork was held one inch away. Further efforts failed to move her.

July 19. — Used the B fork. She fell five times in succession, — only short distances the fourth and fifth times, — after which she would not leave the web.

July 20. — She fell nine times before becoming accustomed to the C fork; the last three times she dropped only two or three inches, and hung at the end of the line.

July 21. — After falling six times she paid no attention to the sound.

July 22. — After falling six times became accustomed to the sound, and would not leave the web.

July 24. — A day having elapsed without a lesson she fell eleven times before becoming accustomed to the sound.

July 25. — Dropped from her web six times as the fork was held near; after that, paid no attention to it.

July 26. — Dropped only five times before becoming accustomed to the vibration.

July 29. — Dropped seven times, and then became indifferent.

July 31. — Dropped eleven times before refusing to move.

August 1. — Dropped seven times, and then remained undisturbed by the sound.

August 2. — She dropped fifteen times, and then refused to move. We left her for fifteen minutes, and, then returning, sounded the fork near her five times without making her move. She probably remembered her former experience and profited by it.

August 3. — Her memory proved short. She dropped eleven times before remaining quiet, as the fork approached. Moreover, she was very slow about returning after each fall, so that it took a much longer time than usual to teach her to pay no heed to the sound.

August 4. — She seems to be in better mood to-day. After the seventh trial she gave no sign of hearing the fork.

August 5. — Education begins to affect her character. When the fork was sounded she seemed startled, and ran up a little way on the band of rubbish, but quickly returned to the centre. This she did a second time, but to nine subsequent trials, the fork being held both behind and in front of her, she gave no response.

August 6. — We could not make her move, though we sounded the fork nine times.

August 7, 4 P.M. — Eight attempts failed to move her in the least. 6.30 P.M. — Made eleven trials, with the same result.

August 8. — Sounded the fork near her fifteen times, but she did not move.

August 9. — Seven trials; the spider remained perfectly quiet.

August 10. — She has spun herself a new web inside the circumferential lines of the old one, preserving the *débris* in its original position. (This is the fourth web she has spun since we found her on July 18.) The fork was sounded ten times, but she paid no attention to it.

August 11. — She seemed more nervous. At the first trial she dropped two inches; at the second and third, she fell about a quarter of an inch, and immediately ran back. Five subsequent efforts failed to move her.

August 12. — In the morning we made nine, and in the afternoon eight, trials with the fork. The spider gave no sign that she heard anything.

August 14. — A day and a half having passed without a lesson, the spider was somewhat startled at the approach of the fork, falling a very short distance the first time it was sounded, but after that remaining imperturbable.

August 15, 10 A.M. — Sounded the fork near the spider ten times. She would not move. 5 P.M. — Made nineteen trials, with the same result.

August 16. — The fork was sounded twenty times in the morning and twenty in the afternoon without disturbing her.

August 17. — While the fork was sounded close to her eleven times she stood immovable in the centre of the web.

August 18. — The web was tenantless. Our little *conica* has probably fallen a prey to some bird or wasp.

As the habit of falling from the web is almost the only safeguard of these spiders in times of danger, the instinct must be of immense importance to them. Taking this into consideration, it seems remarkable that one of them should so soon have learned the sound of the vibrating fork, and should have modified her action accordingly.

In all essentials our results agree with those of Mr. Boys, who says: "If, when a spider has been enticed to the edge of the web, the fork is withdrawn, and then gradually brought near, the spider is aware of its presence and of its direction, and reaches out as far as possible in the direction of the fork; but if a sounding-fork is gradually brought near a spider that has not been disturbed, but which is waiting as usual in the middle of the web, then, instead of reaching out toward the fork, the spider instantly drops — at the end of a thread, of course."[1]

A few experiments were made to determine where the organ of hearing is located, but we can offer nothing positive on this question. It seems probable that the auditory apparatus is but little specialized. Possibly it is spread over a considerable portion of the epidermis.

Finding that *E. strix* and *E. labyrinthea* were very sensitive to the tuning-fork, we removed both palpi from an individual of each of these species. They seemed a good deal disturbed by the operation, and retreated to the tents near their webs. On the next day, when the fork was sounded near them, there was no definite indication that it was heard. On the second day they each responded once; and on the third, they seemed to have entirely recovered, and responded eight or ten times in succession. We afterward removed the palpi from several

[1] *Loc. cit.*, p. 149.

specimens of *strix*, *labyrinthea*, and *insularis*. All seemed able to hear perfectly well without these organs. We also found that the palpi play no essential part in the building of the web, since all these spiders constructed normal webs after their palpi were removed. This confirms, to some extent, the conclusions of Plateau,[1] though his further statement that "these appendages are to be placed in the category of useless organs" seems to be scarcely warranted.

We made an effort to determine how far the first and second pairs of legs subserve the sense of hearing, by removing them, and noting the results. We first removed, at the coxæ, the two anterior legs of a female of *E. insularis*. She soon built a good web, and when, two days later, the B fork was sounded near her, she promptly threw up her second pair of legs in the characteristic way.

Some days later we caught a large female of *A. riparia* that had lost her first pair of legs and also the left leg of the second pair. She was placed in the enclosed porch, and by the next day had built a good web, which lacked, however, the zigzag line down the centre, which is characteristic of the web of this species. (Two other specimens of *A. riparia* that had lost their palpi, also omitted the zigzag.) The remaining leg of the second pair was then removed, leaving the spider with only the posterior two pairs. She was now offered a fly, which she quickly seized and devoured. After her repast the B fork was sounded near her, when she attempted to lift the third pair of legs, but only partly succeeded. Several trials gave similar results. The fork was next held well behind her, when she slowly turned toward the sound.

[*Editor's Note:* In the original, material follows this excerpt.]

[1] *American Naturalist*, April, 1887, p. 384.

Editor's Comments on Papers 3, 4, and 5

Before 1900 most published accounts of learning in animals were anecdotal. Stories of wondrous intellectual feats performed by "dumb brutes" circulated widely. These stories were offered as evidence that features of human mental activity were shared by nonhuman animals, although many of the accounts included feats so spectacular that any human who performed them would be considered unusually gifted.

There is a sense of mission apparent in these early reports of the intelligent activities of animals. Establishing the plausibility of Darwin's theory through evidence of the existence of mind in animals took precedence over the study of animal behavior per se. Early evolutionists wanted to find similarities between the minds of humans and animals, and in their enthusiasm they overinterpreted animal behavior.

Of these accounts those collected by George J. Romanes are without peer. The first systematic attempt to organize the facts of animal psychology, *Animal Intelligence* (1882) is a remarkable catalogue of the capabilities of nonhuman species. The lack of interpretive sophistication in these stories is glaring today, yet

Romane's overall message remains clear. Romanes was a scientist, although his criteria for evaluating information, particularly his reliance on the social status of the observer as a guage of a report's reliability, are obviously unacceptable. Despite its drawbacks, *Animal Intelligence* was an important, seminal contribution. Paper 3, a collection of dog stories, gives a taste of the general flavor of Romanes's approach. Some of the incidents reported are of doubtful reliability, and the murkiness of the division between fact and interpretation is irksome; nevertheless, the observations are fascinating.

Just as the lack of a strong methodology weakened *Animal Intelligence,* so did the lack of a strong conceptual framework for characterizing behavior weaken Romanes's subsequent books, *Mental Evolution in Animals* (1883) and *Mental Evolution in Man* (1888). However, it must be remembered that at the time Romanes was writing these books, there was no serious alternative to anthropomorphism as an interpretational scheme in the study of animal learning. Romanes must be given credit for trying to grapple with a difficult subject matter, using the conceptual means that were at hand.

The classic statement of the conservative reply to blatant anthropomorphism is found in the selection from C. Lloyd Morgan's text (Paper 4). Morgan did not deny the utility of interpreting animal behavior in terms of human experience. His objection was to overinterpreting what animals do. Well aware of the difficulties of comparing any two different behaving organisms, whether they be of different species, of different ages, or of different cultures, Morgan stressed the need to be cautious in inferring psychic faculties.

An important demonstration of how easily animal behavior can be misinterpreted came from the study of a remarkable horse. The apparent learning capacities of a stallion named Hans attracted much interest around the turn of the century. Under the tutelage of Wilhelm von Osten, the horse appeared to learn to count, to read, and to do arithmetic. Observers with impeccable credentials attested to the horse's abilities.

In a series of experiments with Hans, Oskar Pfungst was able to demonstrate the role of minute, unintentional movements of the trainer in controlling the horse's behavior. This demonstration made clear the necessity for caution in imputing reasoning capacities to animals. Pfungst did not show that Hans could not reason, but he showed that a simpler explanation could be applied to the putative feats of intellectual performance. Paper 5 is an excerpt from C. L. Rahn's 1911 translation of Pfungst's 1907 book.

REFERENCES

Pfungst, O., 1907, *Der Kluge Hans,* Johann Umbrofious Barth, Leipzig.
Romanes, G. J., 1883, *Mental Evolution in Animals,* Kegan Paul, Trench, & Co., London.
Romanes, G. J., 1888, *Mental Evolution in Man,* Kegan Paul, Trench, & Co., London.

3

Reprinted from pp. 445–458 of *Animal Intelligence*, Kegan, Paul & Trench, London, 1882, 297 p.

DOG—GENERAL INTELLIGENCE

G. J. Romanes

[*Editor's Note:* In the original, material precedes this excerpt.]

General Intelligence.

I have very definite evidence of the fact that dogs are able to communicate to one another simple ideas. The communication is always effected by gesture or tones of barking, and the ideas are always of such a simple nature as that of a mere 'follow me.' According to my own observations, the dogs must be above the average of canine intelligence, and the gesture they invariably employ is a contact of heads, with a motion between a rub and a butt. It is quite different from anything that occurs in play, and is always followed by a definite course of action. I must add, however, that although the information thus conveyed is always definite, I have never known a case in which it was complex—anything like asking or telling the way, which several writers have said that dogs can do, being, I believe, quite out of the question. One example will suffice. A Skye terrier (not quite pure) was asleep in the room where I was, while his son lay upon a wall which separates the lawn from the high road. The young dog, when alone, would never attack a strange one, but was a keen fighter when in company with his father. Upon the present occasion a large mongrel passed along the road, and shortly afterwards the old dog awoke and went sleepily downstairs. When he arrived upon the door-step his son ran up to him and

[1] *Descent of Man*, p. 71.

made the sign just described. His whole manner immediately altered to that of high animation. Clearing the wall together, the two animals ran down the road as terriers only can when pursuing an enemy. I watched them for a mile and a half, within which distance their speed never abated, although the object of their pursuit had not from the first been in sight.

It is almost superfluous to give cases illustrating the well-known fact that dogs communicate their desires and ideas to man; but as the subject of the communication by signs will afterwards be found of importance in connection with the philosophy of communication by words, I shall here give a few examples of dogs communicating by signs with man, which for my purpose will be the more valuable the less they are recognised as unusual.

Lieutenant-Gen. Sir John H. Lefroy, C.B., K.C.M.G., F.R.S., writes me that he has a terrier which it is the duty of his wife's maid to wash and feed. 'It was her habit after calling her mistress in the morning to go out and milk a goat which was tethered near the house, and give "Button" the milk. One morning, being rather earlier than usual, instead of going out at once she took up some needlework and began to occupy herself. The dog endeavoured in every possible way to attract her attention and draw her forth, and at last pushed aside the curtain of a closet, and never having been taught to fetch or carry, took between his teeth the cup she habitually used, and brought it to her feet. I inquired into every circumstance strictly on the spot, and was shown where he found the cup.'

Similarly I select the following case from a great number of others that I might quote, because it is so closely analogous to the above. It is communicated to me by Mr. A. H. Baines:—

There is a drinking-trough for him in my sitting-room: if at any time it happens to be without water when he goes to drink, he scratches the dish with his fore-paws in order to call attention to his wants, and this is done in an authoritative way, which generally has the desired effect. Another Pomeranian—a member of the same family—when quite young used to soak hard biscuits in water till soft enough to eat. She would carry the biscuit in her mouth to the drinking-trough, drop it in and leave it there for a few minutes, and then fish it out with her paw.

One more instance of the communication of ideas by gestures will no doubt be deemed sufficient. It is one of a kind which has many analogies in the literature of canine intelligence.

Dr. Beattie relates this case of canine sagacity, of which the scene was a place near Aberdeen. The Dee being frozen, a gentleman named Irvine was crossing the ice, which gave way with him about the middle of the river. Having a gun, he was able to keep himself from sinking by placing it across the opening. 'The dog made many fruitless efforts to save his master, and then ran to a neighbouring village, where he saw a man, and with the most significant gestures pulled him by the coat, and prevailed on him to follow. The man arrived on the spot in time to save the gentleman's life.'

Numberless other instances of the same kind might be given, and they display a high degree of intelligence. Even the idea of saving life implies in itself no small amount of intelligence; but in such cases as these we have added the idea of going for help, communicating news of a disaster, and leading the way to its occurrence.

Having thus as briefly as possible considered the emotional and the more ordinary intellectual faculties of the dog, I shall pass on to the statement of cases showing the higher and more exceptional developments of canine sagacity.

Were the purpose of this work that of accumulating anecdotes of animal intelligence, this would be the place to let loose a flood of facts, which might all be well attested, relating to the high intelligence of dogs. But as my aim is rather that of suppressing anecdotes, except in so far as facts are required to prove the presence in animals of the sundry psychological faculties which I believe the different classes to present, I shall here, as elsewhere, follow the method of not multiplying anecdotes further than seems necessary fully to demonstrate the highest level of intelligence to which the animal under consideration can certainly be said to attain. But in order that any who read these pages for the sake of the anecdotes which they necessarily present may not be disappointed by meeting with cases already known to them, I shall draw my material mainly from the facts communicated to me by private correspondents, alluding to previously published facts only as supplementary to those now published for the first time. It may be well to explain to my numerous correspondents that I select the following cases for quoting, not because they are the most sensational that I have received, but rather because they either contain nothing sufficiently exceptional to excite the criticism of incredulity, or because they happen to have been corroborated by the more or less similar cases which I quote from other correspondents.

As showing the high general intelligence of the dog, I shall first begin with the collie. It is certain that many of these dogs can be trusted to gather and drive sheep without supervision. It is enough on this head to refer to the well-known anecdotes of the poet Hogg in his 'Shepherd's Calendar,' concerning his dog 'Sirrah.'

Williams, in his book on 'Dogs and their Ways,' says (p. 124) that a friend of his had a collie which, whenever his master said the words 'Cast, cast,' would run off to seek any sheep that might be cast, and on finding it would at once assist it to rise. He also knew of another dog (p. 102), which would perform the same office even in the absence of his master, going the round of the fields and pastures by himself to right all the sheep that he found to be cast.[1]

One of my correspondents (Mr. Laurie Gentles) sends me an account of a sheep-dog belonging to a friend of his (Mr. Mitchell, of Inverness-shire) which strayed to a neighbouring farm, and took up his residence with the farmer. On the second night after the dog arrived at the farm the farmer 'took the dog down to the meadow to see if the cattle were all right. To his dismay he found that

[1] For many other instances of sheep-dog sagacity, see Watson, *Reasoning Power of Animals*, under 'Shepherd's Dog.'

the fence between his meadow and his neighbour's had got broken down, and that the whole of his neighbour's cattle had got mixed up with his. By the help of the dog the strange cattle were driven back into their proper meadow, and the fence put into temporary repair. The *next* night, at the same hour, the gentleman started off to look after the cattle. The dog, however, was not to be seen. On arriving at the meadow, what was the gentleman's astonishment to find that the dog had preceded him! His astonishment soon changed into delighted approbation when he found the dog sitting on the broken fence between the two meadows, and daring the cattle from either side to cross. The cattle had during the interval between the first and second visits broken down the fence, and had got mixed up with each other. The dog had quietly gone off on his own account to see if all was right, and finding a similar accident to the one the previous evening, had *alone* and *unaided* driven back the *strange* cattle to their proper meadow, and had mounted guard over the broken fence as I have already indicated.'

Colonel Hamilton Smith says that the cattle-dogs of Cuba and Terra Firma are very wise in managing cattle, but require to display different tactics from the cattle-dogs of Europe :—

When vessels with live stock arrive at any of the West India harbours, these animals, some of which are nearly as large as mastiffs, are wonderfully efficient in assisting to land the cargo. The oxen are hoisted out with a sling passing round the base of their horns; and when an ox, thus suspended by the head, is lowered, and allowed to fall into the water, so that it may swim to land, men sometimes swim by the side of it and guide it, but they have often dogs of this breed which will perform the service equally well; for, catching the perplexed animal by the ears, one on each side, they will force it to swim in the direction of the landing-place, and instantly let go their hold when they feel it touch the ground, as the ox will then naturally walk out of the water by itself.[1]

That this sagacity need not be due to special tuition, may be inferred from a closely similar display sponta-

[1] *Naturalist's Library*, vol. x., p. 154 (quoted by Watson).

neously shown in the following case. It is communicated to me by a correspondent, Mr. A. H. Browning. This gentleman was looking at a litter of young pigs in their sty, and when he went away the door of the sty was inadvertently left unfastened. The pigs all escaped into his garden. My correspondent then proceeds:—

My attention was called to my dog appearing in a great state of excitement, *not* barking (he seldom barks), but whining and performing all sorts of antics (in a human subject I should have said 'gesticulating'). The herdmen and myself returned to the sty; we caught but one pig, and put him back; no sooner had we done so than the dog ran after each pig in succession, brought him back to the sty by the ear, and then went after another, until the whole number were again housed.

In Lord Brougham's 'Dialogues on Instinct' (iii.) there is narrated the story told to the author by Lord Truro of a dog that used to worry sheep at night. The animal quietly submitted to be tied up in the evening, but when everybody was asleep he used to slip his collar, worry the sheep, and, returning before dawn, again get into his collar to avoid suspicion. I allude to this remarkable display of sagacity because I am myself able fully to corroborate it by precisely similar cases. A friend of mine (the late Mr. Sutherland Murray) had a dog which was always kept tied up at night, but nevertheless the neighbouring farmers complained of having detected him as the culprit when watching to find what dog it was that committed nightly slaughter among their sheep. My friend, therefore, set a watch upon his dog, and found that when all was still he slipped his collar, and after being absent for some hours, returned and slipped his head in again.

A precisely similar case is given further back, and others are communicated to me by two correspondents (Mr. Goodbehere, of Birmingham, and Mr. Richard Williams, of Buffalo). The latter says:—

And here let me ask if you are aware of the cunning and sagacity of these sheep-killing dogs, that they never kill sheep on the farm to which they belong, or in the immediate vicinity, but often go miles away; that they always return before daylight, and before doing so wash themselves in some stream to get rid of the blood.

In Germany I knew a large dog that was very fond of grapes, and at night used to slip his collar in order to satisfy his propensity; and it was not for some time that the thief was suspected, owing to his returning before daylight and appearing innocently chained up in his kennel.

A closely similar case is recorded in Mr. Duncan's book on 'Instinct' of a dog belonging to the Rev. Mr. Taylor, of Colton. The only difference is that the delinquent dog slipped and afterwards readjusted a muzzle instead of a collar.

In connection with sly sagacity I may also give another story contained in my correspondence, although in this case I am specially requested by my correspondent not to publish his name. I can, therefore, only say that he occupies a high position in the Church, and that the dog (a retriever) was his own property :—

The dog was lying one evening before the kitchen fire where the cook had prepared a turkey for roasting. She left the kitchen for a few moments, when the dog immediately carried away the turkey and placed it in the cleft of a tree close to the house, but which was well concealed by the surrounding laurels. So rapid were his movements that he returned to his post before the cook had come back, and stretching himself before the fire, looked 'as innocent as a child unborn.' Unfortunately for him, however, a man who was in the habit of taking him to shoot, saw him carrying away his prize and watched his progress. On coming into the kitchen the man found the dog in his old place pretending to be asleep. Diver's conduct was all along dictated by a desire to conceal his theft, and if he were a man I should have said that he intended, in case of inquiry, to prove an alibi.

Mr. W. H. Bodley writes me of a retriever dog that belonged to him :—

Before he came to me he lived where another dog of similar size was kept, and on one occasion they fought. Having been chastised for this, on future occasions when they quarrelled they used to swim over a river of some breadth, where they could not be interfered with, and fight out their quarrel on the other side. What seems to me noteworthy in this conduct is the *self-restraint* manifested under the influence of *passion*, and the mutual understanding to defer the fight till they could prosecute it unmolested; like two duellists crossing the Channel to fight in France.

It is, of course, a well-known thing that dogs may easily be taught the use of coin for buying buns, &c. In the 'Scottish Naturalist' for April, 1881, Mr. Japp vouches for the fact that a collie which he knew was in the habit of purchasing cakes with coppers without ever having been taught the use of coin for such purposes. This fact, however, of a dog spontaneously divining the use of money requires corroboration, although it is certain that many dogs have an instinctive idea of giving peace-offerings, and the step from this to the idea of barter may not be large. Thus, to give only two illustrations, Mr. Badcock writes to me that a friend of his had a dog which one day had a quarrel with a companion dog, so that they parted at variance. 'On the next day the friend appeared with a biscuit, which he presented as a peace-offering.' Again, Mr. Thomas D. Smeaton writes to me of his dog that he 'has an amusing practice when he is restored to favour after some slight offence, of immediately picking up and carrying anything that is handiest, stone, stick, paper: it is a deliberate effort to please, a sort of good-will offering, a shaking hands over the past.'

I am indebted for the following to Mr. Goodbehere, of Birmingham; it may be taken as typical of many similar cases:—

My friend (Mr. James Canning, of Birmingham) was acquainted with a small mongrel dog who on being presented with a penny or a halfpenny would run with it in his mouth to a baker's, jump on to the top of the half-door leading into the shop, and ring the bell behind the door until the baker came forward and gave him a bun or a biscuit in exchange for the coin. The dog would accept any small biscuit for a halfpenny, but nothing less than a bun would satisfy him for a penny. On one occasion the baker (being annoyed at the dog's too frequent visits), after receiving the coin, refused to give the dog anything in exchange, and on every future occasion the latter (who declined being *taken in* a second time) would put the coin on the floor, and not permit the baker to pick it up until he had received its equivalent.

Mr. R. O. Backhouse writes to me:—

My dog is a broken-haired rabbit-coursing dog, and is very intelligent. I took him one day to an exhibition of pictures and objects of interest, among which were statues and a bust of Sir Walter Scott. It was a local exhibition, and as there was jewellery, some one had to sit up all night with it as guard. I volunteered, and as we were looking about and sitting on a stand of flowers, my dog suddenly began to bark, and made as if he had found some one hiding. On looking round I found that it was the bust of Sir Walter Scott standing among the flowers, and in which he evidently recognised sufficient likeness to a human being to think the supposed man had no business there at so late an hour.

I adduce this instance because it serves as a sort of introduction to the more remarkable faculty which I cannot have the least doubt is manifested by some dogs—the faculty, namely, of recognising portraits as representing persons, or possibly of mistaking portraits for persons.

Mr. Crehore, writing to 'Nature' (vol. xxi., p. 132), says:—

A Dandie-Dinmont terrier, after the death of his mistress, was playing with some children in a room into which was brought a photograph (large) of her that he had never previously seen. It was placed upon the floor leaning against the wall. In the words of my informant, who witnessed it, the dog, when he suddenly caught sight of the picture, crouched and trembled all over, his whole body quivering. Then he crept along the floor till he reached it, and, seating himself before it, began to bark loudly, as if he would say, 'Why don't you speak to me?' The picture was moved to other parts of the room, and he followed, seating himself before it and repeating his barking.

Mr. Charles W. Peach also gives an account in 'Nature' (vol. xx., p. 196) of a large dog recognising his portrait:—

When it (the portrait) was brought to my house, my old dog was present with the family at the unveiling; nothing was said to him, nor invitation given to him to notice it. We saw that his gaze was steadily fixed on it, and he soon became excited and whined, and tried to lick and scratch it, and was so much taken up with it that we—although so well knowing his intelligence—were all quite surprised—in fact, could scarcely believe that he should know it was my likeness. We, however, had sufficient proof after it was hung up in our parlour. The room was rather low, and under the picture stood a chair: the door was left open, without any thought about the dog; he, however, soon found it out, when a low whining and scratching was heard by the family, and on search being made, he was in the chair trying to get at the picture. After this I put it up higher, so as to prevent its being injured by him. This did not prevent him from paying attention to it, for whenever I was away from home, whether for a short or a long time—sometimes for several days—he spent most of his time gazing on it, and as it appeared to give him comfort the door was always left open for him. When I was long away he made a low whining, as if to draw attention to it. This lasted for years—in fact, as long as he lived.

From this account it appears that when in the first instance the dog's attention was drawn to the picture it was on the floor in the line of the dog's sight; the behaviour of the animal then and subsequently was too marked and peculiar to admit of mistake.

Another correspondent in 'Nature' (vol. xx., p. 220), alluding to the previous letter, writes:—

Having read Mr. Peach's letter on 'Intellect in Brutes,' as shown by the sagacity he witnessed in his dog, I have been asked to send a similar anecdote, which I have often told to friends. Many years ago my husband had his portrait taken by J. Phillips, R.A., and subsequently went to India, leaving the portrait in London to be finished and framed. When it was sent home, about two years after it was taken, it was placed on the floor against the sofa, preparatory to being hung on the wall. We had then a very handsome black-and-tan setter, which was a great pet in the house. As soon as the dog came into the room he recognised his master, though he had not seen him for two years, and went up to the picture and licked the face. When this anecdote was told to Phillips, he said it was the highest compliment that had ever been paid him.

Similarly, in the same periodical (vol. xx., p. 220), Mr. Henry Clark writes:—

Some years ago a fine arts exhibition was held at Derby. A portrait of a Derby artist (Wright) was thus signalised:—'The artist's pet dog distinguished this from a lot of pictures upon the floor of the studio by licking the face of the portrait.'

Again, I learn from Dr. Samuel Wilks, F.R.S., that a friend of his, whom I shall call Mrs. E., has a terrier which recognised her portrait. 'The portrait is now (1881) hanging in the Royal Academy. When it first arrived home the dog barked at it, as it did at strangers; but after a day or two, when Mrs. E. opened the door to show the portrait to some friends, the dog went straight to the picture and licked the hand. The picture is a three-quarter length portrait of a lady with the hand at the bottom of the picture.'

Lastly, my sister, who is a very conscientious and accurate observer, witnessed a most unmistakable recognition of portraits as representative of persons on the part of a small but intelligent terrier of her own. At my request she committed the facts to writing shortly after they occurred. The following is her statement of them:—

I have a small terrier who attained the age of eight months without ever having seen a large picture. One day three nearly life-sized portraits were placed in my room during his absence. Two were hung up, and one left standing against the wall on the floor awaiting the arrival of a picture-rod. When the dog entered the room he appeared much alarmed by the sight of the pictures, barking in a terrified manner first at one and then at another. That is to say, instead of attacking them in an aggressive way with tail erect, as he would have done on thus encountering a strange person, he barked violently and incessantly at some distance from the paintings, with tail down and body elongated, sometimes bolting under the chairs and sofas in the extremity of his fear, and continuing barking from there. Thinking it might be merely the presence of strange objects in the room which excited him, I covered the faces of the portraits with cloths and turned the face of the one on the floor to the wall. The dog soon after emerged from his hiding-place, and having looked intently at the covered pictures and examined the back of the frame on the ground, became quite quiet and contented. I then uncovered one of the pictures, when he immediately flew at it, barking in the same frightened manner as before. I then re-covered that one and took the cover off another. The dog left the covered one and rushed at the one which was exposed. I then turned the face of the one on the floor to the room, and he flew at that with increased fierceness. This I did many times, covering and uncovering each picture alternately, always with the same result. It was only when all three paintings were uncovered at the same time, and he saw one looking at him in whatever direction he turned, that he became utterly terrified. He continued in this state for nearly an hour, at the end of which time, although evidently very nervous and apt to start, he ceased to bark. After that day he never took any more notice of the pictures during the three months he remained in the house. He was then absent from the house for seven months. On his return he went with me into the room where the portraits were hung, immediately on his arrival. He was evidently again much startled on first seeing them, for he rushed at one, barking as he had done on the first occasion, but he only gave three or four barks when he ran back to me with the same apologetic manner as he has when he has barked at a well-known friend by mistake.

It will have been observed that in all these cases the portraits, when first recognised as bearing resemblance to human beings, were placed on the floor, or in the ordinary line of the dog's sight. This is probably an important condition to the success of the recognition. That it certainly was so in the case of my sister's terrier was strikingly proved on a subsequent occasion, when she took the animal into a picture-shop where there were a number of portraits hanging round the walls, and also one of Carlyle standing on the floor. The terrier did not heed those upon the walls, but barked excitedly at the one upon the floor. This case was further interesting from the fact that there were a number of purchasers in the shop who were, of course, strangers to the terrier; yet he took no notice of them, although so much excited by the picture. This shows that the pictorial illusion was not so complete as to make the animal suppose the portrait to be a real person; it was only sufficiently so to make it feel a sense of bewildered uncertainty at the kind of life-in-death appearance of the motionless representation.

If, notwithstanding all this body of mutually corroborative cases, it is still thought incredible that dogs should be able to recognise pictorial representations,[1] we should

[1] Since my MS. went to press I have myself met with a striking

do well to remember that this grade of mental evolution is reached very early in the psychical development of the human child. In my next work I shall adduce evidence to show that children of one year, or even less, are able to distinguish pictures as representations of particular objects, and will point at the proper pictures when asked to show these objects.

Coming now to cases more distinctly indicative of reason in the strict sense of the word, numberless ordinary acts performed by dogs indisputably show that they possess this faculty. Thus, for instance, Livingstone gives the following observation.[1] A dog tracking his master along a road came to a place where three roads diverged. Scenting along two of the roads and not finding the trail, he ran off on the third without waiting to smell. Here, therefore, is a true act of inference. If the track is not on A or B, it must be on C, there being no other alternative.

Again, it is not an unusual thing for intelligent dogs, who know that their masters do not wish to take them out, to leave the house and run a long distance in the direction in which they suppose their masters are about to go, in order that when they are there found the distance may be too great for their masters to return home for the purpose of shutting them up. I have myself known several terriers that would do this, and one of the instances I shall give *in extenso* (quoted from an account which I published at the time in 'Nature'); for I think it displays remarkably complex processes of far-seeing calculation:—

The terrier in question followed a conveyance from the house in which I resided in the country, to a town ten miles distant. *He only did this on one occasion*, and about five months afterwards was taken *by train* to the same town as a present to some friends there. Shortly afterwards I called upon these friends in a different conveyance from the one which the dog had previously followed; but the latter may have known that the two conveyances belonged to the same

display of the recognition of a portrait by a dog. The portrait was one of myself, and the dog a half-bred setter and retreiver of my own.

[1] *Missionary Travels*, chap. i.

house. Anyhow, after I had put up the horses at an inn, I spent the morning with the terrier and his new masters, and in the afternoon was accompanied by them to the inn. I should have mentioned that the inn was the same as that at which the conveyance had been put up on the previous occasion, five months before. Now, the dog evidently remembered this, and, reasoning from analogy, inferred that I was about to return. This is shown by the fact that he stole away from our party—although at what precise moment he did so I cannot say, but it was certainly *after* we had arrived at the inn, for subsequently we all remembered his having entered the coffee-room with us. Now, not only did he infer from a single precedent that I was going home, and make up his mind to go with me, but he also further reasoned thus:—'As my previous master lately sent me to town, it is probable that he does not want me to return to the country; therefore, if I am to seize this opportunity of resuming my poaching life, I must now steal a march upon the conveyance. But not only so, my former master may possibly pick me up and return with me to my proper owners; therefore I must take care only to intercept the conveyance at a point sufficiently far without the town to make sure that he will not think it worth his while to go back with me.'

Complicated as this train of reasoning is, it is the simplest one I can devise to account for the fact that slightly beyond the *third* milestone the terrier was awaiting me, lying right in the middle of the road with his face towards the town. I should add that the second two miles of the road were quite straight, so that I could easily have seen the dog if he had been merely running a comparatively short distance in front of the horses. Why this animal should never have returned to his former home on his own account I cannot suggest, but I think it was merely due to an excessive caution which he also manifested in other things. However, be the explanation of this what it may, as a fact he never did venture to come back upon his own account, although there never was a subsequent occasion upon which any of his former friends went to the town but the terrier was seen to return with them, having always found some way of escape from his intended imprisonment.

[*Editor's Note:* In the original, material follows this excerpt.]

4

Reprinted from pp. 53–59 of *An Introduction to Comparative Psychology*, Walter Scott, London, 1894, 386 p.

OTHER MINDS THAN OURS

C. L. Morgan

[*Editor's Note:* In the original, material precedes this excerpt.]

For in the study of animal psychology as a branch of scientific inquiry, it is necessary that accurate observation, and a sound knowledge of the biological relationships of animals, should go hand in hand with a thorough appreciation of the methods and results of modern psychology. The only fruitful method of procedure is the interpretation of facts observed with due care in the light of sound psychological principles.

What some of these principles are we have considered, or shall consider, in this work. There is one basal principle, however, the brief exposition of which may fitly bring to a close this chapter. It may be thus stated:—*In no case may we interpret an action as the outcome of the exercise of a higher psychical faculty, if it can be interpreted as the outcome of the exercise of one which stands lower in the psychological scale.*

To this principle several objections, none of them however of any real weight, may be raised. First there is the sentimental objection that it is ungenerous to the animal. In dealing with one's fellow-man it is ungenerous to impute to him lower motives for his actions when they may have been dictated by higher motives. Why should we adopt a different course with the poor dumb animal from that which we should adopt with our human neighbour? In the first place, it may be replied, this objection starts by assuming the very point to be proved. The scientific problem is to ascertain the limits of animal psychology. To assume that a given action may be the outcome of the exercise of either a higher or a lower faculty, and that it is more generous to adopt the former alternative, is to assume the existence of the higher faculty, which has to be proved. In the case of our neighbours we have good grounds for knowing that such and such a deed may have been dictated by either a higher or a lower motive. If we had equally good grounds for knowing that the animal was possessed of both higher and lower faculties, the scientific problem would have been solved; and the attribution of the one or the other, in any particular case, would be a purely individual matter of comparatively little general moment. In the second place, this generosity, though eminently desirable in the relations of practical social life, is not precisely the attitude which a critical scientific inquiry demands. Moreover, an ungenerous interpretation of one's neighbour's actions may lead one to express an unjust estimate of his moral character and thus to do him grave social wrong; but an ungenerous interpretation of the faculties of animals can hardly be said to be open to like practical consequences.

A second objection is, that by adopting the principle in question we may be shutting our eyes to the simplest explanation of the phenomena. Is it not simpler to explain the higher activities of animals as the direct outcome of reason or intellectual thought, than to explain them as the complex results of mere intelligence or practical sense-experience? Undoubtedly it may in many cases seem simpler. It is the apparent simplicity of the explanation that leads many people naively to adopt it. But surely the simplicity of an explanation is no necessary criterion of its truth. The explanation of the genesis of the organic world by direct creative fiat, is far simpler than the explanation of its genesis through the indirect method of evolution. The explanation of instinct and early phases of intelligence as due to inherited habit, individually acquired, is undoubtedly simpler than the explanation which Dr Weismann would substitute for it. The formation of the cañon of the Colorado by a sudden rift in the earth's crust, similar to those which opened during the Calabrian earthquakes, is simpler than its formation by the fretting of the stream during long ages under varying meteorological conditions.

In these cases and in many others the simplest explanation is not the one accepted by science. Moreover, the simplicity of the explanation of the phenomena of animal activity as the result of intellectual processes, can only be adopted on the assumption of a correlative complexity in the mental nature of the animal as agent. And to assume this complexity of mental nature on grounds other than those of sound induction, is to depart from the methods of scientific procedure.

But what, it may be asked, is the logical basis upon which this principle is founded? If it be true that the animal mind can only be interpreted in the light of our knowledge of human mind, why should we not use this method of interpretation freely, frankly, and fully? Is there not some contradiction in refusing to do so? For, first, it is contended that we must use the human mind as a key by which to read the brute mind, and then it is contended that this key must be applied with a difference. If we apply the key at all, should we not apply it without reservation?

This criticism might be valid if we were considering the question apart from evolution. Here evolution is postulated. The problem is this: (1) Given a number of divergently ascending grades of organisms, with divergently increasing complexity of organic structure and correlated activities: (2) granted that associated with the increasing organic complexity there is increasing mental or psychical complexity: (3) granted that in man the organic complexity, the complexity of correlated activities, and the associated mental or psychical complexity, has reached the maximum as yet attained: (4) to gauge the psychical level to which any organism has been evolved. As we have already seen, we are forced, as men, to gauge the psychical level of the animal in terms of the only mind of which we have first-hand knowledge, namely the human mind. But how are we to apply the gauge?

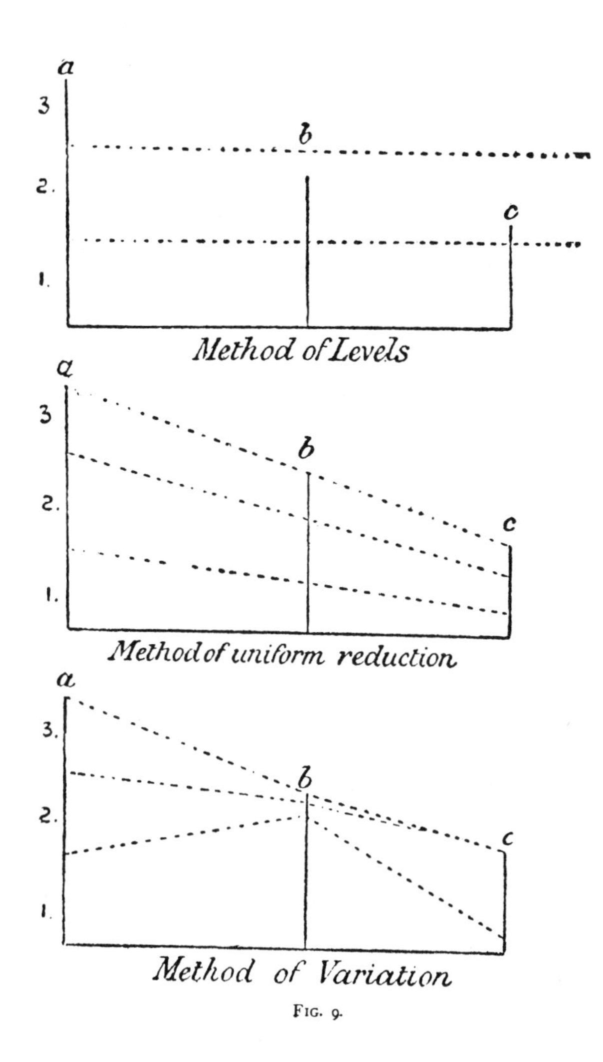
a
3
b
2.
c
1.
Method of Levels
a
3
b
2.
c
1.
Method of uniform reduction
a
3.
b
2.
c
1.
Method of Variation

FIG. 9.

There would appear to be three possible methods, which are exemplified in Fig. 9. Let *a* represent the psychical stature of man, and 1, 2, 3, ascending faculties or stadia in mental development. Let *b c* represent two animals the psychical stature of each of which is to be gauged. It may be gauged first by the "method of levels," according to which the faculties or stadia are of constant value. In the diagram, *b* has not quite reached the level of the beginning of the third or highest faculty, while *c* has only just entered upon the second stadium. Secondly, it may be gauged by the "method of uniform reduction." In both *b* and *c* we have all three faculties represented in the same ratio as in *a*, but all uniformly reduced. And thirdly, it may be gauged by the "method of variation," according to which any one of the faculties 1, 2, or 3, may in *b* and *c* be either increased or reduced relatively to its development in *a*. Let us suppose, for example, that *b* represents the psychical stature of the dog. Then, according to the interpretation on the method of levels, he possesses the lowest faculty (1) in the same degree as man; in the faculty (2) he somewhat falls short of man; while in the highest faculty (3) he is altogether wanting. According to the interpretation on the method of uniform reduction he possesses all the faculties of man but in a reduced degree. And according to the interpretation on the method of variation he excels man in the lowest faculty, while the other two faculties are both reduced but in different degrees. The three "faculties" 1, 2, 3, are not here intended to serve any other purpose than merely to illustrate the three methods of interpretation.

On the principles of evolution we should unquestionably expect that those mental faculties which could give decisive advantage in the struggle for existence would be developed in strict accordance with the divergent conditions of life. Hence it is the third method, which I have termed the method of variation, which we should expect *a priori* to accord most nearly with observed facts. And so far as we can judge from objective observation (the only observation open to us) this would appear to be the case. Presumably there are few observers of animal habit and intelligence who would hesitate in adopting the method of variation as the most probable mode of interpretation. But note that while it is the most probable it is also the most difficult mode of interpretation. According to the method of levels the dog is just like me, without my higher faculties. According to the method of uniform reduction he is just like me, only nowise so highly developed. But according to the method of variation there are many possibilities of error in estimating the amount of such variation. Of the three methods that of variation is the least anthropomorphic, and therefore the most difficult.

In the diagram by which the method of variation is illustrated, the highest faculty 3 is in *c* reduced to zero,—in other words, is absent. It may, however, be objected that this is contrary to the principles of evolution, since the presence of any faculty in higher types involves the germ of this faculty in lower types. This criticism only holds good, however, on the assumption that the evolution of higher faculties out of lower faculties is impossible. Those evolutionists who accept this assumption as valid are logically bound to believe either (1) that all forms of animal life from the amœba upwards have all the faculties of man, only reduced in degree and range, and to interpret all animal psychology on a method of reduction (though not necessarily uniform reduction), or (2) that in the higher forms of life the introduction of the higher faculties has been effected by some means other than that of natural evolution. I am not prepared to accept the assumption as valid; and it will be part of my task in future chapters to consider how the transition from certain lower to certain higher phases of mental development may have been effected.

If this be so it is clear that any animal may be at a stage where certain higher faculties have not yet been evolved from their lower precursors; and hence we are logically bound not to assume the existence of these higher faculties until good reasons shall have been shown for such existence. In other words, we are bound to accept the principle above enunciated: that in no case is an animal activity to be interpreted as the outcome of the exercise of a higher psychical faculty, if it can be fairly interpreted as the outcome of the exercise of one which stands lower in the psychological scale.

5A

Reprinted from pp. 18–22 of *Clever Hans (The Horse of Mr. Von Osten): A Contribution to Experimental Animal and Human Psychology*, C. L. Rahn, trans., Henry Holt, New York, 1911, 274 p.

THE PROBLEM OF ANIMAL CONSCIOUSNESS AND "CLEVER HANS"

O. Pfungst

[*Editor's Note:* In the original, material precedes this excerpt.]

At last the thing so long sought for, was apparently found: A horse that could solve arithmetical problems—an animal which, thanks to long training, mastered not merely rudiments, but seemingly arrived at a power of abstract thought and which surpassed, by far, the highest expectations of the greatest enthusiast.

And now what was it that this wonderful horse could do? The reader may accompany us to an exhibition which was given daily before a select company at about the noon hour in a paved courtyard surrounded by high apartment houses in the northern part of Berlin. No fee was ever taken. The visitor might walk about freely and if he wished, might closely approach the horse and its master, a man between sixty and seventy years of age. His white head was covered with a black, slouch hat. To his left the stately animal, a Russian trotting horse, stood like a docile pupil, managed not by means of the whip, but by gentle encouragement and frequent reward of bread or carrots. He would answer correctly, nearly all of the questions which were put to him in German. If he understood a question, he immediately indicated this by a nod of the head; if he failed to grasp its import, he communicated the fact by a shake of the head. We were told that the questioner had to confine himself to a certain vocabulary, but this was comparatively rich and the horse widened its scope daily without special instruction, but by simple contact with his environment. His master, to be sure, was usually present whenever questions were put to the horse by others, but in the course of time, he gradually responded to a greater and greater number of persons. Even though Hans did not appear as willing and reliable in the case of strangers as in the case of his own master, this might easily be explained by the lack of authoritativeness on their part and of affection on the part of Hans, who for the last four years had had intercourse only with his master.

Our intelligent horse was unable to speak, to be sure. His chief mode of expression was tapping with his right forefoot. A good deal was also expressed by means of movements of the head. Thus "yes" was expressed by a nod, "no" by a deliberate movement from side to side; and "upward," "upper," "downward," "right," "left," were indicated by turning the head in these directions. In this he showed an astonishing ability to put himself in the place of his visitors. Upon being asked which arm was raised by a certain gentleman opposite him, Hans promptly answered by a movement to the right, even though seen from his own side, it would appear to be the left. Hans would also walk toward the persons or things that he was asked to point out, and he would bring from a row of colored cloths, the piece of the particular color demanded. Taking into account his limited means of expression, his master had translated a large number of concepts into numbers; e. g.:—the letters of the alphabet, the tones of the scale, and the names of the playing cards were indicated by taps. In the case of playing cards one tap meant "ace," two taps "king," three "queen," etc.

Let us turn now to some of his specific accomplishments. He had, apparently, completely mastered the cardinal numbers from 1 to 100 and the ordinals to 10, at least. Upon request he would count objects of all sorts, the persons present, even to distinctions of sex. Then hats, umbrellas, and eyeglasses. Even the mechanical activity of tapping seemed to reveal a measure of intelligence. Small numbers were given with a slow tapping of the right foot. With larger numbers he would increase his speed, and would often tap very rapidly right from the start, so that one might have gained the impression that knowing that he had a large number to tap, he desired to hasten the monotonous activity. After the final tap, he would return his right foot—which he used in his counting—to its original position, or he would make the final count with a very energetic tap of the left foot,—to underscore it, as it were. "Zero" was expressed by a shake of the head.

But Hans could not only count, he could also solve problems in arithmetic. The four fundamental processes were entirely familiar to him. Common fractions he changed to decimals, and vice versa; he could solve problems in mensuration—and all with such ease that it was difficult to follow him if one had become somewhat rusty in these branches. The following problems are illustrations of the kind he solved.* "How much is $\frac{2}{5}$ plus $\frac{1}{2}$?" Answer: $\frac{9}{10}$. (In the case of all fractions Hans would first tap the numerator, then the denominator; in

* All examples mentioned are cited from extant works of various observers.

this case, therefore, first 9, then 10). Or again: "I have a number in mind. I subtract 9, and have 3 as a remainder. What is the number I had in mind?"—12. "What are the factors of 28?"—Thereupon Hans tapped consecutively 2, 4, 7, 14, 28. "In the number 365287149 I place a decimal point after the 8. How many are there now in the hundreds place?"—5. "How many in the ten thousandths place?"—9. It will be noticed, therefore, that he was able to operate with numbers far exceeding 100, indeed he could manipulate those of six places. We were told that this, however, was no longer arithmetical computation in the true sense of the term; Hans merely knew after the analogy of 10 and 100 that the thousands take the fourth place, the ten-thousands the fifth, etc. If an error entered into Hans' answer, he could nearly always correct it immediately upon being asked: "By how many units did you go wrong?"

Hans, furthermore, was able to read the German readily, whether written or printed. Mr. von Osten, however, taught him only the small letters, not the capitals. If a series of placards with written words were placed before the horse, he could step up and point with his nose to any of the words required of him. He could even spell some of the words. This was done by the aid of a table devised by Mr. von Osten, in which every letter of the alphabet, as well as a number of diphthongs had an appropriate place which the horse could designate by means of a pair of numbers. Thus in the fifth horizontal row "s" had first place; "sch" second, "ss," third, etc.; so that the horse would indicate the letter "s" by treading first 5, then 1, "sch," by 5 and 2, "ss" by 5 and 3. Upon being asked "What is this woman holding in her hand?" Hans spelled without hesitation: 3, 2; 4, 6; 3, 7; i. e., "Schirm" (parasol). At another time a picture of a horse standing at a manger was shown him and he was asked, "What does this represent?" He promptly spelled "Pferd" (horse) and then "Krippe" (manger).

He, moreover, gave evidence of an excellent memory. In passing we might also mention that he knew the value of all the German coins. But most astonishing of all was the following: Hans carried the entire yearly calendar in his head; he could give you not only the date for each day without having been previously taught anew, but he could give you the date of any day you might mention. He could also answer such inquiries as this: "If the eighth day of a month comes on Tuesday, what is the date for the following Friday?" He could tell the time to the minute by a watch and could answer off-hand the question, "Between what figures is the small hand of a watch at 5 minutes after half-past seven?" or, "How many minutes has the large hand to travel between seven minutes after a quarter past the hour, and three quarters past?" Tasks that were given him but once would be repeated correctly upon request. The sentence: "Brücke und Weg sind vom Feinde besetzt" (The bridge and the road are held by the enemy), was given to Hans one day and upon the following day he tapped consecutively the 58 numbers which were necessary for a correct response. He recognized persons after having seen them but once—yes, even their photographs taken in previous years and bearing but slight resemblance.

[*Editor's Note:* In the original, material follows this excerpt.]

5B

Reprinted from pp. 63–66 of *Clever Hans (The Horse of Mr. Von Osten): A Contribution to Experimental Animal and Human Psychology*, C. L. Rahn, trans., Henry Holt, New York, 1911, 274 p.

EXPERIMENTS AND OBSERVATIONS

O. Pfungst

[*Editor's Note:* In the original, material precedes this excerpt.]

Finally we sought to discover by what movements the horse could be made to cease tapping. We discovered that upward movements served as signals for stopping. The raising of the head was the most effective, though the raising of the eyebrows, or the dilation of the nostrils —as in a sneer—seemed also to be efficacious. However, it was impossible for me to discover whether or not these latter movements were accompanied by some slight, involuntary upward movement of the head. The upward movement of the head was ineffective only when it did not occur as a jerk, but was executed in a circuitous form,— first upward and then back again. Such a movement was occasionally observed in the case of Mr. von Osten. The elevation of the arms or of the elbow nearest the horse, or the elevation of the entire body was also effective. Even if a placard, with which the experimenter tried to cover his face, were raised at a given moment, the horse would make the back-step. On the other hand, head movements to the right and to the left or forward and back, in fine, all horizontal movements, remained ineffective. We also found that all hand movements, including the " wonderfully effective thrust of the hand into the pocket filled with carrots ", brought no response. I might also change my position and walk forward and then backward some distance behind the horse, but the back-step would only occur in response to the characteristic stimulus. After what has been said it is easy to understand how vain were Mr.

Schillings' attempts to disturb the horse and how naturally he might conclude that Hans was not influenced by visual signs. Mr. Schillings simply did not know which signs were effective.

While the horse could thus be interrupted in the process of tapping by movements which were executed at the level of the questioner's head, yet movements below this level had the opposite effect. If Hans showed that he was about to cease tapping before it was desired, it was possible to cause him to continue by simply bending forward a trifle more. The greater angle at which the questioner's trunk was now inclined caused the horse to increase the rate of tapping. The rule may be stated thus: The greater the angle at which the body inclined forward, the greater the horse's rate of tapping, and *vice versa.* It was noticeable that whenever Mr. von Osten asked for a relatively large number—in which case he always bent farther forward than in the case of smaller numbers—Hans would immediately begin to tap very swiftly. Not being entirely satisfied with these observations, the following more exact measurements were taken. I asked the horse to tap 20. From 1 to 10 I held my body at a certain constant angle, at 10 I suddenly bent farther forward and retained this posture until 20 had been reached. If there existed a relationship between the angle of inclination and the rate of tapping, then the time for the last ten taps ought to be less than for the first ten. Of 34 such tests 31 were sucessful. The following are two specimen series.

The first series consisted of ten tests of 15 taps each. In all cases my head was bent at an angle of 30° to the axis of the trunk, but I constantly changed the angle of inclination of the trunk. It was not possible to measure this angle accurately on account of the rapidity with which the whole test had to be made. I was able, however, to differentiate between them with enough accuracy to designate the smallest angle (about 20°) as belonging to Grade I, and the greatest angle (about 100°) as belonging to Grade VII. By fixing certain points in the environment, it was possible to get approximately the same angle repeatedly. The time from the third to the thirteenth tap was, in all cases, taken by Prof. Stumpf by means of a stop-watch. The tests were taken in the following order:

Grade of inclination:	I	VI	II	II	IV	V	VI	VII
Time for 10 taps:	5.2	4.6	5.0	5.0	4.8	4.8	4.6	4.4 sec.

From this series it will be seen that in the case of the same angle of inclination (II and VI were repeated and III was omitted) the same rate obtained in the tapping. In two other tests I constantly increased the angle of inclination during the 15 taps, and Hans gradually increased the rate of tapping accordingly.

In a second series I had the horse tap 14, five times. I myself took the time of the taps up to 7 by means of the stop-watch, while Prof. Stumpf took the time of the taps from 8 to 13. At 8 I suddenly bent forward a little more and retained this position until tap 13. The results were as follows:

Taps 2 to 7 (Pf.):	3.2	2.2–2.4	2.4	2.2–2.4	2.4 seconds.
" 8 to 13 (St.):	2.6	2.0	2.0	2.2	2.2 seconds.

Such good results, however, were possible only after a number of preliminary practice tests had been made. The experiment was especially difficult because the horse was often on the point of stopping in the midst of a test. This was probably due to some unintentional movement on my part. In such cases I could induce him to continue tapping only by bending forward still more, but this effected also, as we have seen, an increase in his rate of tapping. Such tests, of course, could not give unambiguous results.

The rate of tapping was quite independent of my rate of counting. Thus, if I counted aloud rapidly, but bent forward only very slightly, the horse's tapping was slow and lagged behind my count. If I counted slowly but bent far forward, Hans would tap rapidly and advance beyond my count. Thus we see that his rate of tapping was in accordance with the degree of inclination of my body and never in accordance with the rate of my counting, i. e., it was quite independent of every sort of auditory stimulation.

Direct observation and a comparison of the records of the time Hans required in giving to his master responses involving small, medium and large numbers, with the records of the time which he required to respond to my questions when I bent only slightly, moderately or very far forward, proved that the increased rapidity in tapping in the case of large numbers, which many regarded as an evidence of high intelligence, (see page 20), was, as a matter of fact, brought about in the way described. The two series (in each of which the time measured was for 10 taps) are quite in accord. The horse did not tap faster because he had been given a large number by Mr. von Osten, but because the latter had bent farther forward.

[*Editor's Note:* In the original, material follows this excerpt.]

5C

Reprinted from pp. 240–243 of *Clever Hans (The Horse of Mr. Von Osten): A Contribution to Experimental Animal and Human Psychology*, C. L. Rahn, trans., Henry Holt, New York, 1911, 274 p.

CONCLUSION

O. Pfungst

[*Editor's Note:* In the original, material precedes this excerpt.]

If we would make a brief summary of the status of Mr. von Osten's horse in the light of these investigations and try to understand what is the bearing upon the question of animal psychology in general, we may make the following statements.

Hans's accomplishments are founded first upon a one-sided development of the power of perceiving the slightest movements of the questioner, secondly upon the intense and continued, but equally one-sided, power of attention, and lastly upon a rather limited memory, by means of which the animal is able to associate perceptions of movement with a small number of movements of its own which have become thoroughly habitual.

The horse's ability to perceive movements greatly exceeds that of the average man. This superiority is probably due to a different constitution of the retina, and perhaps also of the brain.

Only a diminshingly small number of auditory stimuli are involved.

All conclusions with regard to the presence of emotional reactions, such as stubbornness, etc., have been shown to be without warrant. With regard to the emotional life we are justified in concluding from the behavior of the horse, that the desire for food is the only effective spring to action.

The gradual formation of the associations mentioned above, between the perception of movement and the movements of the horse himself, is in all probability not to be regarded as the result of a training-process, but as an unintentional by-product of an unsuccessful attempt at real education, which, though in no sense a training-process, still produced results equivalent to those of such a process.

All higher psychic processes which find expression in the horse's behavior, are those of the questioner. His relationship to the horse is brought about almost wholly by involuntary movements of the most minute kind. The interrelation existing between ideas having a high degree of affective coloring and the musculature of the body, (which is brought to light in this process), is by no means a novel fact for us. Nevertheless, it is possible that this case may be of no small value, on account of the great difficulties which are usually met in the attempt to establish experimentally the more delicate details in this field.

And, returning to the considerations of the first chapter, if we ask what contributions does this case make toward a solution of the problem of animal consciousness, we may state the following: The proof which was expected by so many, that animals possess the power of thought, was not furnished by Hans. He has served to weaken, rather than strengthen, the position of these enthusiasts. But we must generalize this negative conclusion of ours with care,—for Hans cannot without further qualification be regarded as normal. Hans is a domesticated animal. It is possible (though the opposite is usually assumed), that our animals have suffered in the development of their mental life, as a result of the process of domestication. To be sure, in some respects they have become more specialized than their wild kin, (e. g., our hunting dogs), and in their habits they have become adapted largely to suit our needs. This latter is shown by all the anecdotes concerning "clever" dogs, horses, etc. But with the loss of their freedom they have also gradually been deprived of the urgent need of self-preservation and of the preservation of their species, and thus lack one of the greatest forces that make for psychic development. And often their artificial selection and culture has been with a view to the development of muscle and sinew, fat and wool, all at the expense of brain development.* Our horses are, as a rule, sentenced to an especially dull mode of life. Chained in stalls (and usually dark stalls at that,) during three-fourths of their lives, and more than any other domestic animal, enslaved for thousands of years by reins and whip, they have become estranged from their natural impulses, and owing to continued confinement they may perhaps have suffered even in their sensory life. A gregarious animal, yet kept constantly in isolation, intended by nature to range over vast areas, yet confined to his narrow courtyard, and deprived of opportunity for sexual activity,—he has been forced by a process of education to develop along lines quite opposite to his native characteristics. Nevertheless, I believe that it is very doubtful if it would have been possible by other methods, even, to call forth in the horse the ability to think. Presumably, however, it might be possible, under conditions and with methods of instruction more in accord with the life-needs of the horse, to awaken in a fuller measure those mental activities which would be called into play to meet those needs.

* Buffon,[124] the great naturalist, expresses himself not less pessimistically in his own brilliant manner: "Un animal domestique est un esclave dont on s'amuse, dont on se sert, dont on abuse, qu'on altère, qu'on dépaïse et que l'on dénature."

Though our investigations do not give support to the fantastic acounts of animal intelligence given by Brehms, they by no means warrant a return to Descartes and his theory of the animal-machine (as is advocated by a number of over-critical investigators). We cannot deny the validity of conclusions from analogy without denying at the same time the possibility of an animal psychology—indeed of all psychology. And all such conclusions indicate that the lower forms possess the power of sense-perception, that they, like us, presumably have at their disposal certain images, and that their psychic life is to a large extent also constituted of mere image-associations, and that they too, learn by experience. Also that they are susceptible to feelings of pleasure and of pain and also to emotions, as jealousy, fear, etc., though these may be only of the kind which have a direct relation to their life-needs. We are in no position to deny *a priori* the possibility of traces of conceptual thought in those forms nearest man in the scale—whether living in their natural manner or under artificial conditions. And even less so since the final word has not yet been spoken regarding the nature of conceptual thinking itself. All that is certain is that nothing of the kind has been proven to occur in the lower forms, and that as yet not even a suitable method of discovering its existence has been suggested. But the community of those elementary processes of mental life which we have mentioned above is in itself enough to connect the life of the lower forms with ours, and imposes upon us the duty of regarding them not as objects for exploitation and mistreatment, but as worthy of rational care and affection.

REFERENCE

124. Buffon, Cte de, et L. Daubenton. Histoire naturelle, generale et particuliere. Paris: Imprimerie royale, 1753, vol. 4, p. 169.

Part II

SYSTEMATIC EXPERIMENTAL ANALYSES OF LEARNING PROCESSES IN ANIMALS

Editor's Comments on Papers 6, 7, and 8

6 **THORNDIKE**
Excerpts from *Experimental Study of Associative Processes in Animals*

7 **SMALL**
Experimental Study of the Mental Processes of the Rat. II

8 **PAVLOV**
The Scientific Investigation of the Psychical Faculties or Processes in the Higher Animals

The easy-going anthropomorphism and chatty anecdotes of the early observers of learning in animals gave way only when an alternative approach became available. The three papers reprinted here represent the foundation on which much of the subsequent study of animal learning processes was built. They demonstrated that animal learning could profitably be studied with behavior as the focus of the inquiry.

Edward L. Thorndike began studying animals early in his distinguished career. With the help of William James, who provided cellar space for housing chickens (see Thorndike, 1936), he began to investigate "trial and error" learning. When a fellowship offer induced him to move to Columbia University, he began his famous experiments with cats in puzzle boxes, publishing them in 1898. These experiments led to the formulation of the well-known "law of effect" (see Paper 9). Although Thorndike did not deny consiousness to animals, he emphasized that a great deal can be learned by concentrating on behavior, rather than trying to study animals' minds. Changes that occur in behavior as a result of experience can, he argued, be used as a basis for the study of the formation of associations (Paper 6). Thorndike's subsequent career was a creative and prolific one. He made major contributions to the fields of human learning, lexicography, mental testing, and educational psychology. Throughout his career his thinking was influenced by the results of his early animal experiments.

In 1901 Willard Stanton Small (Paper 7) introduced the maze

apparatus for the study of learning in animals. Thorndike had used rudimentary barriers of books while studying his chickens in William Jame's cellar, and Robert Yerkes (1901) used a maze-like apparatus in a study of learning in the turtle, published the same year as Small's experiments. Small's apparatus was considerably more elegant; it was modeled after a drawing of the Hampton Court Maze that Small had found in the Encyclopedia Brittanica. It was not merely the elegance of the maze design that made it appealing as a means to study learning in rats—Small noted that the requirements of the task conformed to the "psychobiological character" of the rat. By measuring the speed with which the animals traversed the maze and the numbers of errors they made, Small collected numerical data that could serve as a basis for interpreting the "mental aspect" of the learning process in rats. Like Thorndike, Small studied behavior but did not deny mentation to animals. The success he attained with his maze led to its wide use, with a wide range of species, until the maze became something of a cliche in the animal-learning laboratory. Its extensive use (and perhaps overuse) should not obscure the fact that its initial introduction was a valuable contribution: It provided a means of studying behavior that allowed features of behavior that changed with experience to be measured, quantified, recorded, and analyzed.

At about the same time that Small was testing his rats in his maze, Ivan Pavlov began his systematic investigations of conditional salivation with dogs. At the time he began the work, he did not know of the contributions of Thorndike, Yerkes, and Small, but he later acknowledged their priority. Pavlov was slightly over fifty years old when he began the series of experiments that made him world famous (and infamous!). He was already a respected and honored scientist who had completed important work on the regulation of blood circulation and on gastrointestinal secretion, work for which he was awarded the Nobel prize in 1904. His address in Stolkholm on the occasion of accepting the prize was entitled, "The first Sure Steps Along the Path of a New Investigation." It reported some of the first, exploratory work by Pavlov and his colleagues in the investigation of "psychic" secretion. Already Pavlov had had the important insight that such conditional secretion could be studied through the same methods used to study reflexive secretion. In October, 1906, Pavlov delivered the Huxley lecture at Charing Cross Hospital, and the lecture was published the next month in *Science* (Paper 8). It was largely through this brief report that the work became widely known. It was not until

1927 that a full report of the series of experiments was published in English.

One way to note the tremendous impact that Pavlov had on the study of learning is to point out that many of the concepts he introduced, and the names he attached to them, continue to be used extensively. Conditional stimulus (CS), unconditional stimulus (UCS), extinction, spontaneous recovery, conditioned inhibition, induction, generalization—concepts such as these have molded the experimental study of learning.

One of the many reasons for the enormous influence of Pavlov's work is its organized, systematic character. By the time *Conditioned Reflexes* was published in 1927, Pavlov could give a wide-ranging, detailed account of acquired salivation in his dogs, gained through experiments of "physiological" rather than "psychological" design.

REFERENCES

Pavlov, I., 1928, The first sure steps along the path of a new investigation, Nobel Prize Lecture, 1904, in Gantt, W. H. (trans.), *Lectures on Conditioned Reflexes,* W. H. Gantt, trans., International Publishers, New York.

Pavlov, I., 1927, *Conditioned Reflexes,* Oxford University Press, Oxford.

Thorndike, E. L., 1936, Autobiography, in *A History of Psychology in Autobiography,* vol. 3, C. Murchison ed., Clark University Press, Worcestor, Mass. pp. 263–270.

Yerkes, R. M., 1901, The formation of habits in the turtle, *Pop. Sci. Monthly* **58**:519–525.

6

Reprinted from pp. 29–58 of *Animal Intelligence*, Macmillan, New York, 1911, 297 p.

ANIMAL INTELLIGENCE
An Experimental Study of the Associative Processes in Animals

E. L. Thorndike

[*Editor's Note:* In the original, material precedes this excerpt.]

We may now start in with the description of the apparatus and of the behavior of the animals.[1]

Description of Apparatus

The shape and general apparatus of the boxes which were used for the cats is shown by the accompanying drawing of box K. Unless special figures are given, it should be understood that each box is approximately 20 inches long, by 15 broad, by 12 high. Except where mention is made to the contrary, the door was pulled open by a weight attached to a

[1] The experiments now to be described were for the most part made in the Psychological Laboratory of Columbia University during the year '97–'98, but a few of them were made in connection with a general preliminary investigation of animal psychology undertaken at Harvard University in the previous year.

string which ran over a pulley and was fastened to the door, just as soon as the animal loosened the bolt or bar which held it. Especial care was taken not to have the widest openings between the bars at all near the lever, or wire loop, or what not, which governed the bolt on the door.

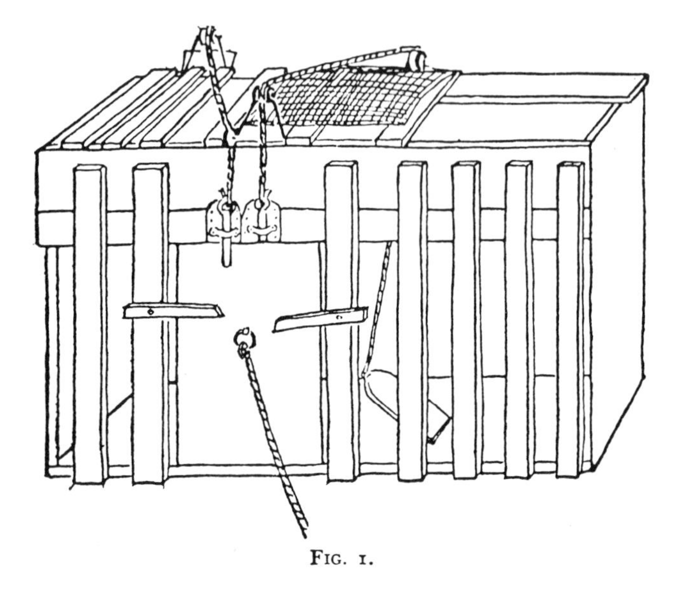

FIG. 1.

For the animal instinctively attacks the large openings first, and if the mechanism which governs the opening of the door is situated near one of them, the animal's task is rendered easier. You do not then get the association-process so free from the helping hand of instinct as you do if you make the box without reference to the position of the mechanism to be set up within it. These various mechanisms are so simple that a verbal description will suffice in most cases. The facts which the reader should note are the nature of the movement which the cat had to make, the nature of the object at which the movement was directed, and the position of the object in the box. In some special cases attention will also be called to the force required. In general, however, that was very slight (20 to 100 grams if applied directly). The various boxes will be designated by capital letters.

A. A string attached to the bolt which held the door ran up over a pulley on the front edge of the box, and was tied to a wire loop (2½ inches in diameter) hanging 6 inches above the floor in front center of box. Clawing or biting it, or rubbing against it even, if in a certain way, opened the door. We may call this box A '*O at front.*'

B. A string attached to the bolt ran up over a pulley on the front edge of the door, then across the box to another pulley screwed into the inside of the back of the box 1¼ inches below the top, and passing over it ended in a wire loop (3 inches in diameter) 6 inches above the floor in back center of box. Force applied to the loop or *to the string* as it ran across the top of the box between two bars would open the door. We may call B '*O at back.*'

B1. In B1 the string ran outside the box, coming down through a hole at the back, and was therefore inaccessible and invisible from within. Only by pulling the loop could the door be opened. B1 may be called '*O at back 2d.*'

C. A door of the usual position and size (as in Fig. 1) was kept closed by a wooden button 3½ inches long, ⅞ inch wide, ½ inch thick. This turned on a nail driven into the box ½ inch above the middle of the top edge of the door. The door would fall inward as soon as the button was turned from its vertical to a horizontal position. A pull of 125 grams would do this if applied sideways at the lowest point of the button 2¼ inches below its pivot. The cats usually clawed the button round by downward pressure on its top edge, which was 1¼ inches above the nail. Then, of course, more force was necessary. C may be called '*Button.*'

D. The door was in the extreme right of the front. A string fastened to the bolt which held it ran up over a pulley on the top edge and back to the top edge of the back side of the box (3 inches in from the right side) and was there firmly fastened. The top of the box was of wire screening and arched over the string ¾ inch above it along its entire length. A slight pull on the string anywhere opened the door. This box was 20 × 16, but a space 7 × 16 was partitioned off at the left by a wire screen. D may be called '*String*.'

D1 was the same box as B, but had the string fastened firmly at the back instead of running over a pulley and ending in a wire loop. We may call it '*String 2d*.'

E. A string ran from the bolt holding the door up over a pulley and down to the floor outside the box, where it was fastened 2 inches in front of the box and 1½ inches to the left of the door (looking from the inside). By poking a paw out between the bars and pulling this string inward the door would be opened. We may call E '*String outside*.'

In F the string was not fastened to the floor but ended in a loop 2½ inches in diameter which could be clawed down so as to open the door. Unless the pull was in just the right direction, the string was likely to catch on the pulley. This loop hung 3 inches above the floor, and 1¾ inches in front of the box. We may call F '*String outside unfastened*.'

G was a box 29 × 20½ × 22½, with a door 29 × 12 hinged on the left side of the box (looking from within), and kept closed by an ordinary thumb latch placed 15 inches from the floor. The remainder of the front of the box was closed in by wooden bars. The door was a wooden frame covered with screening. It was *not* arranged so as to open as soon as the latch was lifted, but required a force of 400 grams, even when applied to the best advantage. The bar of the thumb latch, moreover, would fall back into place again unless the door were pushed out at least a little. The top of this box was not of bars or screening, but solid. We may call G '*Thumb latch.*'

H was, except for the opening where the door was situated, a perfectly solid and dark box. In the front was cut an opening about 9 × 7 inches. A wooden frame covered with wire netting hung in front of this. It was fastened to the box only by a screw through the middle of the frame's top piece, and could therefore be pushed to either side so as to permit escape from the box if it were prevented from swinging back into place.

I was a box 12 × 14 × 23. The door was 8 inches wide, 4 high, and hinged at the left side. It was held closed by a wooden bar which moved easily on a pivot and which could be pushed up by another bar which projected 5 inches into the box. This second bar was pivoted so that downward pressure on it pushed the first bar up and let the door swing open. The second bar entered the box at a point 4 inches above the floor and 2½ inches in from the right side of the box. In its normal position its inner end was 5½ inches above the floor. A depression of 2 inches at that end was necessary to open the door. Of course, nearer the pivot a shorter depression would do. The front of the box was closed by bars, but the rest by solid boards. We may call I '*Lever.*'

J was the same as B except that the door was not opened by a weight as soon as the bolt was pulled up. On the contrary, the door was held closed by a small piece of board (4 × 3½ × ¾ thick) placed against it outside. After pulling the loop at the back the cat had to knock down this support and push the door open. We may call J '*Double.*'

K was a box arranged so that three separate acts were required to open the door, which was held by two bolts at the top and two bars outside. One of the bolts was connected with a platform in the back center of the box so that depressing the platform raised the bolt. The other was raised by a string which ran up over a pulley in the front, across the box 1 inch above the bars, over a pulley near the corner of the box, and down to the floor, where it was fastened. Pulling on this string, either by clawing at it where it was running vertically from the last pulley to the floor, or by putting the paw out between the bars which covered the top of the box, and clawing the string downward, would raise the bolt. If both bolts were raised and *either* bar was pushed up or down far enough to be out of the way, the cat could escape. K, or '*Triple*,' as it may be called, is the box reproduced in Figure 1.

L was a box that also required three acts to open the door. It was a combination of A (O at front), D (string), I (lever). The lever or bar to be depressed was 2 inches to the right of the door, which was in the front center. The string to be clawed or bitten ran from front center to back center 1 inch below the top of the box.

Z was a box with back and sides entirely closed, with front and top closed by bars and screening, with a small opening in the left-hand corner. A box was held in front of this and drawn away when the cats happened to lick themselves. Thus escape and food followed always upon the impulse to lick themselves, and they soon would immediately start doing so as soon as pushed into the box. The same box was used with the impulse changed to that for scratching themselves. The size of this box was 15 × 10 × 16.

Experiments with Cats

In these various boxes were put cats from among the following. I give approximately their ages while under experiment.

No. 1. 8–10 months.	No. 7. 3–5 months.
No. 2. 5–7 months.	No. 8. 6–6½ months.
No. 3. 5–11 months.	No. 10. 4–8 months.
No. 4. 5–8 months.	No. 11. 7–8 months.
No. 5. 5–7 months.	No. 12. 4–6 months.
No. 6. 3–5 months.	No. 13. 18–19 months.

The behavior of all but 11 and 13 was practically the same. When put into the box the cat would show evident signs of discomfort and of an impulse to escape from confinement. It tries to squeeze through any opening; it claws and bites at the bars or wire; it thrusts its paws out through any opening and claws at everything it reaches; it continues its efforts when it strikes anything loose and shaky; it may claw at things within the box. It does not pay very much attention to the food outside, but seems simply to strive instinctively to escape from confinement. The vigor with which it struggles is extraordinary. For eight or ten minutes it will claw and bite and squeeze incessantly. With 13, an old cat, and 11, an uncommonly sluggish cat, the behavior was different. They did not struggle vigorously or continually. On some occasions they did not even struggle at all. It was therefore necessary to let them out of some box a few times, feeding them each time. After they thus associate climbing out of the box with getting food, they will try to get out whenever put in. They do not, even then, struggle so vigorously or get so excited as the rest. In either case, whether the impulse to struggle be due to an instinctive reaction to confinement or to an association, it is likely to succeed in letting the cat out of the box. The cat that is clawing all over the box in her impulsive struggle will probably claw the string or loop or button so as to open the door. And gradually all the other non-successful impulses will be stamped out and the particular impulse leading to the successful act will be stamped in by the resulting pleasure, until, after many trials, the cat will, when put in the box, immediately claw the button or loop in a definite way.

The starting point for the formation of any association in these cases, then, is the set of instinctive activities which are aroused when a cat feels discomfort in the box either because of confinement or a desire for food. This discomfort, plus the sense-impression of a surrounding, confining wall, expresses itself, prior to any experience, in squeezings, clawings, bitings, etc. From among these movements one is selected by success. But this is the starting point only in the case of the first box experienced. After that the cat has associated with the feeling of confinement certain impulses which have led to success more than others and are thereby strengthened. A cat that has learned to escape from A by clawing has, when put into C or G, a greater tendency to claw at things than it instinctively had at the start, and a less tendency to squeeze through holes. A very pleasant form of this decrease in instinctive impulses was noticed in the gradual cessation of howling and mewing. However, the useless instinctive impulses die out slowly, and often play an important part even after the cat has had experience with six or eight boxes. And what is important in our previous statement, namely, that the activity of an animal when first put into a new box is not directed by any appreciation of *that* box's character, but by certain general impulses to act, is not affected by this modification. Most of this activity is determined by heredity; some of it, by previous experience.

My use of the words *instinctive* and *impulse* may cause some misunderstanding unless explained here. Let us, throughout this book, understand by instinct any reaction which an animal makes to a situation *without experience*. It thus includes unconscious as well as conscious acts. Any reaction, then, to totally new phenomena, when first experienced, will be called instinctive. Any impulse then felt will be called an instinctive impulse. Instincts include whatever the nervous system of an animal, as far as inherited, is capable of. My use of the word will, I hope, everywhere make clear what fact I mean. If the reader gets the fact meant in mind it does not in the least matter whether he would himself call such a fact instinct or not. Any one who objects to the word may substitute 'hocus-pocus' for it wherever it occurs. The definition here made will not be used to prove or disprove any theory, but simply as a signal for the reader to imagine a certain sort of fact.

The word *impulse* is used against the writer's will, but there is no better. Its meaning will probably become clear as the reader finds it in actual use, but to avoid misconception at any time I will state now that *impulse* means the consciousness accompanying a muscular innervation *apart from that feeling of the act which comes from seeing oneself move, from feeling one's body in a different position, etc.* It is the *direct feeling of the doing* as distinguished from the *idea of the act done* gained through eye, etc. For this reason I say 'impulse *and* act' instead of simply 'act.' Above all, it must be borne in mind that by impulse I never mean the *motive* to the act. In popular speech you may say that hunger is the impulse which makes the cat claw. That will never be the use here. The word *motive* will always denote that sort of consciousness. Any one who thinks that the act ought not to be thus subdivided into impulse and deed may feel free to use the word *act* for *impulse* or *impulse and act* throughout, if he will remember that the act in this aspect of being felt as to be done or as doing is in animals the important thing, is the thing which gets associated, while the act as done, as viewed from outside, is a secondary affair. I prefer to have a separate word, *impulse*, for the former, and keep the word *act* for the latter, which it commonly means.

Starting, then, with its store of instinctive impulses, the cat hits upon the successful movement, and gradually associates it with the sense-impression of the interior of the box until the connection is perfect, so that it performs the act as soon as confronted with the sense-impression. The formation of each association may be represented graphically by a time-curve. In these curves lengths of one millimeter along the abscissa represent successive experiences in the box, and heights of one millimeter above it each represent ten seconds of time. The curve is formed by joining the tops of perpendiculars erected along the abscissa 1 mm. apart (the first perpendicular coinciding with the y line), each perpendicular representing the time the cat was in the box before escaping. Thus, in Fig. 2 on page 39 the curve marked *12 in A* shows that, in 24 experiences or trials in box A, cat 12 took the following times to perform the act, 160 sec., 30 sec., 90 sec., 60, 15, 28, 20, 30, 22, 11, 15, 20, 12, 10, 14, 10, 8, 8, 5, 10, 8, 6, 6, 7. A short vertical line below the abscissa denotes that an interval of approximately 24 hours elapsed before the next trial. Where the interval was longer it is designated by a figure 2 for two days, 3 for three days, etc. If the interval was shorter, the number of

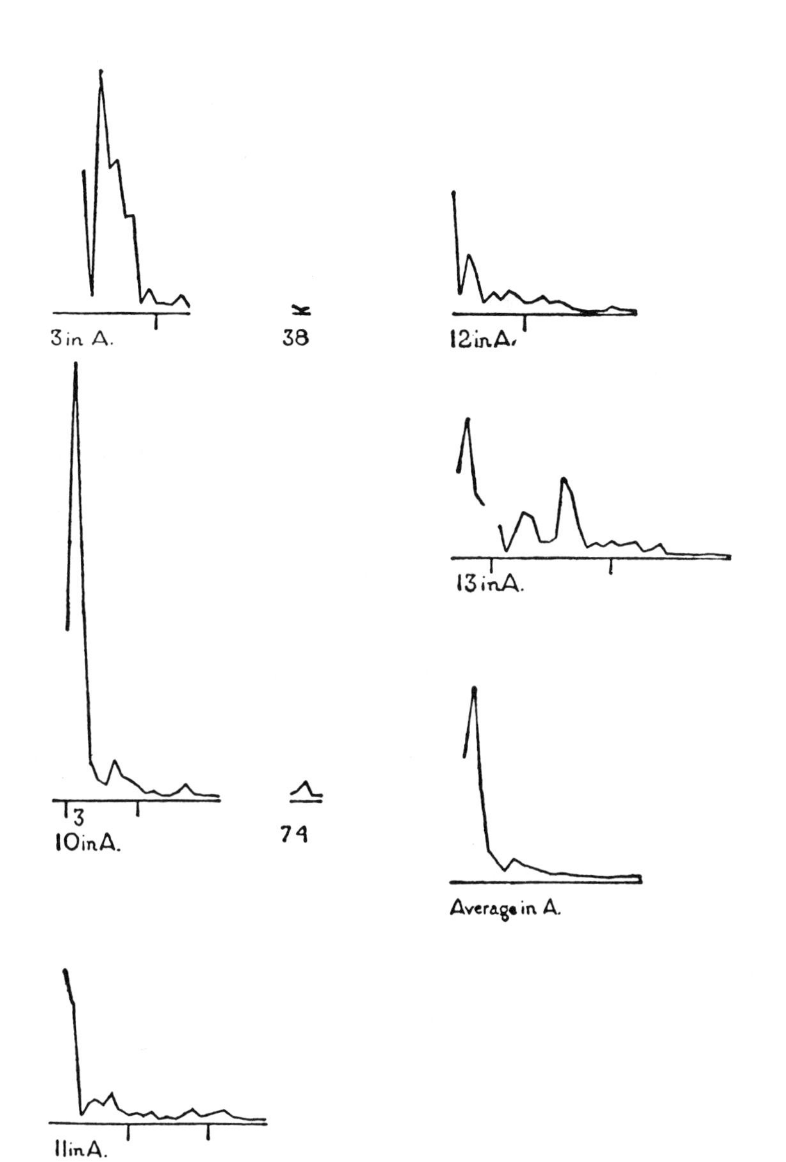

Fig. 2.

hours is specified by 1 hr., 2 hrs., etc. In many cases the animal failed in some trial to perform the act in ten or fifteen minutes and was then taken out by me. Such failures are denoted by a break in the curve either at its start or along its course. In some cases there are short curves after the main ones. These, as shown by the figures beneath, represent the animal's mastery of the association after a very long interval of time, and may be called memory-curves. A discussion of them will come in the last part of the chapter.

The time-curve is obviously a fair representation of the progress of the formation of the association, for the two essential factors in the latter are the disappearance of all activity save the particular sort which brings success with it, and perfection of that particular sort of act so that it is done precisely and at will. Of these the second is, on deeper analysis, found to be a part of the first; any clawing at a loop except the particular claw which depresses it is theoretically a useless activity. If we stick to the looser phraseology, however, no harm will be done. The combination of these two factors is inversely proportional to the time taken, provided the animal surely wants to get out at once. This was rendered almost certain by the degree of hunger. Theoretically a perfect association is formed when both factors are perfect, — when the animal, for example, does nothing but claw at the loop, and claws at it in the most useful way for the purpose. In some cases (*e.g.* 2 in K on page 53) neither factor ever gets perfected in a great many trials. In some cases the first factor does but the second does not, and the cat goes at the thing not always in the desirable way. In all cases there is a fraction of the time which represents getting oneself together after being dropped in the box, and realizing where one is. But for

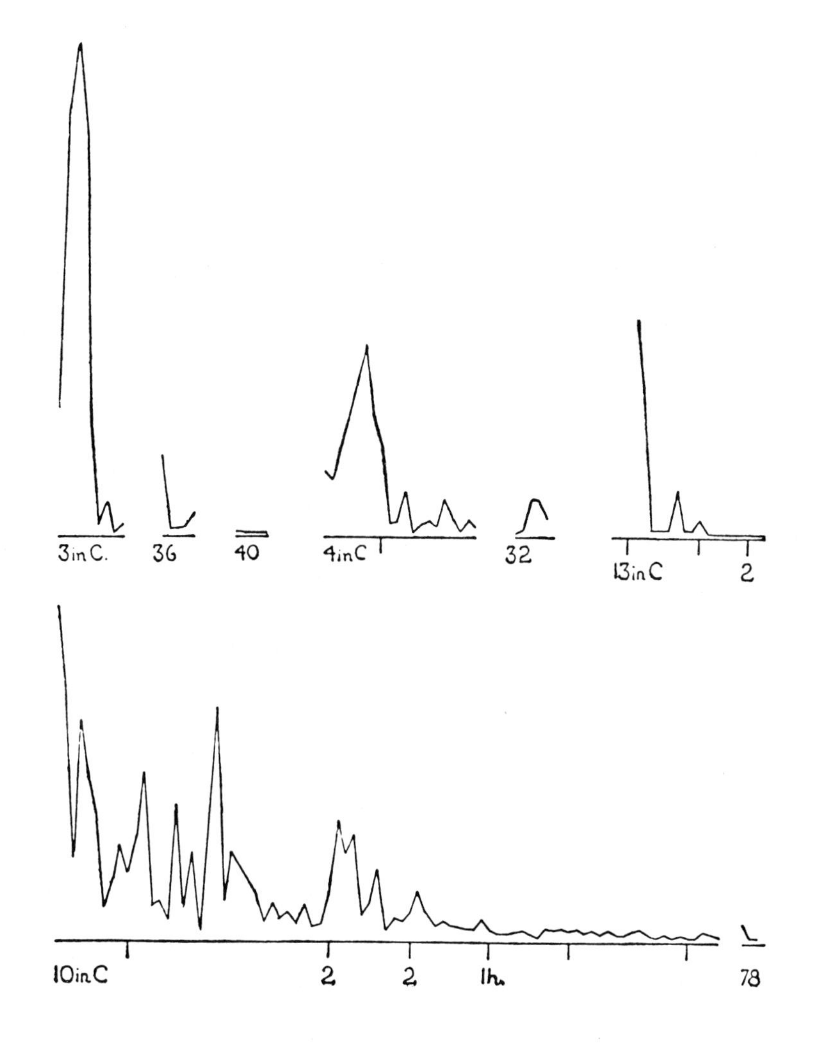

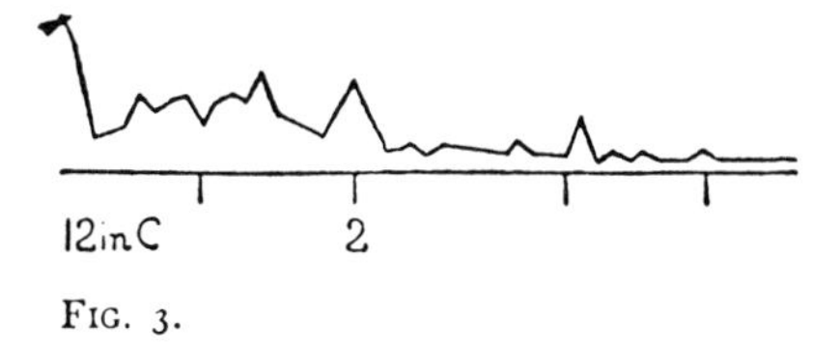

FIG. 3.

our purpose all these matters count little, and we may take the general slope of the curve as representing very fairly the progress of the association. The slope of any particular part of it may be due to accident. Thus, very often the second experience may have a higher time-point than the first, because the first few successes may all be entirely due to accidentally hitting the loop, or whatever it is, and whether the accident will happen sooner in one trial than another is then a matter of chance. Considering the general slope, it is, of course, apparent that a gradual descent — say, from initial times of 300 sec. to a constant time of 6 or 8 sec. in the course of 20 to 30 trials —represents a difficult association; while an abrupt descent, say in 5 trials, from a similar initial height, represents a very easy association. Thus, 2 in Z, on page 57, is a hard, and 1 in I, on page 49, an easy association.

In boxes A, C, D, E, I, 100 per cent of the cats given a chance to do so, hit upon the movement and formed the association. The following table shows the results where some cats failed: —

TABLE I

	No. Cats Tried	No. Cats Failed
F	5	4
G	8	5
H	9	2
J	5	2
K	5	2

The time-curves follow. By referring to the description of apparatus they will be easily understood. Each mm. along the abscissa represents one trial. Each mm. above it represents 10 seconds.

These time-curves show, in the first place, what associa-

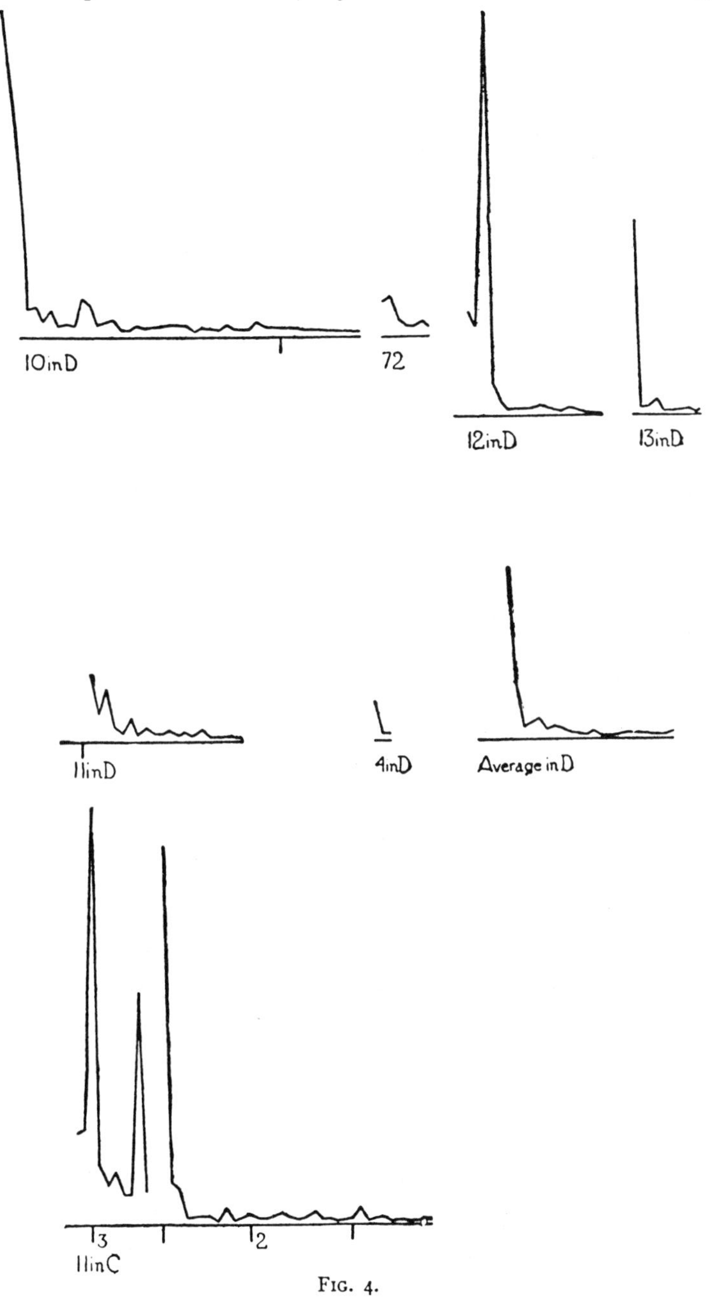

FIG. 4.

tions are easy for an animal to form, and what are hard. The act must be one which the animal will perform in the course of the activity which its inherited equipment incites or its previous experience has connected with the sense-impression of a box's interior. The oftener the act nat-

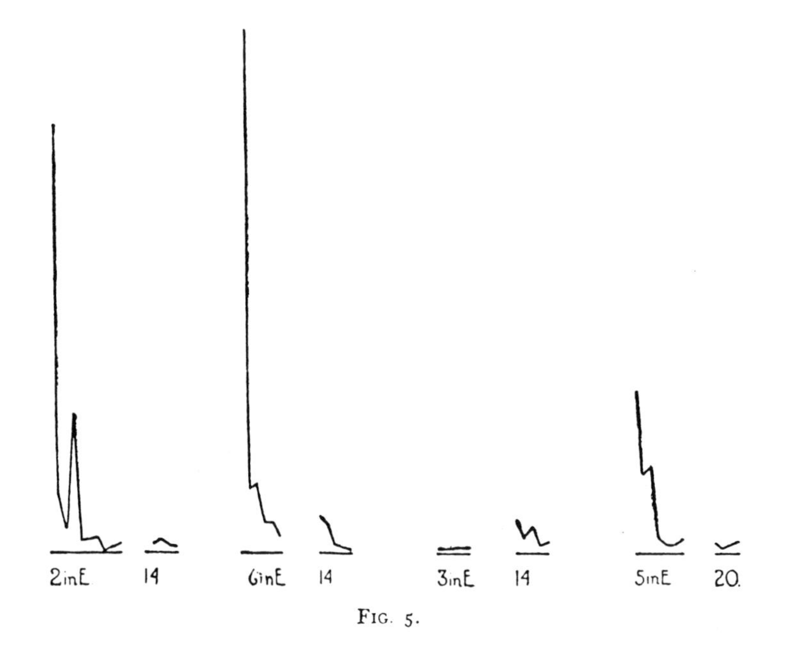

Fig. 5.

urally occurs in the course of such activity, the sooner it will be performed in the first trial or so, and this is one condition, sometimes, of the ease of forming the association. For if the first few successes are five minutes apart, the influence of one may nearly wear off before the next, while if they are forty seconds apart the influences may get summated. But this is not the only or the main condition of the celerity with which an association may be formed. It depends also on the amount of attention given to the act. An act of the sort likely to be well attended to will be learned

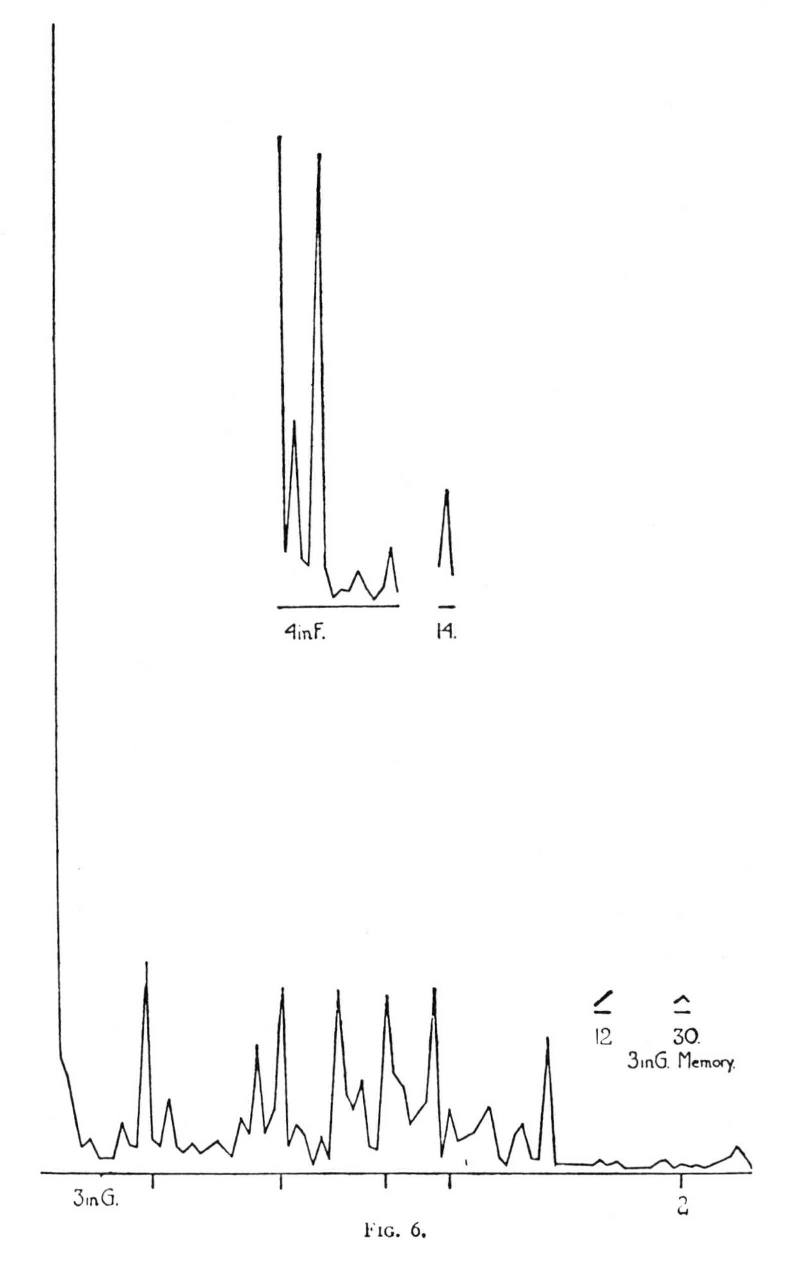

Fig. 6.

more quickly. Here, too, accident may play a part, for a cat may merely happen to be attending to its paw when it claws. The kind of acts which insure attention are those where the movement which works the mechanism is one which the cat makes definitely to get out. Thus A (O at front) is easier to learn than C (button), because the cat does A in trying to claw down the front of the box and so is attending to what it does; whereas it does C generally in a vague scramble along the front or while trying to claw outside with the other paw, and so does not attend to the little unimportant part of its act which turns the button round. Above all, *simplicity* and *definiteness* in the act make the association easy. G (thumb latch), J (double) and K and L (triples) are hard, because complex. E is easy, because directly in the line of the instinctive impulse to try to pull oneself out of the box by clawing at anything outside. It is thus very closely attended to. The extreme of ease is reached when a single experience stamps the association in so completely that ever after the act is done at once. This is approached in I and E.

In these experiments the sense-impressions offered no difficulty one more than the other.

Vigor, abundance of movements, was observed to make differences between individuals in the same association. It works by shortening the first times, the times when the cat still does the act largely by accident. Nos. 3 and 4 show this throughout. Attention, often correlated with lack of vigor, makes a cat form an association more quickly after he gets started. No. 13 shows this somewhat. The absence of a fury of activity let him be more conscious of what he did do.

The curves on pages 57 and 58, showing the history of cats 1, 5, 13 and 3, which were let out of the box Z when

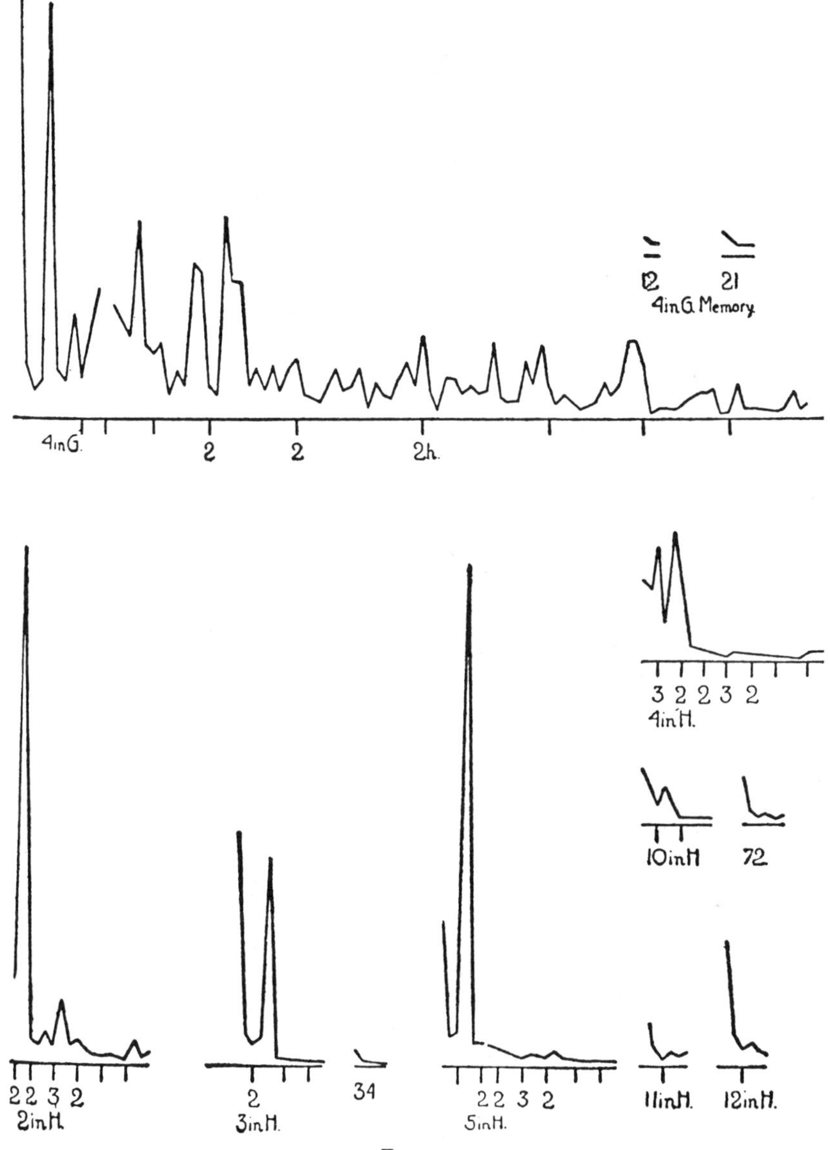
12
21
4in G. Memory
4in G.
2
2
2h.
3 2 2 3 2
4in H.
10in H
72
2 2 3 2
2in H.
2
3in H.
34
2 2 3 2
5in H.
11in H.
12in H.

Fig. 7.

they licked themselves, and of cats 6, 2 and 4, which were let out when they scratched themselves, are interesting because they show associations where there is no congruity (no more to a cat than to a man) between the act and the result. One chick, too, was thus freed whenever he pecked at his feathers to dress them. He formed the association, and would whirl his head round and poke it into his feathers as soon as dropped in the box. There is in all these cases a noticeable tendency, of the cause of which I am ignorant, to diminish the act until it becomes a mere vestige of a lick or scratch. After the cat gets so that it performs the act soon after being put in, it begins to do it less and less vigorously. The licking degenerates into a mere quick turn of the head with one or two motions up and down with tongue extended. Instead of a hearty scratch, the cat waves its paw up and down rapidly for an instant. Moreover, if sometimes you do not let the cat out after this feeble reaction, it does not at once repeat the movement, as it would do if it depressed a thumb piece, for instance, without success in getting the door open. Of the reason for this difference I am again ignorant.

Previous experience makes a difference in the quickness with which the cat forms the associations. After getting out of six or eight boxes by different sorts of acts the cat's general tendency to claw at loose objects within the box is strengthened and its tendency to squeeze through holes and bite bars is weakened; accordingly it will learn associations along the general line of the old more quickly. Further, its tendency to pay attention to what it is doing gets strengthened, and this is something which may properly be called a change in degree of intelligence. A test was made of the influence of experience in this latter way by putting two groups of cats through I (lever), one group

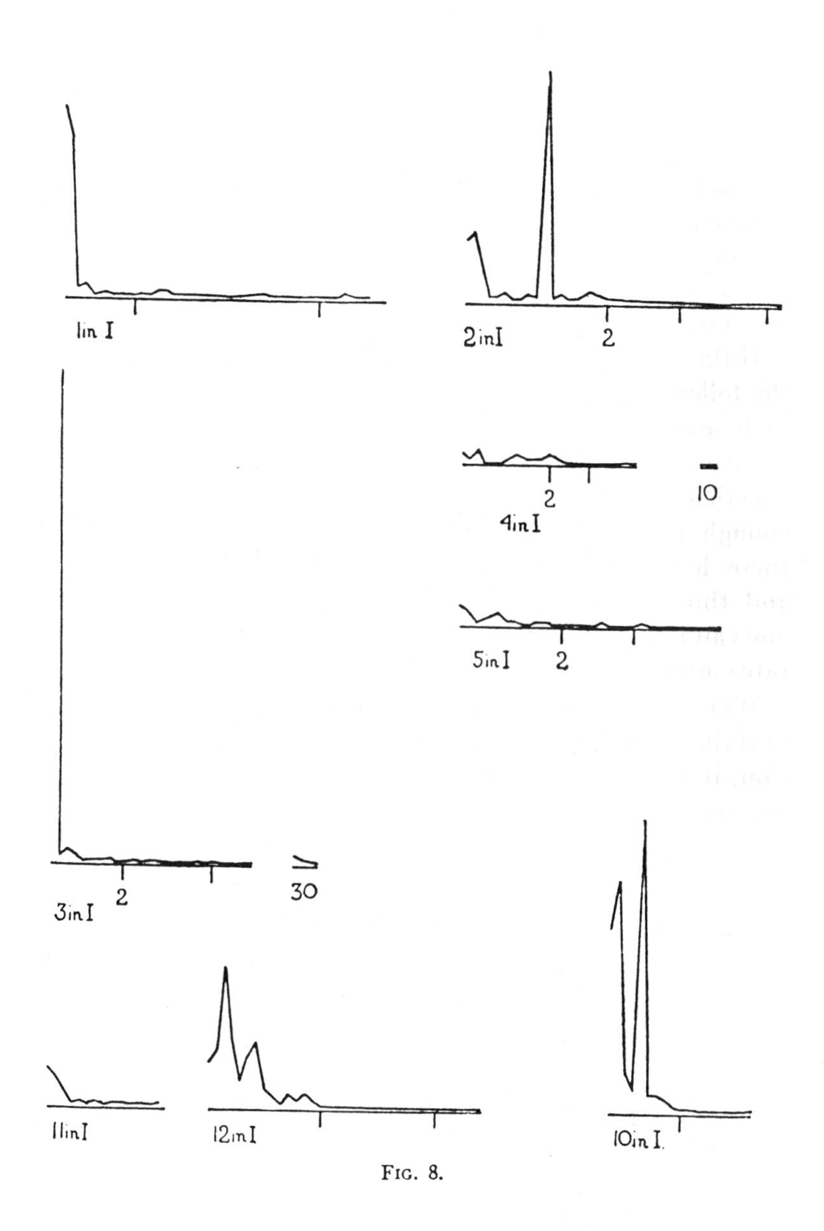

Fig. 8.

(1, 2, 3, 4, 5) after considerable experience, the other (10, 11, 12) after experience with only one box. As the act in I was not along the line of the acts in previous boxes, and as a decrease in the squeezings and bitings would be of little use in the box as arranged, the influence of experience in the former way was of little account. The curves of all are shown on page 49.

If the whole set of curves are examined in connection with the following table, which gives the general order in which each animal took up the different associations which he eventually formed, many suggestions of the influence of experience will be met with. The results are not exhaustive enough to justify more than the general conclusion that there is such an influence. By taking more individuals and thus eliminating all other factors besides experience, one can easily show just how and how far experience facilitates association.

When, in this table, the letters designating the boxes are in italics it means that, though the cat formed the association, it was in connection with other experiments and so is not recorded in the curves.

TABLE II

Cat 1	*A B C* D_1 *D* Z I
Cat 2	*C* D_1 *D* E Z H J I K
Cat 3	A C E G H J Z I K
Cat 4	C F G D Z H J I K
Cat 5	*C* E Z H I
Cat 6	*A C* E Z
Cat 7	*A C*
Cat 10	C I A H D L
Cat 11	C I A H D L
Cat 12	C I A H D L
Cat 13	A C D G Z

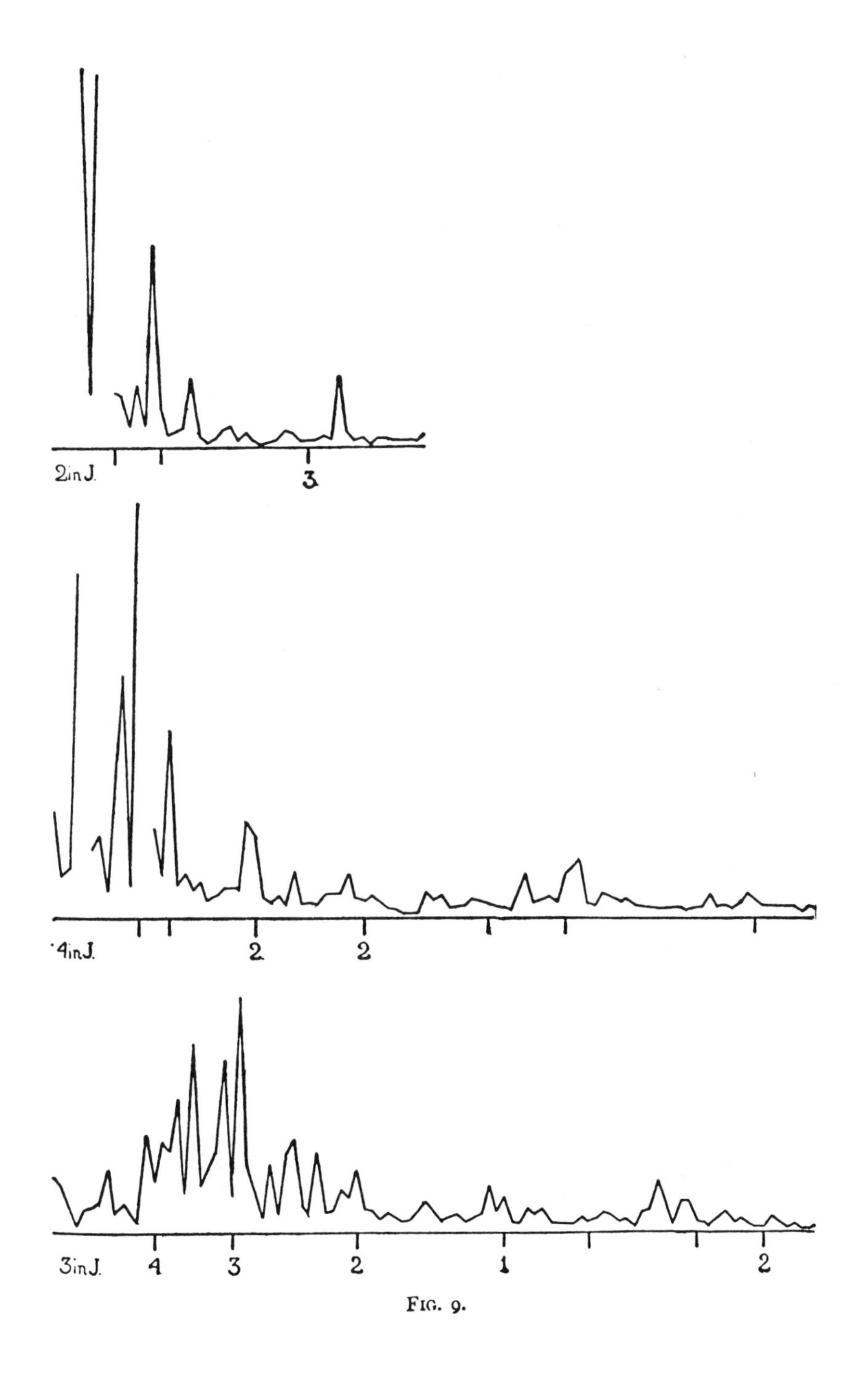

FIG. 9.

The advantage due to experience in our experiments is not, however, the same as ordinarily in the case of trained animals. With them the associations are with the acts or voice of man or with sense-impressions to which they naturally do not attend (*e.g.* figures on a blackboard, ringing of a bell, some act of another animal). Here the advantage of experience is mainly due to the fact that by such experience the animals gain the habit of attending to the master's face and voice and acts and to sense-impressions in general.

I made no attempt to find the differences in ability to acquire associations due to age or sex or fatigue or circumstances of any sort. By simply finding the average slope in the different cases to be compared, one can easily demonstrate any such differences that exist. So far as this discovery is profitable, investigation along this line ought now to go on without delay, the method being made clear. Of differences due to differences in the species, genus, etc., of the animals I will speak after reviewing the time-curves of dogs and chicks.

In the present state of animal psychology there is another value to these results which was especially aimed at by the investigator from the start. They furnish a quantitative estimate of what the average cat can do, so that if any one has an animal which he thinks has shown superior intelligence or perhaps reasoning power, he may test his observations and opinion by taking the time-curves of the animal in such boxes as I have described.

If his animal in a number of cases forms the associations very much more quickly, or deals with the situation in a more intelligent fashion than my cats did, then he may have ground for claiming in his individual a variation toward greater intelligence and, possibly, intelligence of a different

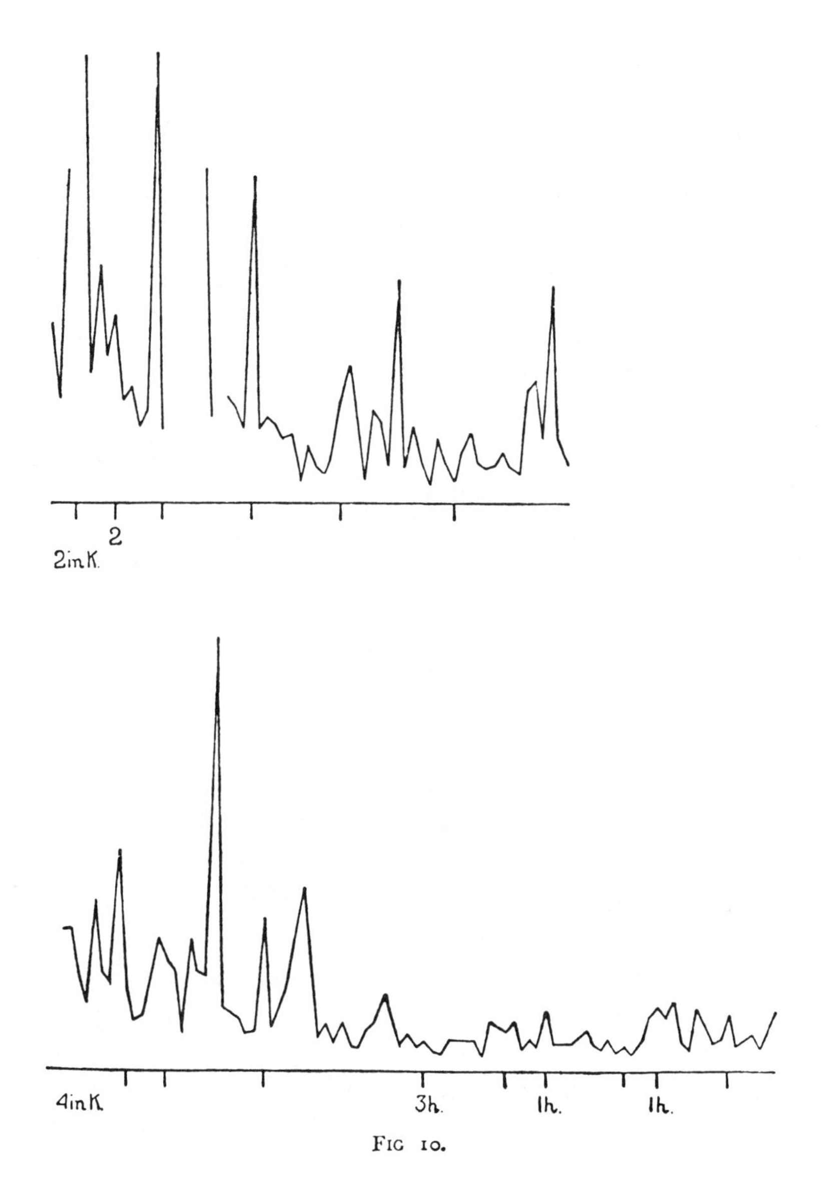

FIG 10.

order. On the other hand, if the animal fails to rise above the type in his dealings with the boxes, the observer should confess that his opinion of the animal's intelligence may have been at fault and should look for a correction of it.

We have in these time-curves a fairly adequate measure of what the ordinary cat can do, and how it does it, and in similar curves soon to be presented a less adequate measure of what a dog may do. If other investigators, especially all amateurs who are interested in animal intelligence, will take other cats and dogs, especially those supposed by owners to be extraordinarily intelligent, and experiment with them in this way, we shall soon get a notion of how much variation there is among animals in the direction of more or superior intelligence. The beginning here made is meager but solid. The knowledge it gives needs to be much extended. The variations found in individuals should be correlated, not merely with supposed superiority in intelligence, a factor too vague to be very serviceable, but with observed differences in vigor, attention, memory and muscular skill. No phenomena are more capable of exact and thorough investigation by experiment than the associations of animal consciousness. Never will you get a better psychological subject than a hungry cat. When the crude beginnings of this research have been improved and replaced by more ingenious and adroit experimenters, the results ought to be very valuable.

Surely every one must agree that no man now has a right to advance theories about what is in animals' minds or to deny previous theories unless he supports his thesis by systematic and extended experiments. My own theories, soon to be proclaimed, will doubtless be opposed by many. I sincerely hope they will, provided the denial is accompanied by actual experimental work. In fact, I shall be tempted

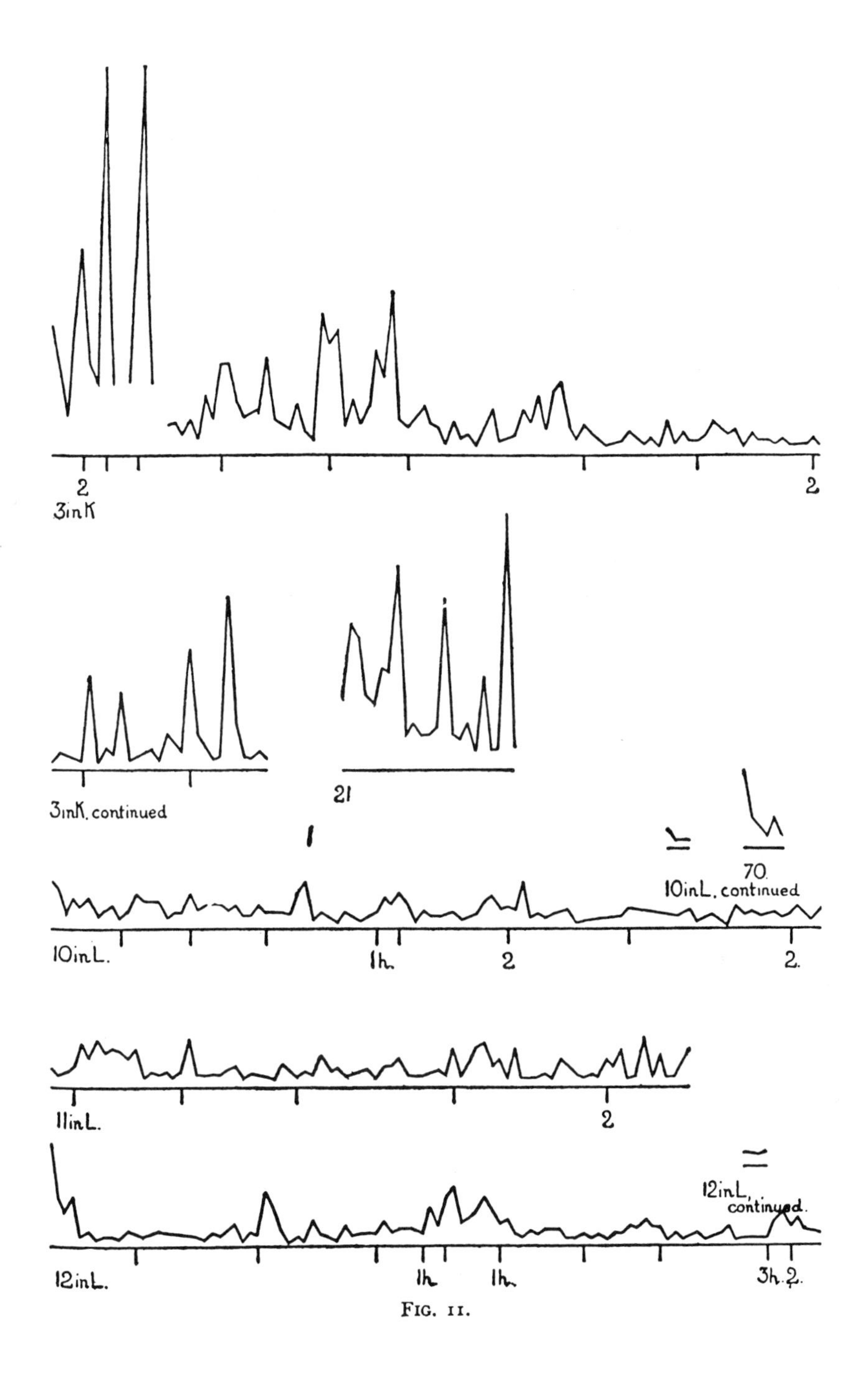
2
2
3inK
3inK. continued
21
70.
10inL. continued
10inL.
1h.
2
2.
11inL.
2
12inL, continued.
12inL.
1h.
1h.
3h.2.

FIG. 11.

again and again in the course of this book to defend some theory, dubious enough to my own mind, in the hope of thereby inducing some one to oppose me and in opposing me to make the experiments I have myself had no opportunity to make yet. Probably there will be enough opposition if I confine myself to the theories I feel sure of.

[*Editor's Note:* In the original, material follows this excerpt.]

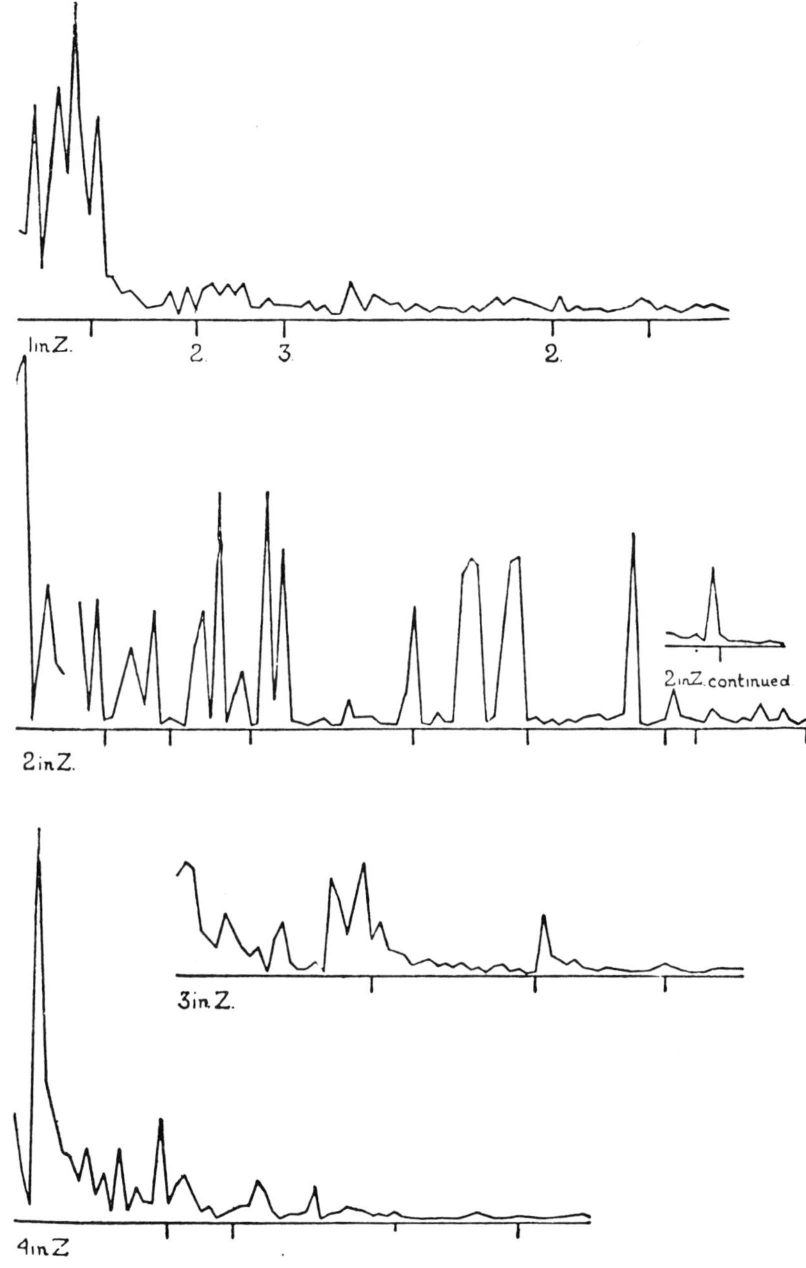

FIG. 12.

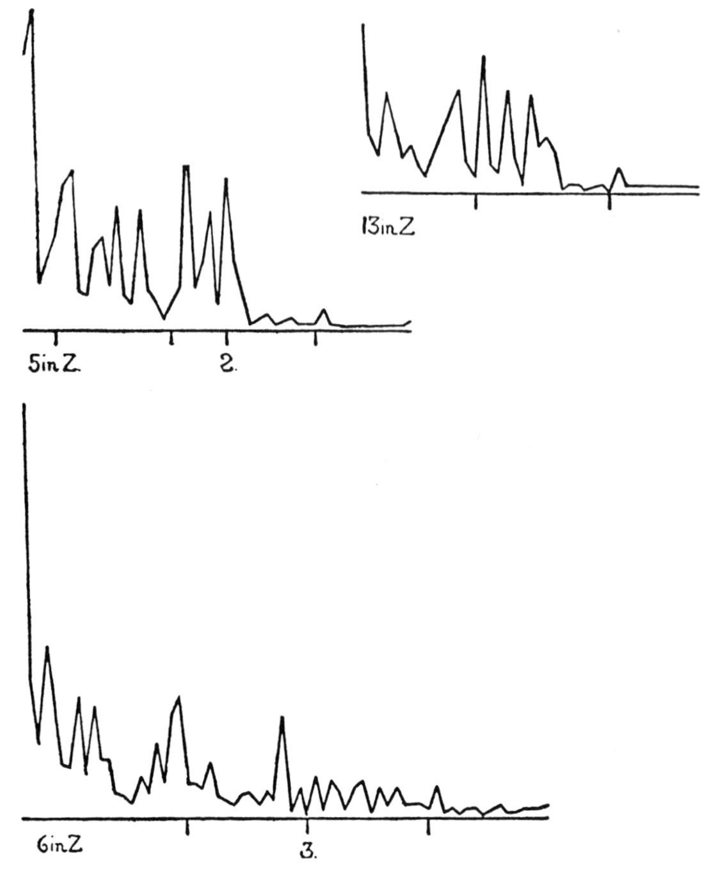

FIG. 13.

7

Reprinted from *Am. J. Psychol.* **12**:206–239 (1901)

EXPERIMENTAL STUDY OF THE MENTAL PROCESSES OF THE RAT. II.

By WILLARD S. SMALL,
Honorary Fellow in Psychology, Clark University.

The present paper, which supplements, in a measure, two former papers, (*Am. Jour. Psy.* Vol. XI, Nos. 1, 2), presents in detail the results of some further experimental studies upon the mental life of the rat. Primarily a study in method, an attempt to observe this animal under approximately experimental conditions, the methods and devices have worked well enough to warrant considerable confidence in the results, and the inferences therefrom. The paper describes the apparatus used and the conditions of experiment, gives a detailed account of a typical series of experiments, compares the intelligence of wild rats and the tame white rats (same variety) upon the basis of these experiments, discusses briefly the general form or character of animal intelligence, and makes some suggestions in regard to the mental facts involved in solving the problems set in the tests. It is not in any sense a systematic rat-psychology.

APPARATUS AND CONDITIONS OF EXPERIMENTS.

The aim in these experiments, as indicated above, was to make observations upon the free expression of the animal's mental processes, under as definitely controlled conditions as possible; and, at the same time, to minimize the inhibitive influence of restraint, confinement and unfamiliar or unnatural circumstances. Fear, which in lack of a better term, may be used to include the three influences just noted, and too great difficulties are the things most rigorously to be guarded against. On the positive side, the experiments must conform to the psycho-biological character of an animal if sane results are to be obtained. This is not the same thing as guarding against "too great difficulties." The difficulty of two tasks, judged by their complexity and the quality of intelligence necessary for their performance, may be identical; yet the problem involved in the one may be so different from that in the other, so remote from the animal's racial experience and life habits as to be absolutely outside his capabilities. A human being has to know how the elements of an action feel—or better, has to know the *feeling* of the elements—in order to perform the action. The importance of these characteristic differences is obvious enough where structural differences provide the basis: no one would

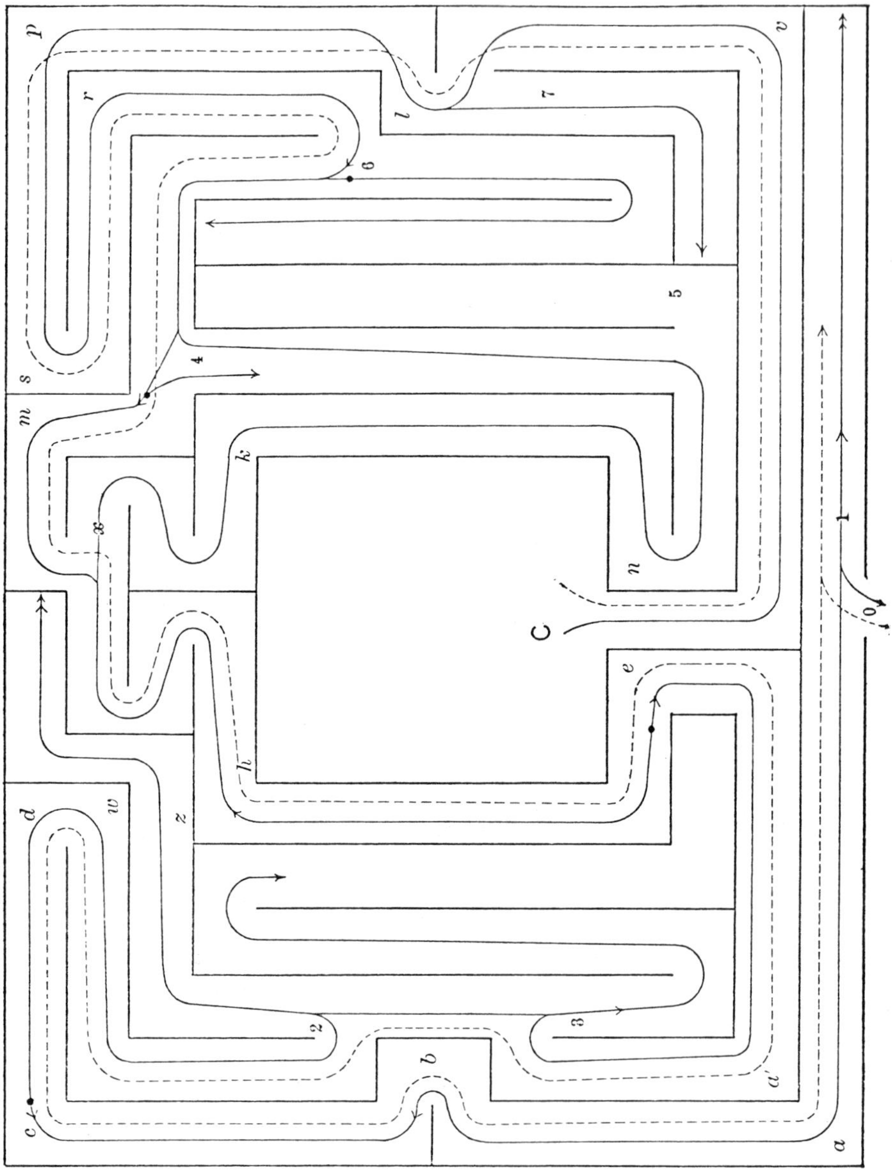

expect a buffalo and a rabbit to do the same things in the same way. No less important, though less easy to define and designate, are the specifically psychic traits of an animal which constitute its character, and which depend in the main upon its biological conditions. The cat and the rat are not antithetical in structure, nor do they differ very widely in degree and kind of intelligence; but their life habits, both on the emotional and the intellectual side, present an effective contrast. The cat, primarily a hunter, is bold, independent, and aggressively open; the rat, on the contrary, primarily the hunted, and only secondarily a hunter, is timid and furtive. His timidity is comparable to an intellectual obsession. His boldness when displayed is impudent and half apologetic, never self-contained and unconscious like that of the cat. I daily see this matter exemplified in the case of a pet rat and a young cat. Though the rat has compelled the cat to respect his rights, the characteristic difference of mental attitude is not greatly changed.[1]

Conforming with such considerations, appeal was made to the rat's propensity for winding passages. A recent magazine article upon the Kangaroo Rat, by Mr. Ernest Seton Thompson[2] illustrates well the radical character of this rodent trait. Mr. Thompson gives a diagram of the Kangaroo Rat's home-burrow, the outline of which bears a striking resemblance to that of the apparatus used in these experiments. It suggests that the experiments were couched in a familiar language. Not only do they conform to the sensori-motor experience of the animals, but they also fall in with their constructive instinct relative to home building.

The Hampton Court Maze[3] served as model for the apparatus. The diagram given in the Encyclopedia Britannica was corrected to a rectangular form, as being easier of construction. The character of the problem was not affected. Three mazes were made. The first was as follows. The dimensions were 6 by 8 feet. The bottom was of wood, the boards being fastened together so as to make a portable whole of the apparatus. All the rest: top, sides, and partitions between galleries were of wire netting, ¼ in. mesh. The height of the sides was 4 inches; the width of galleries, the same. In the center was a large open space. The accompanying diagram gives the ground plan of the maze. The entrance is marked by the figure *O*. Figures *1* to *7* indicate seven blind alleys—seven

[1] The character of the white rat is modified somewhat by its immemorial life of captivity, but less than might be supposed. All the primitive traits emerge under the appropriate conditions.

[2] Scribner's Magazine, April, 1900.

[3] *Cf.* Ency. Brit. Art. "Labyrinth."

possibilities of error. It will be observed that *4* does not lead necessarily into the *cul-de-sac 5*, but does inevitably furnish a chance for error. The letter *x* marks a dividing of the ways either of which, however, may be followed without completely losing the trail. Certain other points are indicated by letters *a*, *b*, *c*, etc., for convenience in description. C stands for center. A glance at the maze will be sufficient to convince one of the difficulty of the problem. This maze is designated as Maze I. The two other mazes were identical with this, except that they were made throughout of wire netting, bottom as well as top and sides. · The apparatus could thus be reversed and a mirror reproduction of the original form could be obtained. In use they were fastened temporarily to wooden floors. These mazes will be spoken of as II and III. In all cases the mazes were placed upon tables 2½ feet above the floor. The floor of the maze was covered with sawdust. This was renewed whenever a new rat was introduced.

To obviate the affective disturbance which would have resulted from change of conditions just prior to experiment, as little difference as possible between the conditions of experiment and of ordinary experience was allowed. The event frequently justified the precaution. The rats were kept in an ordinary observation cage,[1] a well ventilated and commodious apartment, standing flush with the maze at *O*. Back of the cage, upon the same table, was a screen completely concealing the observer during experimentation, but permitting him to look down upon the maze and mark every movement of the rats. When ready for experimenting, the sliding glass door which closed the passage into the maze was lifted by means of a pulley connection running over the screen. This door always opened with slight noise and friction. Food was placed in the central enclosure. To obtain this, the rats had to find their way through the tangled maze. They were kept hungry enough so that they would set about the task vigorously. Experimenting was done in the evening to minimize the inflnence of distracting noises.

In spite of these precautions there was, as might be expected, some lack of uniformity in the results—so much has to be allowed to the internal conditions and to individual variation.

As a beginning, two series of experiments were made; one with wild brown rats, *mus decumanus;* the other with white rats, albinos of the same variety. The original purpose was to use the wild rats exclusively; but the difficulties were considerable, and the white rats were found to serve the purposes of experiment quite as well in every way. In all the experiments after the first series white rats were used. The most of the later

[1] *Cf. Am. Jour. Psy.*, X, 3, p. 135.

experiments were made to determine special points, and will be described in their immediate connections.

As before stated, the rats were let into the maze at the time of experiment and were left until the following morning. The entire night was theirs for investigating the maze. This method excludes the possibility of a quantitative study, and intentionally, for reasons given within.

Description of Experiment.

In presenting the results of these experiments I give a detailed account of a typical series, following this with explanatory suggestions. The conduct of the rat is described as it was minutely recorded at the time of observation. This is supplemented in passing by comparison with other experiments and by details taken from them. This plan of presentation has the advantage of exhibiting in detail the method of animal intelligence in actual operation,—it is analogous with vital staining in histology—rather than in figuring its general form as could be done by the method of graphic representation in curves. Again, in view of the rôle one's own prepossessions and limitations are bound to play in interpretation, it will be an advantage for the reader to know the observed facts upon which the writer's interpretations are based. I readily concede that my facts may be found more valuable than my inferences. Be that as it may, all who are interested in comparative psychology will agree that careful and impartial description of the objective facts of animal intelligence is a desideratum.

I have selected for detailed description my second series of experiments. The subjects were two young male white rats, about 3½ months old. They were perfectly tame—pets from birth.[1] Nine experiments were made at intervals of two days; and a tenth after a lapse of three weeks. The conditions were the same as with the wild rats of the first series.

Before giving this detailed description I wish to present a brief summary of the observations upon the wild rats in series 1, in order to make clear the passing comparison, given within, between the wild and the tame rats. This comparison is purely a by-product of the study—it was not contemplated at all in the beginning—and it takes us a little out of the straight line of exposition; but the results have a suggestive value sufficient to warrant the delay.

The subjects of Series I were three adult wild rats—a male and two females. They had been in captivity about three months, but were in perfect physical condition. No particular

---

[1] Rats A and B of Group V, reported in *Amer. Jour. Psy.*, Vol. XI, No. 2, pp. 156 ff.

effort had been made to tame them except by excluding as far as possible occasion of fright. They remained so wild, however, as to necessitate the extreme precautions described above.

The first two experiments failed of definite results because of imperfect conditions. The rats were frightened by external disturbances, so that they failed to move (Exp. 1,) for more than 10 m., and did not reach C in two hours. This they accomplished during the night in each case. It was not until Exp. 3 that the conditions were made to conform perfectly to the description above. After that, observation was made and results recorded more satisfactorily. The influence of the first two trials is apparent in Exp. 3, the journey being made in 15 m., and with 8 errors.[1]

In this experiment and in all succeeding experiments of the series the male showed striking superiority to the females. In this case they did not reach C until 15 m. after the male—indeed they did not leave the cage till after he had reached C and was well on his way back to the cage. He met one of the females at *c*. Under the same stress of hunger, they exhibited throughout the series very much less boldness and initiative. This humbler character comes out again in the deference the females showed the male. Whenever he was monopolizing the food they would make furtive attemps to steal the prize, or approach with deprecating squeaks as if begging to be allowed to share. On several occasions I saw him chastise a female severely for trying to get the food away. The females, too, were less apt in familiarizing themselves with the maze. In the fifth experiment, I observed that they showed little familiarity or assurance, and seemed much more disconcerted than the male when they got into a blind alley. The male died after Exp. 7, but even after that the improvement of the females was relatively slow. Four more experiments showed slight gain. Their timidity seemed to abate but little, and prevented them from giving full attention to the problem.

Later tests with male and female white rats gave the same general result, though the difference was not so marked. In only one case out of ten or twelve pairs observed did the female show equal initiative. The significance of this is not, I think, that the intelligence of the males is higher or more refined, but, rather, that it is more effectual by reason of being less subject to affective disturbances and inhibitions. The case is analogous with a point made later in comparing the white and the gray rats: the appearance of greater intelligence really may mean

[1] By errors I mean the returns—*e. g.*, going back from *d* to *a*—as well as the positive errors of following wrong paths.

nothing more than that one animal is somewhat more energetic than another, moves more and faster, and consequently has more chances of succeeding in a trial and error series.

It is interesting to note, also, that domestication tends to reduce this disparity between the sexes. It suggests the very live question whether civilization, which is a kind of domestication, operates similarly with the human species. Popular opinion seems to lean towards the affirmative; and we find expression of this movement of the social mind in the tendency to identify the aims, conquests, and education of the sexes. On the other hand anthropologists recognize a progressive differentiation of the sex types, physiologically and biologically. Possibly a point in development has been reached where psychological development has attained a greater freedom than heretofore from physiological and biological determinants.

But to return from this digression. The male (m) showed the influence of the previous experience not only in the reduction of time and errors, but also manifestly by avoiding several errors and by hesitation at certain points. He even returned to the cage without any mistakes. In some parts of the maze the movement had some definiteness, but generally had a groping appearance. This in itself is significant, suggesting that already he passed beyond the stage of purely accidental selection, that he had already acquired some kind of recognition, however vague, of rights and wrongs. The selection of paths begins to be purposive. In the four succeeding trials, in which this rat was a subject, there was observable a constant increase in the purposiveness of the movements. This takes the form of definiteness and speed where no error was possible, and of doubt and indecision where there was choice of path. The indecision tended to fade away, but did not disappear completely in this (too) brief series.

If the matter were illustrated graphically by curves, we should have the curve of speed and definiteness beginning with a maximum of slowness and complete indefiniteness, and sweeping down gradually to relatively perfect definiteness and a minimum of time. The curve of choice, on the contrary, would begin very low, rise gradually as the recognition of critical points became more acute, and would then fall gradually as the right association became habitual.

As further illustrative of this increasing purposiveness I note from the records the following pertinent observations. In Exp. 4, "one could not fail to see that the rat was trying to select his path—there seemed to be some kind of an image in his mind that he was trying to follow. The first time he came to *4* he hesitated, then went wrong; each succeeding time he seemed to recognize this place, for he went by confidently.

Indeed after the first time around he seemed sure of his path beyond this point, and at *s* visibly accelerated his pace each time." In Exp. 5, he made fewer errors, and recoiled more quickly from an error when made. Movements were all rapid, yet cautious. Again made "the error at *4*, and again quickened his pace as he passed it the second time. This acceleration was accompanied by a flick of the tail and a general abandon that said, 'I've struck the right trail.'" The movements in Exp. 6 "were purposive. He apparently knew when he was on the right and when on the wrong road. The right (latter) half of the maze seemed better in mind than the left; before reaching *m* the movements were cautious and uncertain—several mistakes—but beyond *m*, they were rapid and secure—no mistakes." Exp. 7 gave similar results.[1] There is little reason to doubt that a few more trials would have enabled him to learn the way perfectly as did the white rats in many later experiments.

The time and error factors of the five experiments, which were accurately recorded, were as follows:[2] III—15, 8′. IV—10, 2′. V—1¾, 4′. VI—3, 4′. VII—⅚, 3′. It will be seen that the relative decrease in time is more nearly constant than that in number of errors—*e. g.*, Exp. *4*, which required 10 m., gave only 2 errors; Exp. 7, requiring only 50 seconds, gave 3 errors. Probably the small number of errors in Exp. *4* is largely accidental. There is doubtless a somewhat constant relation between knowledge of the path (indicated by relative number of errors) and the time required to traverse it. It must be remembered, however, that organic conditions vary so as to make this relation practically indemonstrable.

In this series of experiments, the writer failed to observe any thing that would indicate that the sense of smell played an important rôle in the process of solving the problem, at least so far as following a trail is concerned. In no case could I make out that the male followed his own latest track or the females followed the track of the male, which often would have led them almost directly to the goal. All later experiments confirmed this view. The evidence is given in detail in its proper place *infra*.

It should be remarked, also, that the females gave no evidence of intelligently following the male, when by doing so they might have gone directly to C. The rats passed and repassed each other, each going his or her own way. Occasionally one would be deflected from his path to follow another, but seldom for any considerable distance. This same independence was observed

[1] Reason for discontinuance was death of the rat—disease unknown.

[2] Roman numerals indicate the experiment number; the plain Arabic, the time in minutes; the accented Arabic, the number of errors.

in all later experiments, so that I think imitation even in its simplest form[1] cuts a very insignificant figure in this matter.

One other matter of interest is the general conduct of the rats in the maze. I have spoken already of the initial timidity. The manner soon became more confident, affective tension was relaxed, and curiosity and the play instinct were unloosed. In Exp. *3*, after the rat[2] had eaten and drunk a little, he seemed to become thoroughly happy, and, for the nonce, quite oblivious of the world of traps and snares. He alternately ate, drank, and ran curiously about the maze, investigating all the passages in a sprightly and eager manner. In Exp. *4*, I note "the restless curiosity they all manifest after the first pangs of hunger are appeased, carefully exploring all the passages, feeling and sniffing along the sides and tops and into all the angles and corners. Their conduct impresses one strongly as being the expression of a free curiosity—a fundamental and irreducible desire to know all this new environment. It has not at all the appearance of a further search for food, but rather a seeking for the feeling of security and at-homeness attendant upon knowledge of surroundings.

I pass on now to describe in detail Series II, with white rats.

Exp. 1. *A.* 13 m. *A* came out of cage 2 m. after the door was opened. A few steps into *1*—turned and forward to *b*—back, end of *1*—into cage—end of *1*—to end of *2* without pause—into *3*, half to end—turned and forward to *4*—a few steps into *4*—turned, foward to *6*—a pause, then to end of *6*—out and back, via *4*, *n*, *k*, *x*, as far as *c*. A slight pause, then forward to end of *2* again—out, and a few steps into *3*—back and thence to *h*, pausing a moment at *e* to sniff and dig—back to *e*—forward continuously to end of *7* (very slowly by *4*, as if having some recognition of the place, sniffing cautiously from side to side of gallery)—back to *6*—thence forward to C.

Rat *B* up to this time had made little progress. *A* soon came out of C, and for 15 m. the two rats ran about the maze, most of the time in the circuit immediately around the center, *i. e.*, *z-h-x-k-n*, fagging back and forth, digging at the base of the wall and biting at the wires—the depth of stupidity, one would say. *B* reached C 15 m. later, and *A*, a second time, a minute later than *B*.

It is well briefly to notice the points of interest presented in this first experiment. (1) In regard to timidity they presented a contrast with the wild rats of the preceding series. They spent little or no time looking and listening. Their movements were free and unconstrained. (2) Their movements were a trifle less vigorous than those of the brown male. They go about the business more deliberately; their movements appear less automatic. (3) At the turns in the road, the rats frequently stopped, as if in doubt which way to go. (4) The meaning of the sudden stops and turns is not very intelligible. A rat going at a good pace through an unobstructed gallery may stop suddenly or wheel about and go in the opposite direction. This indecisiveness is a constant trait of the rat character, as had been remarked in all my preceding observations and experiments. Other than this

[1] *Amer. Jour. Psy.*, Vol. XI, No. 2, p. 162.
[2] I refer solely to the male in these illustrative examples.

flightiness or unstable attention, no explanation of these movements offers itself. I do not regard this as ultimate; but for the present, at least, it must serve, as the conditions of attention are not apparent. (5) The fortuitousness of the first success is evident. The great number of errors, the repetition of the same errors, the marching and countermarching, and the general appearance of *lostness* amounted to a demonstration of the accidental nature of this first success. (6) It appears also that the very first experience was little profited by. It was noted that *A*, in course of his first wandering journey to C, did seem vaguely to recognize the locality of *4* when he approached it the second time—his slow and cautious movements seeming to indicate that—and there was a suspicion of a similar recognition at 3; but these were the only suggestions of having 'profited by experience,' and after leaving C, *A* wandered just as blindly as *B*, when trying to return.

Most of these points will come up for consideration in the fuller discussion later on.

Exp. 2. A, 3 m. The two rats kept together until *a'*, having gone to the end of *1*, but having avoided *2* and *3*. *B* paused to pick up a crumb. *A*, forward without pausing, by direct path, *i. e.*, *x-m*, to end of *6*—turned instantly and went like a flash to *7*—thence more slowly to C, pausing but once (to eat a crumb) on the way. *B*, after eating his crumb, went forward slowly and carefully to end of *6*. He looked as though he were following *A's* trail by scent. Here an insane suggestion seemed to grip him (there was no external occasion for fright); he wheeled, made a wild rush, and bit up at *h*. He then ran about foolishly for 15 m, twice going again to the end of *6*. The last time, he flashed out much as *A* had done, not pausing until *r*—thence slowly, as if following a trail. At *s*, *A* and *B* met, stopped a minute to play, then each went his own way, *B* still appearing to follow the trail.

In this second experiment the noticeable things are the immense improvement shown by *A*, and the appearance of "trailing" by *B*. Three possible explanations of *A's* improvement suggest themselves: (1) the right path is selected by sense of smell; (2) by lucky accident; (3) the path is identified by other means than smell. A fourth explanation might be one that included these three as variable factors. I defer discussion of this point, merely indicating that I do not regard the first factor (smell) as important, in spite of the presupposition in its favor, and apparent following of the trail by *B*, noted above.

Exp. 3. A, 4 m. Distraction on part of the rat prevented a full record. The time of actual performance was not more than 1½ m. For the other 2½ m., *A's* attention was directed to a foreign matter. On the whole, however, he showed more familiarity with the path, making fewer pauses, and moving more rapidly and securely.

Exp. 4. A, 1¾ m. *B*, a few seconds longer. *A's* course: End of *1*—forward quickly and into *2*—stopped at last turn and back, seeming to recognize his error here—paused at *3*, then to end of that *cul de sac*—turned instantly, and ran swiftly and continuously to *4*—entered *4*, going slowly around circuit *n*, *k*, *x*—paused meditatively at *x*—then suddenly started on a quick gallop, accelerating at *m* and *4*, and not pausing till he reached C. *B* followed about the same course, but went less quickly. He showed less the appearance of "trailing." In the case of both rats, something very like disgust was manifest when they found themselves at the end of a blind alley. The instant recoil, the swift retracing of their steps, and the decisiveness with which they turned from the blind alley into the right path, seldom going *back* now beyond the entrance to the *cul de sac*, seemed to indicate something more of mental content than the mere recognition of the impossibility of getting further that way. Another noticeable fact was the increased security and confidence of all their movements. The slow, blundering *modus operandi* of the first experiment had given place to rapid, definite, purposive movements. This was evident alike in respect to the right moves and in connection with the errors. It has been noted already that the errors were retraced with great rapidity. It remains to remark that in nearly all cases, the entrance into the blind alleys was marked by hesitation, and the journey to the end was made slowly and doubtfully. It was also noted in passing that *A*, after passing *x* the second time accelerated his pace, and increased the acceleration at *m* and *4*, indicating thus a greatly increased familiarity with the path from *x* on.

Exp. 5. *B*, 1 m. *A*, 1½ m. Started together. Omitted *1* for the first time. At *2*, separated, *A* entering *2*, *B* going on after a momentary pause. *B* entered *3*—proceeded only a few steps—turned confidently—out and forward to *n*. Delayed here and was overtaken by *A*.[1] Both remained here a little: *B* first recovered his wits, and suddenly dashed back through *4*, and forward to C. Less than 10 sec. from *n* to C; and went with full assurance, hesitating at no point. The appearance of the action was as if some kind of an image of the path to C had flashed in upon the creature's mind touching of simultaneously the motor discharge. At this point I do not wish to discuss the possible nature of the process involved further than to say that the term image does not mean here visual image. Representation, perhaps, would be a less objectionable term to describe this hypothetical mental correlative of the action.

In this experiment *B* for the first time equalled *A's* performance. In all the former tests, both with this apparatus and with that of preceding experiments (*Am. Jour. Psy.*, Vol. XI, No. 2, p. 156 ff.), he had shown a slight inferiority. In the succeeding experiments he was generally first to *C*, though there was slight difference.[2]

Exp. 6. *A*, 1½ m. A variation introduced. The entrance into the center, C was barred by Exp. Box II (*Loc. cit.* above) used in a former series of experiments with these rats. The door of this box, which was held closed by strips of paper pasted upon the door and the sill, was opened inward by a spring when the papers were removed. Food inside the box. The reappearance of this problem so unexpectedly and under such different circumstances after the lapse of 27 days constituted an interesting test of the permanence and distinctness of the memory.[3]

B first out. Directly to *2*, pause—a very human-like indecision. After 5 or 6 abortive starts each way, finally entered *2* and proceeded slowly to end. Turned and swiftly retraced his steps. At mouth of *2* joined by *A*. Together they proceeded placidly to end of *3*. Turned instantly and galloped back swiftly out of *3*, not slowing up until *e*. Here *B* charmed by the odor from C stopped to dig. *A*, forward soberly, hesitated at *x* turning now right, now left, but finally on to *n*.

---

[1] Apropos of the possible rôle played by smell in selecting the path, it is significant that *A*, after following the path traversed by *B* from 3 to 5, *here* diverged, turning into *5* before going to *n*.

[2] Later in this same experiment I observed a striking instance of the difference between the affective character of these rats and of their wild brothers. The laboratory cat jumped upon the maze and showed a friendly interest in the rats. They were startled at first, but soon were trying to bite her toes through the meshes. Under the same circumstances, the wild rats were stiff with terror for 15 minutes.

[3] The term memory is used in this paper in its generic sense. In the present instance it implies: in the presence of accustomed conditions, a mental state leading to the performance of the task incident to the situation. It does not simply "actual and distinct recall;" much less, reference of the reproduction to a former time in the subject's experience. Nor does it imply self-consciousness. It does imply a feeling of recognition—'at-homeness.' The phrase 'permanence of association' might be used, but that this recognition-element is not properly a part of its connotation.

He delayed there a moment—then hastened on to C, via *4*, without further delay or mistake. (*B*, reached *n* before *A* left, but seemed to 'lose his head'—ran back to *e*—to *n* again—then *5*—reached C more than a minute after *A*.) When *A* came to the barricading door of the box, he attacked the paper without hesitation, almost automatically. After pulling off the first paper, he ran back a little way in the characteristic half-frightened manner of the rat, but returned immediately and removed the other paper. All accomplished in a few seconds. The memory was perfect.

This experiment illustrated also the fact of the increasingly definite recognition of critical points, as evidenced by the hesitation and indecision at those points. This was decidedly more striking than in the preceding experiments, especially in the case of *B* at the entrance of *2*, where, as stated, he exhibited a quite human hesitation, turning now one way, now another. The conduct of *A*, at *x*, was hardly less noticeable. *B's* hesitation continued all the way to the end of *2*, as indicated by his slow and doubtful progress; and his disgust when he reached the end, was as manifest as had been his indecision at the beginning.

The rats were exceedingly active during the evening, traversing curiously all the galleries of the maze, investigating every angle and nook; and returning frequently to C for a drink of milk—they had carried the bread out into the maze at once. As an illustration of the rapidity and accuracy of their movements, I saw both rats go from *n* to C without a suspicion of a pause in less than 5 seconds. The distance is approximately 16 ft. The path includes 16 right angles and 4 chances for error.[1]

Exp. 7. B, 3 m. Slow in starting. *B* went to end of *1*—then slowly forward as if feeling his way—eliminated errors *2* and *3*, hesitating only slightly—blundered at *x*, going to *n*. After running back and to between *n* and *k* for a moment, he passed on *via x, m*, etc., to C, without further hesitation or error. Tore off the papers instantly on reaching C. Memory on this point quite as perfect as *A's*. After taking a drink of milk, *B* seized a piece of bread and started back—paused at *h* to eat—met here by *A*, and hastened on to *c*—stopped again to eat. *A* followed here in a few moments, and *B* went to the cage. Almost immediately he went again to C and brought back the other piece of bread. The rats ate their supper in the cage, making frequent trips to C for milk. *B* made the journey twice very swiftly without error or delay; and both of them, several times with but one error. They take a run to C for a drink as naturally as they would turn around, or as a man would go out between the acts. The ease of it all indicates a pretty definite knowledge of the way through the maze, as well as of the hedonistic end to be attained by traversing that path.

Exp. 8. B, 30 sec. *B* made the journey with only one error—went half to end of *1*—and without indecision at any point. (He paused 6 or 7 seconds at one point, but because his attention was caught by something outside of the maze). Tore off the papers instantly. The total time between entering the maze and entering the box was just 30 sec. Deducting the time of the pause (above noted) and that for pulling off the papers—about 5 sec.—we find the actual time in traversing the maze less than 20 sec. This rapidity of movement illustrates—measures roughly—the almost automatic definiteness of the mental pro-

[1] The next morning, I found the rats had carefully deposited the remnants of bread and the *bits of paper* in their cage. A few days later, after I had given them abundant excelsior for a bed, they ceased to save the bits of paper from the box. The collecting impulse seems to be inhibited by the feeling *enough-at-home*.

cess—not quite automatic, as appears from the fact that the animal will again make mistakes and show indecision at critical points. The movements are habitual but not secondary automatic. Attention and discrimination are not wholly shelved. *A* was a few seconds later. Both went back to the cage with their bread without mistake. In making another trip to C a few seconds later, *A* went right to *4*; here hesitated as if "scratching his head," then entered this gallery slowly and doubtfully—only a few steps however; then with a sudden turn and a triumphant flick of his tail he returned to the correct path. This is peculiarly interestiug from the fact that he made this error complete, the first time, going completely around the circuit *4*, *n*, *k*, *x*, *m*. I watched *A* make two more journeys to C. The fourth time he did not pause or hesitate or slacken his pace at *4*. As in Exp. 7, frequent journeys were made to C for milk. These averaged one every 4 or 5 minutes during the half hour I watched. They were made generally without error and in as brief time as 15 or 20 sec.

After the edge of the appetite is worn off a little, the rats tend to let loose the play instinct in the fullest degree. In all their journeys they 'play by the way,' strolling nonchalantly into the blind alleys, now sniffing listlessly, now with half-eager curiosity in all the corners, and angles. That they *know* their way pretty well, however, is evident from the manner in which they take a sudden start from any place in the maze and 'flash' to the end—either end. To one who is familiar with the ways of the white rat, or who is able to imagine his action from the descriptions and figures above, the term 'flash' in this connection will be appreciated as realistically literal.

Exp. 9. *B*, 1 m. (Exp. Box II not used). Each rat made 2 errors: both entered *1*; *A* took long way at *x*—*i. e.*, *k*, *n*, *4*—*B* turned into *4*, went as far as *n*, then retraced his steps. It is noteworthy that these are the most persistent errors. Neither rat hesitated at *2* or *3*. It is not apparent why the error at *1* is so persistent—it was fully eliminated only once. The persistence of the errors at *x* and *4* is a simpler matter. A glance at the diagram will show that these mistakes may be rectified without a return—indeed they are not, properly speaking, mistakes at all, but rather failures to select the shorter path. In each case the rat has but to push ahead in order to recover the right trail. It is probable that these errors would be fully reduced with a sufficient number of trials—in fact other series of experiments confirm this view, though, as will be seen, they are the last errors to be sloughed off; and, with some rats, absolute certainty is not reached in a great number of trials. In one case, forty did not suffice.

The series really concludes with this experiment, the tenth serving merely to test the memory of the maze experience. It adds little or nothing to the data for explaining the formation of the associative nexus, and the mental material and powers involved in that process.

It was made 22 days after Exp. 9. The results were: *B* reached C in about two m., making errors at *1*, *2*, and *3*; *A*, a few seconds later, errors at *1*, *3* and *x*. They clearly recognized the maze, but some of the details had slipped away. *B* hesitated at both *2* and *3*, as did *A* at *3* and *x*, showing that there was recognition of these critical points, but that the memory was indistinct. After returning from *3*, *B* proceeded more securely—his gait had been slow and doubtful before—hesitating only at *x* and but a second there.

The same experiment in several other cases gave similar results—a constant tendency towards a partial lapse of the association.

ANALYSES OF RESULTS.

In appreciating the results of this series of experiments,

about the same facts come into view, only more distinctly, as in the case of the wild gray rats; the initial indefiniteness of movement and the fortuitousness of success; the just observable profit from the first experiences; the gradually increasing certainty of knowledge indicated by increase of speed and definiteness, and the recognition of critical points indicated by hesitation and indecision; the lack of imitation and the improbability of following by scent; the outbreak of the instincts of play and curiosity after the edge of appetite is dulled. In addition are to be noted the further observations upon the contrast between the slow and cautious entrance into, and the rapid exit from the blind alleys, after the first few trials; the appearance of disgust on reaching the end of a blind alley; the clear indication of centrally excited sensation (images) of some kind; memory (as I have used the term); the persistence of certain errors; and the almost automatic character of the movements in the later experiments. Viewed objectively, these observations all converge towards one central consideration; the continuous and rapid improvement of the rats in threading the maze, amounting to almost perfect accuracy in the last experiments. No qualification of this view was found necessary in the light of many later experiments. Rather they all confirm it. The mental aspect is considerably more complex, the mental factors, much more difficult of analysis and evaluation; but the central fact in the process seems to be the recognition by the rats of particular parts of the maze. Deferring consideration of this side of the matter, and looking now only at the objective side the important points are, as in case of the wild rats: (1) the increase in speed, and (2) the decrease in the number of errors and in uncertainty. Comparison of the two points is as follows:[1] I-13, 13′. II-3, 2′. III-4, —. IV-1¾, 4′. V-1, 2′. VI-1½, 3′. VII-3, 2′. VIII-½, 1′. IX-1, 2′. (X-2, 2′). As with the wild rats. we find a fairly constant decrease in time and in number of errors. Similarly, there are fluctuations both in time and number of errors, *e. g.*, Exp. 9 shows an increase over Exp. 8 in both respects. This is to be expected from the character of the animal and the confessed impossibility of completely controlling even the external conditions, not to mention the particular internal conditions in each case. Allowing for such variable factors, the relative time required and number of mistakes made furnish a fairly accurate index of the progressive acquaintance of the rats with the problem. The contrast between the first slow, blundering, accidental success and the definitely foreseen success of Exp. 8 (taken as the best) is striking. This is brought out even more saliently by the

[1] Significance of the numerals is the same as *supra*.

graphic representation on p. 201 than by the figures collocated above. The solid line indicates the course followed in Exp. 1; the dotted line that followed in Exp. 8.[1] The arrows mark the point where the rat stopped and turned about, and are pointed in the direction he was headed when he stopped. The dots (•) indicate points where considerable pauses were made. The arrow and dot between *4* and *m* indicate that the rat, when returning from an abortive essay into *4*, went as far as that arrow, paused, then turned and went forward.

Turning now for a brief consideration of the relative results of experiments with the two kinds of rats, some interesting and rather unexpected facts crop out. Comparison of the time and error tables of the two series discloses no considerable superiority on either side, although in this comparison I disregarded entirely the first two experiments of Series I, comparing the 5 trials recorded of Series I with the first 5 of Series II. The average of times required and errors made gives the white rats the advantage in regard to time: the brown rats, in regard to errors.[2] In view of the handicap of two experiments, it will be seen that the advantage lies with the white rats throughout. The brown rats had gained comparatively little from the first two experiments, imperfectly conditioned to be sure, but in which they had the usual freedom of the maze during the entire night. I am of the opinion that even if the conditions of experiment had been identical, the balance still would have tipped in favor of the white rats, so potent is the inhibitive influence of fear with the wild rats. The rational conclusion seems to be that there is little difference between the two in actual intelligence. The wild rat is somewhat more vigorous and active, and consequently this excess of activity increases his chances of accidentally hitting upon the right path. This might secure him a slight advantage *in time* in perfecting his knowledge of the path. It would not signify, however, any advantage in quality or degree of intelligence. It might ensure a larger number of associations (supposing a free life for both animals) but would not make any difference with their delicacy or complexity. In short, the results of superior activity are not intelligence, though they seem often not to be discriminated, in accounts of human as well as animal doings. This superior activity, however, is fully balanced by the wild rats' greater susceptibility to fear under these strange conditions.

[1] That Exp. 1 gives the record of Rat *A* and Exp. 8, of Rat *B* is inconsequential. The attainments of the two were practically equal. (The course was marked on a diagram at the time of observation.)

[2] Average time: Series I, 5 m. 19 s.; Series II, 4 m. 33 s. Average of errors: Series I, 4 1-5; Series II, 5 1-5. In Exp. 3 of Series II, I was not sure whether there were 4 or 5 errors. The maximum is taken.

In ability to profit by experience—in this case, to learn a definite route involving possibilities of frequent error—the two are not far different.

This conclusion does not tally very well with the general opinion that animals suffer mental deterioration under domestication. However that may be with other animals, it evidently is doubtful in this instance. The white rat in comparison with his wild congener is somewhat less vigorous and hardy (especially does not endure cold or hunger so well), and has sloughed off some of the timidity and suspiciousness of the wild rat; on the other hand, his senses with the exception of sight are as keen, his characteristic rat traits are as persistent, and his mental adaptability is as considerable. The *modus operandi* of the two kinds in the maze shows little variation. Likewise there is no difference in the curiosity manifested, either in kind or degree. In view of the many generations of luxurious idleness [1] of the white rat, this profound and enduring nature of specific psychic traits is striking. A pertinent illustration was furnished by a young rat that escaped from his cage and was loose about the laboratory for several days. He had just been weaned when the accident occurred. Food was rather scarce and he got pretty hungry. Finally one morning he found his way into the chicken pen, and in less than two minutes had killed two chickens, and was upon a third when discovered. The chickens were three times as large as himself. The killing was done by biting through the throat of the victim, and was as neatly and deftly executed as if the executioner were an old hand. The importance of this illustration lies in the fact, that this is exactly the method of killing employed by the wild rats. The only possible preparation in his own experience this pigmy could have had for such serious business must have been in play with his fellows. That, however, was general rather than specific; and, at best, was of slight importance, as he had reached the playing age but few days before. Another typical illustration of the persistence of specific traits is furnished by a perfectly tame pet rat that exhibited the greatest fondness for a hole in the base-board of a room where he was allowed to run, making for the hole every time he was set free in the room, and dodging in and out at every sound; yet so tame that he would come out and allow himself to be caught as often as I went to the hole and called. Such cases give some suggestion of the tenacity of those fundamental specific traits which "persist with undiminished vigor" long after the conditions of life which

[1] I cannot find that the white rat is known to exist in the wild state. There is a tradition that it was brought to the Occident from an immemorial existence in China. I am not able to verify this tradition—or disprove it.

called them into being have changed radically. In these cases non-use certainly has had but little effect in reducing the potential force of specific instincts.

What is true of such relatively superficial specific traits is doubly true of generic instinct-feelings. Curiosity is a good example, frequently coming into evidence as it does in these experiments. Its intensity is not diminished with the long domesticated white rats. Nor is it greatly changed. Ribot, rightly regards curiosity as the basis of the 'intellectual sentiment.' "This primitive craving—the craving for knowledge—under its instinctive form is called curiosity. It exists in all degrees, from the animal which touches or smells an unknown object, to the all-examining, all-embracing scrutiny of a Gœthe; it always remains identical with itself."[1] I am persuaded that Ribot is right in regarding this affection as primitive and as a primitive craving for knowledge; not merely a reaction to hunger or sex stimulus. The desire for familiar acquaintance with environment, concomitant with fear and uneasiness in strange surroundings, is about as fundamental as hunger. Observers of wild animals in their native haunts tell the same tale. The astronomer who orients himself with respect to infinite worlds satisfies the same craving for knowledge and calms the same uneasiness in strangeness as does the animal which seeks all the knowledge possible to him of his universe. The reduction of chaos to cosmos begins there.

Such considerations as these suggest the question whether zoölogical psychology may not profitably turn from its almost exclusive search for variation, to a search for the relatively invariable factors in the animal mind. In order to do this the ideals of structural psychology must be departed from somewhat, and attention directed to the study of the instinctive traits and tendencies, out of which, in higher differentiations, human nature is made. From the point of view of psychic statics these are composite and analyzable; but, from the point of view of psychic dynamics, they are themselves primitive and elementary. By this method, if by any, will be gathered the material for a natural history of mind. From this source light may be expected upon many obscure problems in individual and anthropological psychology.

General Form of Animal Intelligence.

The amount and variety of fact brought out in the preceding description and analysis must serve as excuse for this somewhat tedious presentation of details. It justifies the initial assumption that the results of such a study as this are qualitative

[1] Ribot: Psy. of Emotions. (Eng. Tr.) p. 368.

rather than quantitative; and that a generalized statement of results or any graphic representation whatever of the data by curves, would indicate only the most general form or tendency of animal intelligence, which was decidedly less what I wished to exhibit than the details of the performance. My conviction of the importance of this aspect of the case is strengthened by M. Hachet-Souplet's brilliant suggestion of a new method of classification of species from the psychological point of view.[1] M. Hachet-Souplet shows clearly that there is very good reason for cutting loose from the trammels of morphological classification in our psychological investigations of animals. If, however,.such a desideratum is to be realized, it must be by studying the mentality of the different kinds of animals with the same minuteness that morphology employs in its domain.

A few words, however, as to the 'general form of animal intelligence' and a more adequate appreciation of the value and limitations of the 'curve' in connection with the same may not be *mal apropos*. In a former paper,[2] I have noted that a time curve would be an insufficient index of the definiteness and certainty of an animal's mental processes, on account of the inconstancy of internal conditions. In this paper I have pointed out that a more adequate representation might be made (at least for these experiments) by compounding the time and error curves, and also a curve representing the indecision at critical points—if such could be extracted. Such a compound curve, however, would still be far from telling the tale fully and precisely.

The one extensive and important study in comparative psychology in which the graphic method of presenting results is largely employed is that of Dr. Edward Thorndike upon cats, dogs, and chickens.[3] Dr. Thorndike's methods with his cats and dogs was to confine the hungry animals in boxes, small enough to be uncomfortable, from which they might escape by "some simple act, such as pulling at a loop of cord, pressing a lever, or stepping on a platform." Food was exhibited outside

[1] P. Hachet-Souplet: Examen psychologique des animaux, Paris, 1900. Schleicher Frères.

[2] *Am. Jour. Psy.*, Vol. XI, No. 2, p. 136.

[3] Edward Thorndike: Animal Intelligence. Psy. Rev. Monograph, Vol. II, No. 4, June, 1898. Dr. Thorndike's methods have been criticised by: W. Mills (Psy. Rev., May, 1899); C. Lloyd Morgan (Nature, July 14, 1898); and Kline (*Am. Jour. Psy.*, Vol. X, Nos. 1, 2, 3). Kline's criticisms are fragmentary, but valuable. Morgan's Review in Nature is discriminating and sympathetic. Mills raises some objections to Dr. Thordike's work, but is blinded at times to the other's meaning by his polemical ardor. Thorndike has replied to Mills (Psy. Rev., July, 1899).

the enclosure as a special incitement to vigorous effort; and escape was rewarded with the food. The time required to escape in successive experiments was recorded, and the results represented by curves. These time curves are regarded by Dr. Thorndike as exact indices of the progress of the animal in the formation of the required associations. Though these temporal data do not seem so significant to me, as to Dr. Thorndike, for the reasons given above, yet they do exhibit, I think, what I have called the general form of animal intelligence; or as Professor Morgan expresses it in an analysis of Dr. Thorndike's paper (v. note above), "they bear out the contention that the method of animal intelligence is to profit by chance experience, and depends upon the gradual establishment of direct associations." I suppose Höffding means essentially the same thing when he says: "The simple primitive consciousness does not feel the need of concepts but goes passive from disappointment to disappointment." Professor Morgan reports in the same paper, having attempted to extract from some of Mr. Thorndike's carefully plotted data, a mean curve for the method of trial and error. The attempt was not very successful he admits; but he thinks the resultant curve does indicate the gradualness of the process which theoretically would be expected. On the whole, this conclusion seems to me essentially sound. At least, my own observations with rats, under what seem to me more natural, and sympathetic conditions, give confirmatory results. I have plotted tentatively a number of time curves, both from the results of my experiments with the maze and with other devices; and they show no radical variation from the general form of Dr. Thorndike's curves. I think Morgan is right in asserting that "the form of his curves affords no particle of evidenee for reasoned behavior." No more do mine. It must be remembered that Morgan rigidly limits the term reasoning or rational procedure to the process of drawing logical inference. The relation between experiences must be perceived as such. The transitive moment between focal points in consciousness must itself be capable of becoming focal. The reasoning creature must be able to 'focus the therefore, think the why.'[1] In the broader sense of practical adaptation to varying conditions by direct association; or as Binet[2] defines it: "an organization of images determined by the properties of the images themselves, so that the images have merely to be brought together for them to become organized, and reasoning follow with the inevitable necessity of a reflex"—Morgan readily admits reasoning in animals.

[1] Morgan: Comparative Psychology. *In locis.*
[2] Psychology of Reasoning. (Eng. Tr., p. 3.)

Morgan further made a comparison of his mean trial and error curve, with a mean curve of rational procedure, which he plotted from Dr. Lindley's data compiled in his "Study of Puzzles." (*Am. Jour. Psy.*, Vol. VIII, No. 4.) These present a clear contrast, he thinks. In the latter case he finds "a sudden leap from failure to success when the trick of the puzzle was discovered and understood;" as opposed to a "gradual sweep towards rapid and assured success" with the former.

There is one fact, however, in connection with the trial and error curves that Morgan does not remark. Nearly all of Dr. Thorndike's curves show a sudden fall after the first success. My own experiments with various devices gave similar results. This fall is analogous with the fall in the curve of rational procedure after the 'trick was understood.' This fact does not affect the distinction between the two processes. The sudden fall in the trial and error curve indicates only effectiveness for reproduction of the first right association. In this method, learning the trick or the task, depends in the first instance upon performing it fortuitously so far as previsioned end is concerned; and requires then time and repetition for perfecting the knowledge. Improvement depends upon memory of previous performances, or forgetting useless details. The relation of the acts does not become focal. On the other hand, in rational procedure, the trick may be understood before it is performed. The entire plan of solution may be envisaged before a move is made. As a matter of fact, however, in the great majority of cases tested by Dr. Lindley, in which rational procedure was employed, the understanding at the beginning was only partial, the plan was vague and hazy. The understanding became full and definite through abortive attempts at practical solution. The younger children, Dr. Lindley found, as would be expected, used almost exclusively the trial and error method, chance success and direct association. With older children rational procedure based upon a considered plan came progressively into evidence. "The younger children succeed through a long series of slight variations. Occasional lapses into useless movements occur; but the trend is by a slow and primitive method of exclusion towards the goal." Among older persons the inhibitive influence of failure is stronger. The memory of the failure takes its place as a substantive element in consciousness and constrains the subject to reconsider, to deliberate. The variations, too, are wider and more far-reaching, indicating a larger and more complex grasp of attention. Yet comparatively few, even of adults, pursue a strictly rational procedure. The close similarity between animals and children is obvious.

The fact that the trial and error method plays so large a part in human mentality, and especially that it predominates with children, still further supports the view taken in the beginning of this paper that animal intelligence works almost exclusively by this method. Although this hypothesis was assumed for the present case only, it certainly covers adequately a very large part of animal activity. Most anecdotal cases of animal reasoning are explainable upon this ground. So far as the narrowly experimental studies of animals go, they point in the same direction. As yet, however, these are too limited in number and too restricted in scope to be very conclusive.

On the other hand, however, there is yet something to be said. The analogy between children and animals is so close that the manifestation of reason in some very young children suggests that some animals may have the same power. Preyer cites the case of a two-year-old child getting a cricket and standing upon it in order to reach a desired object.[1] M. Hachet-Souplet[2] shows wide variations of psychic faculty among animals of closely related morphological groups. Individual variations, too, are indubitable. M. Hachet-Souplet has no doubt that some animals reason. His opinion is important since he is an experimenter in a large and fruitful way, and is at the same time in the main sharply critical of his facts. He describes a striking case of reasoning in a coati, as follows: "Un de nos amis nous ayant rapporté un merveilleux trait d'intelligence de la part d'un coati, nous avons résolu de provoquer artificiellement autour d'un autre coati, des circonstances analogue à celles dans lesquelles le premier s'était trouvé, quand il donna une si grande preuve de sagacité. On sait que l'espèce est très friande d'œufs de poule; nous en plaçâmes un sur une haute cheminée de façon à ce qu'l pût être vu du coati et, après avoir éloigné légèrement les sièges, nous quittâmes la pièce; en nous arrangeant toutefois de manière à ne rien perdre de ce que ferait notre sujet.

"Il s'agita d'abord, saute deux ou trois fois; mais, voyant que son élan ne le portait qu'à mi-hauteur de la tablette, il semblait réfléchir un instant. Il se dirigea ensuite vers une chaise en chêne ciré qu'il essaya d'attirer du côté de la cheminée, mais ses pattes glissait sur le bois et il renonça à son enterprise; il semblait désespéré. Cependant il aperçut dans un coin un paquet de vieux chiffon et parut frappé d'une véritable *idée. Ayant pris une des bandelettes, il en entoura le pied de la chaise et il se mit à l'attirer à reculons.* Quand le siège fut contre la cheminée, en deux bonds, mon coati monta sur celle-ci et s'empara de l'œuf." Manifestly the coati in this case perceives relations, more complicated relations, indeed, than the

[1] Preyer: Infant Mind. (Eng. Tr., p. 185.)
[2] *Loc. cit.*

relation perceived by the child cited above. Such a mental expression goes beyond the explanatory possibilities of 'direct association by trial and error methods.'

Perhaps the real question after all is not whether animals perceive relations, but, rather, *what animals perceive what kind of relations.* This would seem to be the logic of M. Hachet-Souplet's position; and, what is more important, it is the logical method of psychogenetic study. Granting that animals may perceive relations, we have, then, definite and feasible problems to investigate: the kind and complexity of the relations perceived. This method should give data for a better understanding of the nature of animal reasoning,[1] judgment, inference, and the consciousness attending these processes. It would not be surprising if degrees of complication and of symbolism were found to constitute the differentiæ.

On the whole, the modern studies of comparative psychology confirm Hume's acute observations on the nature of animal intelligence. "Animals, as well as men, learn many things from experience, and infer that the same events will always follow the same causes. By this principle they become acquainted with the more obvious properties of external objects, and gradually, from their birth, treasure up a knowledge of the nature of fire, water, earth, stones, heights, depths, etc., and of the effects which result from their operation. The ignorance and inexperience of the young are here plainly distinguishable from the cunning and sagacity of the old who have learned from long experience to avoid what hurt them, and to pursue what gave ease or pleasure. . . . An old greyhound will trust the more fatiguing part of a chase to the younger, and will place himself so as to meet the hare in her doubles; *nor are the conjectures which he forms on this occasion founded in anything but his observation and experience.* Animals, therefore, are not guided in these inferences by reasoning, neither are children, neither are the generality of mankind in their ordinary actions and conclusions, neither are the philosophers themselves. Animals undoubtedly owe a large part of their knowledge to what we call instinct. *But the experimental reasoning itself, which we possess in common with beasts is nothing but a species of instinct or mechanical power that acts in us unknown to ourselves.*" This statement of Hume's[2] as to the general form of animal intelligence has not been improved upon. His expression 'experi-

[1] Benn (Gk. Philosophers, I, p. 381,) makes the interesting suggestion that animals reason disjunctively ("after a canine fashion"). It is from the disjunctive form that all other forms of reasoning are successively evolved.

[2] Hume: An Enquiry concerning Human Understanding. Section IX, pp. 85 ff.

mental reasoning' seems to me singularly happy and accurate, guarded, as it is, by the suggestion of its subconscious character —especially happy in comparison with the clumsy expressions of some modern comparative psychologists whose abhorrence of anthropomorphism leads to the opposite extreme both in thought and expression; and causes them to inveigh against that vice *ad nauseam*. I suppose Hume would be dealt with summarily on this score by these writers. But is not a certain amount of chastened anthropomorphism a wholesome specific, a kind of saving grace against the scientific pedantry that thinks to create a new science of comparative psychology with the imperfect instruments of experiment and the law of parsimony. The law of parsimony is important, no doubt, but it may be employed too rigorously. The real difficulty lies not in the tendency to interpret animal intelligence in the terms of human experience, for we have no other way; but in the faulty and imperfect analysis of human experience. That is the real vice of Romanes' work. His analysis did not go much deeper than the discursive adult human understanding. This difficulty is intensified, not only by the fact that human consciousness is permeated, and, as it were, recreated by self-consciousness; but also by the fact, not always heeded, that the more elementary and obscure phases of human experience, as yet, have not been fully and definitely analyzed.

Another difficulty, not less real and important, but not sufficiently remarked, is found in psychophysical limitations. This difficulty is frequently met with in human psychology. Galton's Academicians who regarded mental imagery as 'moonshine' illustrate the point. They experienced no mental images; therefore, mental images did not exist. A similar limitation, Ribot thinks, leads psychologists to cavil at memory of emotions. Now, doubtless, the psychophysical disparities between man and brute are inconceivably greater than between man and man. The immensely greater rôle played by smell, for example, or motor experience, in the economy of the animal mind cannot be fully appreciated. We can hardly have any idea of the radically different tone and feeling of such consciousness; and we cannot say with any precision what modifications of intelligent processes are concealed from our view in this way.

The Process of Learning.

Whether the process of learning the way through this maze is adequately described as a gradual establishment of direct associations by profiting by chance experience depends upon the meaning attached to the phrase 'profiting by chance experience.' I wish in the remainder of this paper to attempt an analysis of the mental factors involved in the animal's solution of the problem; and to offer some suggestions upon the character of the perfected knowledge.

It will be well to résumé the facts by following the rats *A* and *B* of Series II from their introduction to the maze, throughout their experience of getting acquainted with their new environment, to the time when they have perfect mastery of the situation. In their first trial, after a lapse of 13 minutes and after many errors, returns, and delays, they find their way into C. Here they are rewarded for their labors by all the pleasure possible to a rat from the satisfaction of a keen-edged appetite by a good meal of bread and milk. This first success is assumed to be accidental; its realization does not depend at all upon previsioning intelligence. The animal does not foresee the end and set to work to attain that end. There is no reflection. The determining conditions in the rat's mind are more immediate in their effect. The most obvious of these are: hunger, perception of the odor of the food, curiosity, normal activity (the obverse of curiosity) and the instinctive special trait of following out tortuous passages—a definite rat-hole consciousness that acts, as it were, thygmotactically. These factors, with the inhibitive balance-spring of timidity, firmly rooted and deeply toned emotionally, constitute the relatively stable background of consciousness over which play the lights of the perceptive and discriminative processes as the animal proceeds with the task. The rat, when he enters the maze, is psychically a confused complexus of these factors. No one of them looms high above the others in the wave of consciousness. Attention is dispersed; perhaps, better, distraction prevails. Nevertheless, 'experimental reasoning' begins at once. The animal keeps constantly moving; but his activity at this stage is evidently sensori-motor (or organo-motor). Motive in the sense of ideated end is absent. The nearest approach to this is the possible idea of a definite kind of food incident to the perception of the familiar food-odor. This is not impossible. On the other hand the effect on consciousness may be only an intensification of the hunger psychosis resulting in an increase of motor activity. However this may be it probably is the animal's instinctive fondness for following out devious ways his, thygmotactic rat-hole psychosis, rather than the smell of the food that gives determinateness to his movements at first. Were this trait less imperative, the rat, when he comes near the food (*e. g.*, at *e*) would become the victim of his hunger and his perception of the position of the food—for the food at this point could be located by smell—and would spend himself stupidly in endeavoring to force an entrance. In general, however, he soon passes on, going directly away from the Canaan of his desire.[1] Failure to get through by gnawing and digging,

[1] Occasionally one does just this stupid thing. Cases were noted in passing.

quickly results in a wavering and dispersion of attention. Concomitantly the perception of the odor relapses into the margin of consciousness, and the instinctive motor tendency at this juncture reasserts itself as the focal and directive influence.[1] The first success then may be set down as the accidental issue of a trial and error series, motivated by hunger and curiosity, mediated by the sense of smell and, more largely, by this instinctive motor trait, and consummated by the pleasure of hunger satisfied. And yet the term *accidental* must be used with reservations. The rats of this series, and of all others in their first trial, seemed to profit at once by experience. By this I mean that after they had made an error once or twice, though they had not yet succeeded in reaching C, they would hesitate or even avoid the error when going over the ground a second time. For example, rat *A* went only a few steps into *3* the second time he reached that place, and avoided *6* and *7* completely. A glance at the record of the first experiment will make this perfectly clear. Such cases may be attributed to pure chance; the conduct of the rat, his hesitation more than his avoidance of the error, indicates, rather, recognition and selection.

It will be remembered that the rats have the entire night each time for exploration of the maze. This results in remarkable improvement in the second trial. In the succeeding experiments the improvement is continuous in the elimination of errors and in the increase in definiteness and speed. The rats soon acquire a practically perfect knowledge of the maze, so that they can make the journey quickly and accurately when they want to do so, or stroll about as they list.

How explain this improvement? What does 'profiting by chance experience' mean in this instance? how is it assimilated and how utilized?

Doubtless one factor in the process is the memory of the pleasant experience at the end. In addition to the undirected and undifferentiated motive of hunger and the motor trait of the first trial, there is, in the second, a dimly ideated end which probably becomes progressively clearer in the subsequent experience. But the essential point is certainly the recognition of the critical points along the way and the discrimination of

[1] With several subjects the odor stimulus was done away with entirely to see whether they would make the first journey to C as well as the rats that had that stimulus. No appreciable difference was shown either in time or number of errors. The rats followed out the maze to the end, C, just as perseveringly as if the food had been there dispensing its savory solicitations. The expected did happen, however, in respect to learning the direct way to C. They made little progress in five or six trials. As there was no pleasant association at the end of the journey there was nothing to determine the building up of this definite association train.

the divergent paths at these points, leading to purposive selection of the right path. The memory of the pleasant experience at the end would be of slight avail, if the rats did not recognize the critical points and discriminate and select their paths. The animal begins by going right and wrong wholly by chance. After a few trials he comes to recognize the doubtful places, and hesitates when he comes to them, undecided which way to take. The external signs of indecision vary between standing still as if trying to think which way to go, and abortive starts each way. Sometimes to these is added standing up and sniffing in the usual manner of orientation. This movement seldom was observed after the first two or three experiments, *i. e.*, after the dilemma began to be clearly felt. At this stage, the choice of path is still about as often wrong as right. The distinguishing accidentia are acquired gradually. Progress in discrimination is marked by decrease of hesitation and in more frequent choice of the right path. The path chosen often is pursued doubtfully. If the wrong one is chosen, the error frequently is retrieved after a few steps; if it is followed to the end the return is made swiftly and the right path is taken confidently. In the final stage, errors and hesitation drop out entirely. The right path is followed from start to finish without attention to specific points *en route*.

It should be noted that the learning was slowest in connection with *o*, *x*, and *4*. The persistent confusion at *o* is attributable probably to its being at the entrance of the maze. At this point there is a maximum of affective excitement. The momentum of association has to be gathered as the animal goes along in the familiar path. In a remote way it may be likened to the stumbling and groping of an orator at the beginning of a familiar theme. The suggestion that the rats might have a penchant for right as opposed to left was found baseless by the use of the reversed maze. A definite memory of direction seems to be required. The slower discrimination at *x* and *4* was due doubtless to the fact that wrong choice at these points consisted in taking the roundabout, rather than the direct path. Strictly speaking there was no error. In all the other cases, taking one path was associated ultimately with success; taking the other, with failure—disappointment. In these cases the association would seem to be between path and distance.

In such cases profiting by experience manifestly involves the processes of recognition, discrimination and choice. If the problem set were merely the selection of one effective movement out of several haphazard movements, as was the case with the puzzle-box experiments reported by Dr. Thorndike[1] and my-

[1] *Loc. cit.*

self,[1] then the profiting by experience could be accounted for by the fading away of the useless movements. They would drop off like dead branches from a tree, of their own weight. They would be associated with nothing—either positively or negatively. The right movement would be selected *naturally*. In the present case, however, two direct associations are formed and discriminated between, and the advantageous one selected. Recognition of the critical places is equivalent to doubt as to the right path. This doubt is the correlative in consciousness of the struggle between the two associations or 'constructs.'[2] The positively useless or the less advantageous association does not fall away mechanically, but only in virtue of discrimination between the two constructs, and, finally, the conscious selection of the right one. In such a case as that of choice at x if the animal did not consciously select, there could never be any fixed association; consequently never any habitual reaction. Both ways lead to success. In a sufficient number of trials the theory of probabilities would require an equal number of selections of each path. But the short road is soon habitually selected, just as is the right path at other critical points. There is involved an elementary form of comparison and judgment; for comparison, judgment and reflection, even, are present in embryo. They all take their rise in the struggle of ideas and images, and lower down of 'constructs,' which "gives in animal, as in man, the illusion of choice and free intelligence."

Modalities of Sensation.

This section is an attempt to appreciate the rôle of the different sense-modalities in learning the task.

The conditions of the experiment were such as to exclude any very direct influence of taste and hearing. Taste gives only a pleasurably-toned experience at the end, the significance of which has been noted. The influence of hearing is limited to occasioning affective variations. Neither gives any data for solving particular difficulties.

Smell. The sense of smell might be supposed, *à priori*, to play the leading rôle, but in the present case its claims to primacy are doubtful.

In the preceding section it has been shown that the location of the food by odor, and hence the end to be reached, was an unimportant factor. In fact, it is improbable that olfactory sensation *per se* has much greater spatial significance with animals than with man. In general, animals perceive direction of odors only with the aid of air currents. The perception is quite as much tactual as olfactory.

---

[1] *Loc. cit.*

[2] *Cf.* Morgan: Comparative Psychology.

It is even clearer that the trail of the first accidental success was not followed subsequently by scent. In the first trial the rats invariably traversed practically all the galleries; and, after appeasing their hunger a little, carefully investigated the entire maze. It would be impossible, therefore, for them to select the right path by scenting the trail. Again, the second rat frequently turned aside from the route marked out by his immediate predecessor; either he was not following the trail or he could not discriminate the fresh trail from one a day old. Further, the recognition of critical points, and the fact that the rats frequently ran long distances with heads up—*e. g.*, when carrying food—are evidence against the supposition. These facts together are sufficient to throw the theory out of court.

The conclusion is drawn for the present case only. It is perfectly apparent that animals of this class do follow trails by scent in the right circumstances. These facts point to the complexity and variety of the animal mind, and are a warning against naïvely accepting 'simplest explanations.'

Another possibility in regard to smell is that particular points in the maze may have been associated with definite peculiarities of odor. The constant sniffing and extensive olfactory investigations of the rats lend color to this thought. The experience thus acquired, may, however, influence only the affective tone—connect directly with the emotional tendencies which determine the animal's conduct. Such a relation is indubitable. I found, for example, that putting rats perfectly familiar with the task, into a new maze, differing from the one learned only in *newness*, threw them into extreme emotional excitement. They acted as though the task were absolutely new to them. They were curious, timid, and hesitant; errors were as frequent as in their introductory trial. After finding the food, they continued eagerly exploring and re-exploring the maze. As soon, however, as they had become familiar with the new odors, their former facility returned; they made the journey as quickly and accurately as before. This would not have occurred, had it been necessary for them to establish a new series of smell-position-direction associations. The inference is clear that the effect of smell sensations is general and emotional, rather than that delicate and discrete associations of odors with special positions are set up. The point, however, is not absolutely secure. Probably more conclusive evidence might be obtained by testing rats with olfactory nerves paralyzed.

Sight. Sight is much less relied upon, and, relatively, much less acute than smell and hearing—the psychic organs respectively of food-getting and defence. This corresponds with the poor development of the eye and optic nerve.[1]

[1] The eye of the wild brown rat is better developed than that of the white rat, but the two rats varied slightly in their conduct in the maze.

Several tests were made, the results of which indicated that visual perception played no part in the processes of recognition and discrimination.

1. It was suggested to me that the direction of light, by analogy with Lubbock's[1] experience with ants, might be a factor in the chance of path. Lubbock found that "when the direction of the light was changed, but everything else left as before, out of seven ants, five were deceived and went in the wrong direction." (This was after they had learned their path perfectly, of course.) Fortified by further experimentation, Lubbock concludes that "in determining their course the ants are greatly influenced by the direction of the light."

As the rats did most of their exploring in the dark, and as the brightness element is only one factor in the visual datum, not the total datum as with the insect, it was improbable that this factor should be very influential. Nevertheless, it was made a matter of experiment. Tests were made by having the rats learn the path perfectly with the direction of the light constant. The light was then transferred to the opposite side for a few trials; after which, it was alternated at unequal, though frequent, intervals. The results were: (1) In most cases, change of direction of the light seemed to produce a very slight effect upon certainty and celerity of movement; but hardly more than might occur as normal variations under constant conditions. (2) Some subjects showed absolutely no effect. (3) After the first change the alternation produced no effect. This shows that the effect when it occurred was merely a slight affective disturbance—a retardation, not a change of the cognitive process. Plainly this is a very minor factor.

2. A partial test of the part played by sight in the recognition and discrimination necessary to the formation of special associations at critical points was made as follows: At all such points, bright red posts, ¼ inch diameter, were placed in the middle of the right path a few inches beyond the dividing of the ways. When the rats had learned the path perfectly the posts were removed. Two rats only were tried, the results being *nil*. These rats did not learn the path more quickly; nor did they exhibit the slightest variation in conduct after the posts were removed. It is tolerably clear that visual data, if effective, must be of a more general character. The animal does not hang his association upon a gross and obvious object.[2]

3. Another method of partial experiment suggesting itself was to blindfold the rat after he had learned the path. The evident

---

[1] Lubbock: Ants, Bees, and Wasps, p. 267 ff. (6th Ed.).

[2] It is improbable that the action had become so habitual as to dispense with what was at first a determining factor in the formation of the association.

objection to this plan is that it would change completely the conditions of attention and emotion. If the subject blundered and failed nothing would be proved. If he did his work about as well as before there would be a negative demonstration of the slight importance of visual perceptions; *not, however, of visual sensation*, for its effect might still be present as visual images co-operating in the mental process.

Fortunately, nature stepped in and performed a conclusive experiment for me. A number of my rats came to me with diseased eyes. Before I discovered this, two of them, an adult male, *X*, and a young female (about 10 weeks old), Y, had become blind. I had already started them learning the maze, with two others, when I noticed their blindness. After the fifth experiment they were totally blind. In the first two experiments distinct impressions—if white rats have such—may have been possible to *X*; and brightness sensation until the fifth. Rat *Y* may have had brightness sensations in the first two experiments, but not later. At this time the general health, vigor and temperament of these rats were unaffected by their malady.

The results of the experiments with these blind rats were so striking that I give them somewhat fully. Until after the ninth trial, the two normal rats were continued with the blind ones. They were then removed, and thirteen more experiments were made with the blind ones. Following that, the latter were tried in the reversed maze. The blind rats learned the original task as well as the normals—all the normals experimented with. Rat *X* in this case learned the path before either of his normal companions. In Exps. 5 and 6 he was first to C, and made fewest errors. In Exp. 7 he made the round in 50 seconds, without error, and with slight hesitation at two points only. In the succeeding 15 experiments he showed practically perfect acquaintance, though occasionally making errors. His conduct in the maze did not differ materially from that of normal rats. He ran in the middle of the galleries, rounded the corners quickly and precisely, and carried on the usual investigations. At critical points there were the same hesitation and indecision manifested as with the normals, by alternately turning each way as if stayed in the grasp of conflicting images. Occasionally he would nose along the several sides before starting on again. This probably was not a direct means of ascertaining the way, for, later, I cut off his feelers, also those of some normals, without any effect upon their ability to find their way.

The results with Rat *Y* were even more interesting, as she was certainly totally blind after Exp. 2. She was somewhat longer in learning the way than *X* and the normals. At first she was slow and diffident in starting, and less facile in getting about; she ran somewhat gropingly, and frequently almost bumped into the ends of the galleries. These defects soon wore off, and she kept her path and rounded the corners as nicely as the others. Certain errors, however, clung persistently, notably at *2* and *3*. After Exp. 3, she had little difficulty beyond *3*. Until Exp. 11, she went each time to the end of *2* and *3* as mechanically as if these were essential stages of the journey. I began to wonder whether the habit was so firmly fixed as to defy the benefit of chance experience. In Exp. 12, however, *2* was dropped out; in Exp. 13, *3* was dropped, but *2* reinstated. In Exp. 15, *2* and *3* were finally eliminated, and did not reappear. In the last 7 experiments she made no errors and seldom hesitated or showed indecision. This blind rat thus eliminated errors that had become almost automatically habitual. The experiments with the reversed maze[1] gave the only suggestion of importance of the visual factor. The blind rats when first put into the reversed maze were more disoriented and confused than the normals. Not until Exp. 3 did *X* succeed in getting to C.[2] The normals, on the contrary, seemed to have profited by their experience with the other maze. They made better time, fewer errors, and showed less indecision than in the first experiment in Maze I. They fell off badly, however, in the second trial. The discrepancy between the blind and the normals quickly disappeared; in Exp. 5 the blind rat did as well as the normals. So also, in the successive alternations of the mazes, the blind rat perfected his distinct knowledges of the two as quickly as the normals.

This slight superiority of the normal rats, however, does not seem to me to mean that visual data exerted a determining influence. Rather, with the blind rats the motor element was so exclusive in the reproductive process as to make readaptation proportionately more difficult. With the normals sight, though probably contributing no determining data, served to distract attention from the established reproduction. Sight certainly is not a *sine qua non* in the process of experimental reasoning incident to these experiments. Its service is superficial, and may be dispensed with almost without loss. Its office in the essential processes of recognition and discrimination is hardly

[1] The method of experimentation was to transfer the subject, after he had mastered the original maze, to the reversed form—*i. e.*, right and left interchanged. After this reversed form was learned the rat was returned to Maze I. The mazes then were alternated till both were learned perfectly.

[2] Rat *Y* died in Exp. 3, so these data are from observations of Rat *X*.

appreciable. Its forces are deployed in the background of consciousness; they do not get into the forefront of action.

By this process of elimination the conclusion emerges that the tactual motor sensations furnish the essential data for the recognition and discrimination involved in forming the special associations at critical points. How the animal recognizes critical points it is impossible to say. The most reasonable supposition is that in the gradual formation of the motor memory of the entire course, at the established distance-intervals, the conflicting images of turning in one direction or the other spontaneously arise, resulting in indecision—the sign of recognition. There is some positive evidence in support of a distance quality in the animal's image to be remarked below. The machinery of discrimination seems tolerably clear. In any given case it consists of direct association of *the motor image of turning in one direction* with success, and the motor image of turning in the other direction, with failure. This is the explanation in the cases where one alternative is a blind alley. In the cases where the alternatives are longer and shorter we have to suppose an association between direction and distance; between turning in one direction and the distance traversed or *the time consumed.* Perhaps the quality is temporal rather than spatial; or indeed it may be that the temporal and the spatial *qualia* of this modality of sense are undifferentiated in the lower animals. The positive evidence of the distance quality noted above appears in the fact that the rats quickly adopt the shorter road where there is choice of longer or shorter. It has been pointed out that all the rats experimented with learned, sooner or later, to take the shorter road at *x*. Inasmuch as both roads lead to Rome it is difficult to see why the shorter one invariably should be selected, unless it is known as shorter, or, in other words, unless quicker satisfaction is associated with this path than with the other. It is equally difficult to see in what terms this association could be mediated other than those suggested—a distance or temporal idea in tactual-motor terms.

The fact of the invariable adoption of the 'shorter circuit' was brought out more clearly and forcibly by a special test. A normal and the blind rat *X* were used. Both had been familiar with both mazes (direct and reversed) for weeks. Their knowledge was as nearly automatic as possible. A path was opened, then, between *d* and *h*, by cutting the walls at *w* and *z*. A large part of the first (left) half of the journey thus was cut out.

In the first trial (normal rat) the rat went automatically *via* the old route—paid no attention to the new one. He likewise went directly by the opening at *z* when he came up from *e*. In the next trial, however, he took the new path unhesitatingly through *w* and *z*, and turned correctly—*i. e.*, to the right—at *h*. In the third trial he took the new path, turned correctly at *h*, then paused and went half way to *e*; paused here again, made several abortive movements each way, but finally turned correctly and went forward to C with confidence and speed. In the next two trials the right association became pretty well fixed; the new route was learned perfectly and the old abandoned.

The conduct of the blind rat was really striking. In the first trial he did not notice the new path. In the second, however, he selected it after brief hesitation. The experience of the preceding night was thus strikingly effective. Singularly enough, he perfectly acquired the new association more quickly than the normals.[1] After the second trial he rarely went astray. In all three cases, however, the old habit was quickly broken, and a new, more advantageous one established. This preference for the shorter path is difficult to explain except upon the supposition that the path is known as shorter. To charge it up to the animal's "short-circuiting tendency," or his "tendency to eliminate useless movements" is to beg the question. Unless the advantage of the new path over the old is known in some way the old habit would persist simply in virtue of its own inertia. Again we cannot speak of a direct connection between *turning* in one direction and satisfaction; no such association is formed directly. A direct association is admitted; but it is formed only after experience with both paths, and deliberation often repeated between them. It is difficult to see what the association turns upon, if not upon distance-direction ideas of motor origin.

Why should not the repetition of the same motor expenditure establish in the psychophysical organism a path, the conscious concomitant of which is capable of being discriminated from similar feelings of other expenditures, and capable of reproduction under appropriate conditions?

The chief obstacle in the way of realizing to ourselves the existence and character of such representations as those suggested above is our lack of experience with pure tactual-motor ideas. It is a pertinent illustration of the influence of psychophysical limitation. We recognize that tactual-motor experience is fundamental in our own spatial perception; but it is so grown over and obscured by visual experience that it is next to impossible for us to realize in ourselves a pure tactual-motor image. No fusion of elements is more complete than this. The imperfect isolation we are able to give the tactual-motor element in

[1] A second normal was tried with results essentially the same as obtained with the other.

our own representations helps us, however, to imagine the pure idea of this kind. I observe that my own representation of the course through the maze is strongly motor. I find it just as impossible to see the course as a static visual image as to abstract the motor image from the interfused visual elements. I can alternately make one or the other focal, though the motor seems less *real* than the visual. This imperfect and mongrel tactual-motor experience of ours gives, however, a remote suggestion of the quality and feeling of such ideas in their pure condition. When we consider the clearness and definiteness of the spatial ideas of the blind deaf-mutes it ought not to be very difficult to conceive that an animal of poor visual endowment, or of semi-subterranean habits of life, may do a large part of his thinking in tactual-motor terms; and that the content may have a clearness and fullness of meaning hard for our visually over-slaughed minds to appreciate. With no feature of this study have I been so impressed as with the possibilities it reveals of thinking in motor or tactual-motor terms.

In carrying on this study I have profited greatly by aid and suggestions from many persons, and also from many books. I wish especially to express my indebtedness to Dr. E. C. Sanford for the initial suggestion, for ample laboratory facilities, and for continued interest and helpful criticism; to Dr. L. W. Kline for practical suggestions; to President Hall, and the members of his seminary, for salutary and stimulating criticism. The published works which I have found most helpful are those of Professors Lloyd Morgan, Wesley Mills, and Edward Thorndike, and the *Examen psychologique des animaux* of M. Pierre Hachet-Souplet.

8

Reprinted from *Science* **24**:613–619 (1906)

THE SCIENTIFIC INVESTIGATION OF THE PSYCHICAL FACULTIES OR PROCESSES IN THE HIGHER ANIMALS

I. P. Pavlov

For a consistent investigator there is in the higher animals only one thing to be considered—namely, the response of the animal to external impressions. This response may be extremely complicated in comparison with the reaction of animals of a lower class. Strictly speaking, natural science is under an obligation to determine

[1] The Huxley lecture on recent advances in science and their bearing on medicine and surgery. Delivered by Professor Ivan P. Pavlov at Charing Cross Hospital on October 1, 1906. From the report in the *British Medical Journal.*

only the precise connection which exists between the given natural phenomenon and the responsive faculty of the living organism with respect to this phenomenon—or, in other words, to ascertain completely how the given living object maintains itself in constant relation with its environment. The question is simply whether this law is now applicable to the examination of the higher functions of the higher quadrupeds. I and my colleagues in the laboratory began this work some years ago, and we have recently devoted ourselves to it almost completely. All our experiments were made on dogs. The only response of the animals to external impressions was a physiologically unimportant process—namely, the excretion of saliva. The experimenter always used perfectly normal animals, the meaning of this expression being that the animals were not subjected to any abnormal influence during the experiments. By means of a systematic procedure easy of manipulation it was possible to obtain an exact observation of the work of the salivary glands at any desired time.

It is already well known that there is always a flow of saliva in the dog when something to eat is given to it or when anything is forcibly introduced into its mouth. In these circumstances the escaping saliva varies both in quality and quantity very closely in accordance with the nature of the substances thus brought into the dog's mouth. Here we have before us a well-known physiological process—namely, reflex action. It is the response of the animal to external influences, a response which is accomplished by the aid of the nervous system. The force exerted from without is transformed into a nervous impression, which is transmitted by a circuitous route from the peripheral extremity of the centripetal nerve through the centripetal nerve, the central nervous system, and the centrifugal nerve, ultimately arriving at the particular organ concerned and exciting its activity. This response is specific and permanent. Its specificity is a manifestation of a close and peculiar action of the external phenomena to physiological action, and is founded on the specific sensibility of the peripheral nerve-endings in the given nervous chain. These specific reflex actions are constant under normal vital conditions, or, to speak more properly, during the absence of abnormal vital conditions.

The responses of the salivary glands to external influences are, however, not exhausted by the above-mentioned ordinary reflex actions. We all know that the salivary glands begin to secrete, not only when the stimulus of appropriate substances is impressed on the interior surface of the mouth, but that they also often begin to secrete when other receptive surfaces, including the eye and the ear, are similarly stimulated. The actions last mentioned are, however, generally considered apart from physiology and receive the name of psychical stimuli. We will take another course, and will endeavor to restore to physiology what properly belongs to it. These exceptional manifestations unquestionably have much in common with ordinary reflex action. Every time that there is a flow of saliva attributable to this cause, the occurrence of some special stimulus among the external influences may be recognized. On very careful exercise of his attention, the observer perceives that the number of spontaneous flows of saliva forms a rapidly diminishing series, and it is in the highest degree probable that those extremely infrequent flows of saliva for which no particular cause is at first sight apparent are in reality the result of some stimulus invisible to the eye of the observer. From this it follows that the centripetal paths are always stimulated primarily and the centrifugal paths secondarily, of course, with the interposition of the central nervous system.

In the first place, they arise from all the bodily surfaces which are sensitive to stimulation, even from such regions as the eye and the ear, from which an ordinary reflex action affecting the salivary glands is never known to proceed.

It must be observed that ordinary salivary reflexes may originate not only from the cavity of the mouth, but also from the skin and the nasal cavity. In the second place, a conspicuous feature of these reflexes is that they are in the highest degree inconstant. All stimuli introduced into the mouth of the dog unfailingly give a positive result in reference to the secretion of saliva, but the same objects when presented to the eye, the ear, etc., may be sometimes efficient and sometimes not. In consequence of the last-mentioned fact, we have provisionally called the new reflexes 'conditioned reflexes,' and for the sake of distinction we have called the old ones 'unconditioned.' Every conditioned stimulus becomes totally ineffective on repetition, the explanation being that the reflex action ceases. The shorter the interval between the separate repetitions of the conditioned reflex the more quickly is this reflex obliterated. The obliteration of one conditioned reflex does not affect the operation of the others. Spontaneous restoration of the obliterated conditioned reflexes does not occur until after the lapse of one, two or more hours, but there is a way in which our reflex may be restored immediately. All that is necessary is to obtain a repetition of the unconditioned reflex—as, for instance, by pouring vinegar into the dog's mouth and then either showing it to him or letting him smell it. The action of the last-mentioned stimuli, which was previously quite obliterated, is now restored in its full extent. If for a somewhat long time—such as days or weeks continuously—a certain kind of food is shown to the animal without being given to him to eat, it loses its power of imparting a stimulus from a distance—that is, its power of acting on the eye, the nose, etc.

We may, therefore, say that the conditioned reflex is in some way dependent on the unconditioned reflex. At the same time we see also the mechanism which is necessary for the production of our conditioned reflex. When an object is placed in the mouth, some of its properties exercise an action on the simple reflex apparatus of the salivary glands, and for the production of our conditioned reflex that action must synchronize with the action of other properties of the same object when the last-mentioned action, after influencing other superficial parts of the body that are sensitive to such stimuli, arrives in other parts of the central nervous system. Just as the stimulant effects due to certain properties of an object placed in the mouth may be associated as regards time with a number of stimuli arising from other objects, so all these manifold stimuli may by frequent repetition be turned into conditioned stimuli for the salivary glands. It must be remembered that in feeding a dog or forcing something into its mouth each separate movement and each variation of a movement may by itself represent a conditioned stimulus. If that is the case, and if our hypothesis as to the origin of the conditioned reflex is correct, it follows that any natural phenomenon chosen at will may, if required, be converted into a conditioned stimulus. Any ocular stimulus, any desired sound, any odor that might be selected, and the stimulation of any portion of the skin, either by mechanical means or by the application of heat or cold, have in our hands never failed to stimulate the salivary glands, although they were all of them at one time supposed to be ineffective for such a purpose. This was accomplished by applying these stimuli simultaneously with the action of the salivary glands, this action having been evoked by the giving of certain kinds of food, or by forcing certain substances into the dog's mouth. These artificial conditioned reflexes, the product of our training, showed exactly the same properties as the natural conditioned reflexes previously described. As regards their obliteration and restoration, they followed essentially the same laws as the natural conditioned reflexes.

Up to the present time the stimuli with which we had to do were comparatively few in number, but were constant in action. Now, however, in another more complicated portion of the nervous system we encounter a new phenomenon—namely, the conditioned reflex. On the one hand, the nervous apparatus is responsive in the highest degree—that is, it is susceptible to the most varied external stimuli, but, on the other hand, these stimuli are not constant in their operation and are not uniformly associated with a definite physiological effect. The introduction of the idea of conditioned reflexes into physiology seems to me to be justified because it corresponds to the facts that have been adduced, since it represents a direct inference from them. It is in agreement with the general mechanical hypotheses of natural science. It is completely covered by the ideas of paths and inhibition, ideas which have been sufficiently worked up in the physiological material of the present day. Finally, in these conditioned stimuli, looked at from the point of view of general biology, there is nothing but a very complete mechanism of accommodation or, which amounts to the same thing, a very delicate apparatus for maintaining the natural equilibrium. There are reasons for considering the process of the conditioned reflex to be an elementary process—namely, a process which really consists in the coincidence of any one of the innumerable vague external stimuli with a stimulated condition of any point in a certain portion of the central nervous system. In this way for the time being a path is made by which the stimulus may reach the given point.

Although there are differences in the time required for the establishing of the conditioned reflexes, some proportionality may be perceived. From our experiments it is very evident that the intensity of the stimulation is of essential importance. In contradistinction to this we must state with regard to acoustic impressions that very powerful stimuli, such as the violent ringing of a bell, were not, in comparison with weaker stimuli, quick to produce conditioned increase of function in the salivary glands. It must be supposed that powerful acoustic stimuli produce in the body some other important reaction which hinders the development of the salivary reaction.

What is it that the nervous system of the dog recognizes as individual phenomena of external origin? or, in other words, what are the elements of a stimulus? If the application of cold to a definite area of the skin acts as a conditioned stimulus of the salivary glands, the application of cold to another portion of the skin causes secretion of saliva on the very first occasion. This shows that the stimulus of cold generalizes itself over a considerable portion of the skin, or perhaps even over the whole of it.

Stimulation by musical sounds or by noise in general is remarkably convenient for determining the discriminating or analytical faculty of the nervous system of the dog. In this respect the precision of our reaction goes a great way. If a certain note of an instrument is employed as a conditioned stimulus, it often happens that not only all the notes adjoining it, but even those differing from it by a quarter of a tone, fail to produce any effect. Musical *timbre* is recognized with similar or even much greater precision.

We have hitherto spoken of the analytical

faculty of the nervous system as it presents itself to us in, so to say, the finished condition. We have now accumulated material which contains evidence of a continuous and great increase of this faculty if the experimenter perseveres in subdividing and varying the conditioned stimulus, and thereby makes it coincide with the unconditioned stimulus. Here, again, is a new field of enormous extent. In this material relative to the conditioned stimuli there are not a few cases in which an evident connection between the effect and the intensity of a stimulus can be seen. As soon as a temperature of 50° C. had begun to induce a flow of saliva it was found that even a temperature of 30° C. had a similar effect but in a much less degree. Trial was then made of combinations consisting of stimuli of the same kind and also of stimuli of different kinds. The simplest example is a combination of different musical notes, such as a harmonic chord, which consists of three notes. When this is employed as a conditioned stimulus each two notes together and each separate note of the chord produce an effect, but the notes played two and two together accomplish less than the whole, and the notes played separately accomplish less than those played in pairs. The case becomes more complicated when we employ as a conditioned stimulus a combination of stimuli of different kinds, that is, of stimuli acting upon different kinds of susceptible surfaces. Only a few of such combinations have been provisionally experimented with. In these cases for the most part one of the stimuli was a conditioned stimulus. In a combination in which rubbing and cold were employed the former was preponderant as a conditioned stimulus while the application of cold taken by itself produces a hardly perceptible effect. But if an attempt is made to convert the weaker stimulus separately into a conditioned stimulus it soon becomes an energetic conditioned stimulus. If we now apply the two stimuli together we have before us an evident case of them acting in combination. The following problem had for its object to explain what happens to an active-conditioned stimulus when a new stimulus is added to it. In the cases that were examined, the action of the preexisting conditioned stimulus was hindered when a new stimulus of a like kind was added to it. A new odor of a like kind hindered the operation of another odor which was already acting as a conditioned stimulus; a new musical note similarly hindered the operation of the note previously employed which was a conditioned stimulus. After a conditioned stimulus had been applied, together with another one which inhibited its action, the action of the first one alone was greatly weakened and sometimes even stopped altogether. This is either an after-effect of the inhibiting stimulus which was added or it is the obliteration of the conditioned reflex, because in the experiment of the added stimulus the conditioned reflex is not strengthened by the unconditioned reflex. The inhibition of the conditioned reflex is also observed in the converse case. When you have a combination of stimuli acting as a conditioned stimulus—in which, as has been already stated, one of the stimuli by itself produces almost no effect—frequent repetition of the powerful stimulus by itself without the other one leads to a powerful inhibition of its action, even to the extent of its action being almost destroyed. The relative magnitudes of all these manifestations of stimulation and inhibition have a very close connection with their dependence on the conditions under which they originate.

Experiments have been made in the production of conditioned reflexes by traces or latent remnants both of a conditioned and of an unconditioned stimulus. The method was that a conditioned stimulus was either allowed to act for one minute immediately in advance of an unconditioned stimulus or it was even applied two minutes earlier. Conversely, also, the conditioned stimulus was not brought into action until the unconditioned reflex was at an end. In all these cases the conditioned reflex developed itself; but in the cases in which the conditioned stimulus was applied three minutes before the unconditioned one, and was separated from the latter by an interval of two minutes, we obtained a condition which was quite unexpected and extremely peculiar, but was always repeated. When scratching was applied to a particular spot—for instance, as a conditioned stimulus—after it began to produce an effect it was found that scratching of any other place also produced an effect, just as in the case of cold or heat applied to the skin, new musical sounds, optical stimuli and odors. The unusually copious secretion of saliva, and the extremely expressive movements of the animal attracted our attention. It may appear that this manifestation is of a different kind from those with which we have hitherto been occupied. The fact was that in the earlier experiments at least one coincidence of the conditioned stimulus with the unconditioned one was necessary, but on the present occasion manifestations which had never occurred simultaneously with an unconditioned reflex were acting as conditioned stimuli. Here an unquestionable point of difference naturally comes to light, but at the same time there is also to be seen another essential property of these manifestations which they have in common with the former ones—that is, the existence of a very sensitive point in the central nervous system, and in consequence of its position this point becomes the destination of all the important stimuli coming from the external world to make impressions on the receptive cells of the higher regions of the brain.

Three characteristic features of this subject make a deep impression upon him who works at it. In the first place, these manifestations present great facilities for exact investigation. I am here referring to the ease with which they may be repeated, to their character of uniformity under similar conditions of environment, and to the fact that they are capable of further subdivision experimentally. In the second place, it is inevitable that opinions formed on this subject must be objective only. In the third place, the subject involves an unusual abundance of questions. To what departments of physiology does it correspond? It corresponds partly to what was in former days the physiology of the organs of special sense and partly to the physiology of the central nervous system.

Up to the present time the physiology of the eye, the ear and other superficial organs which are of importance as recipients of impressions has been regarded almost exclusively in its subjective aspect; this presented some advantages, but at the same time, of course, limited the range of inquiry. In the investigation of the conditioned stimuli in the higher animals, this limitation is got rid of and a number of important questions in this field of research can be at once examined with the aid of all the immense resources which experiments on animals place in the hand of the physiologist. The investigation of the conditioned reflexes is of very great importance for the physiology of the higher parts of the central nervous system. Hitherto this department of physiology has throughout most of its extent availed itself of ideas not its own, ideas borrowed from psychology, but now there is a possibility of its being liberated from such evil influences. The conditioned reflexes lead us to the consideration of the position of animals in nature; this is a subject of immense extent and one that must be treated objectively.

Broadly regarded, physiology and medicine are inseparable. Since the medical man's object is to remedy the various ills to which the human body is liable, every fresh discovery in physiology will sooner or later be serviceable to him in the preservation and repair of that wonderful structure. It is an extreme satisfaction to me that in honoring the memory of a great physiologist and man of science I am able to make use of ideas and facts which from a unique standpoint affording every prospect of success throw light upon the highest and most complicated portion of the animal mechanism.

Editor's Comments on Papers 9 and 10

9 THORNDIKE
Excerpt from *Laws and Hypotheses for Behavior*

10 SKINNER
Superstition in the Pigeon

Among the most fascinating instances of animal behavior are those that appear to be "purposive." Few topics generate as much controversy among students of animal learning as does the question of how apparently purposive behavior is to be accounted for and characterized. There is still no universally accepted answer to this question, although it has been a major focus of the field of animal learning for years. Of the various answers that have been proposed (and often defended with religious fervor), none has been more influential than the law of effect in its various forms.

Before Thorndike proposed the law of effect, behavior that appeared willful or purposeful seemed to be determined by what the animal foresaw the consequences of its behavior would be. The directedness of much learned behavior, such as that of a rat in a maze, seemed to demand interpretation as "volition," behavior with a psychological purpose guiding it.

Thorndike's law of effect provided an alternative interpretation. Voluntary behavior could be accounted for in terms of an animal's past experience with "satisfying" and "annoying" events. The law of effect accounts for the directedness of behavior in terms of its past consequences, rather than its future ones. The statement of the law of effect and of a second principle, the law of exercise, reprinted here as Paper 9, is from Thorndike's 1911 book, *Animal Intelligence*. This version of the law of effect includes the assertion that responses closely followed by discomfort will "have their connections with that situation weakened." Thorndike later (1931) revised this part of his law, arguing: "Annoyers do not act on learning in general by weakening whatever connection they follow. If they do anything to learning they do it indirectly."

The person most responsible for elaborations and extensions of Thorndike's law of effect is B. F. Skinner. An articulate defender of a sophisticated version of behaviorism (see Skinner, 1974), Skinner has made many important empirical contributions to experimental psychology. Behavioral techniques he developed are widely employed by behavioral scientists. Although there is considerable controversy about Skinner's version of behaviorism, his typology of behavior, and his view of human nature, the manipulative power of the behavioral techniques he developed and studied is acknowledged even by his most severe critics.

Paper 10 reports behavior in pigeons that appears to be analogous in some ways to human superstition. The paper demonstrates how Skinner's interpretational scheme can be applied to a rather unusual behavioral observation. Further analysis of the behavior of pigeons in situations similar to that employed by Skinner suggests that his interpretation of the behavior is not entirely adequate (Staddon and Simmelhag, 1971), and there are some internal conceptual difficulties with the interpretational edifice (Schick, 1971). Nevertheless, this report describes a striking class of behaviors and provides a provocative interpretation.

REFERENCES

Schick, K., 1971, Operants. *J. Exp. Anal. Behav.* **19**:413–423.

Skinner, B. F., 1974, *About Behaviorism,* Alfred A. Knopf, New York.

Staddon, J. E. R., and V. L. Simmelhag, 1971, The "superstition" experiment: A re-examination of its implications for the principles of adaptive behavior, *Psychol. Rev.* **78**:3–43.

Thorndike, E. L., 1931, *Human Learning,* Century, New York.

9

Reprinted from pp. 244–245 of *Animal Intelligence*, Macmillan, New York, 1911, 297 p.

LAWS AND HYPOTHESES FOR BEHAVIOR

E. L. Thorndike

[*Editor's Note:* In the original, material precedes this excerpt.]

The Law of Effect is that: *Of several responses made to the same situation, those which are accompanied or closely followed by satisfaction to the animal will, other things being equal, be more firmly connected with the situation, so that, when it recurs, they will be more likely to recur; those which are accompanied or closely followed by discomfort to the animal will, other things being equal, have their connections with that situation weakened, so that, when it recurs, they will be less likely to occur. The greater the satisfaction or discomfort, the greater the strengthening or weakening of the bond.*

The Law of Exercise is that: *Any response to a situation will, other things being equal, be more strongly connected with the situation in proportion to the number of times it has been connected with that situation and to the average vigor and duration of the connections.*

These two laws stand out clearly in every series of experiments on animal learning and in the entire history of the management of human affairs. They give an account of learning that is satisfactory over a wide range of experience, so long as all that is demanded is a rough and general means of prophecy. We can, as a rule, get an animal to learn a given accomplishment by getting him to accomplish it, rewarding him when he does, and punishing him when he does not; or, if reward or punishment are kept indifferent, by getting him to accomplish it much oftener than he does any other response to the situation in question.

For more detailed and perfect prophecy, the phrases 'result in satisfaction' and 'result in discomfort' need further definition, and the other things that are to be equal need comment.

By a satisfying state of affairs is meant one which the animal does nothing to avoid, often doing such things as attain and preserve it. By a discomforting or annoying state of affairs is meant one which the animal commonly avoids and abandons.

The satisfiers for any animal in any given condition cannot be determined with precision and surety save by observation. Food when hungry, society when lonesome, sleep when fatigued, relief from pain, are samples of the common occurrence that what favors the life of the species satisfies its individual members. But this does not furnish a completely valid rule.

The satisfying and annoying are not synonymous with favorable and unfavorable to the life of either the individual or the species. Many animals are satisfied by deleterious conditions. Excitement, overeating, and alcoholic intoxication are, for instance, three very common and very potent satisfiers of man. Conditions useful to the life of the species in moderation are often satisfying far beyond their useful point: many conditions of great utility to the life of the species do not satisfy and may even annoy its members.

[*Editor's Note:* In the original, material follows this excerpt.]

10

Reprinted by permission from *J. Exp. Psychol.* **38**:168–172 (1948)

'SUPERSTITION' IN THE PIGEON

BY B. F. SKINNER

Indiana University

To say that a reinforcement is contingent upon a response may mean nothing more than that it follows the response. It may follow because of some mechanical connection or because of the mediation of another organism; but conditioning takes place presumably because of the temporal relation only, expressed in terms of the order and proximity of response and reinforcement. Whenever we present a state of affairs which is known to be reinforcing at a given drive, we must suppose that conditioning takes place, even though we have paid no attention to the behavior of the organism in making the presentation. A simple experiment demonstrates this to be the case.

A pigeon is brought to a stable state of hunger by reducing it to 75 percent of its weight when well fed. It is put into an experimental cage for a few minutes each day. A food hopper attached to the cage may be swung into place so that the pigeon can eat from it. A solenoid and a timing relay hold the hopper in place for five sec. at each reinforcement.

If a clock is now arranged to present the food hopper at regular intervals *with no reference whatsoever to the bird's behavior*, operant conditioning usually takes place. In six out of eight cases the resulting responses were so clearly defined that two observers could agree perfectly in counting instances. One bird was conditioned to turn counter-clockwise about the cage, making two or three turns between reinforcements. Another repeatedly thrust its head into one of the upper corners of the cage. A third developed a 'tossing' response, as if placing its head beneath an invisible bar and lifting it repeatedly. Two birds developed a pendulum motion of the head and body, in which the head was extended forward and swung from right to left with a sharp movement followed by a somewhat slower return. The body generally followed the movement and a few steps might be taken when it was extensive. Another bird was conditioned to make incomplete pecking or brushing movements directed toward but not touching the floor. None of these responses appeared in any noticeable strength during adaptation to the cage or until the food hopper was periodically presented. In the remaining two cases, conditioned responses were not clearly marked.

The conditioning process is usually obvious. The bird happens to be executing some response as the hopper appears; as a result it tends to repeat this response. If the interval before the next presentation is not so great that extinction takes place, a second 'contingency' is probable. This strengthens the response still further and subsequent reinforcement becomes more probable. It is true that some responses go unreinforced and some reinforcements appear when the response has not just been made, but the net result is the development of a considerable state of strength.

With the exception of the counter-clockwise turn, each response was almost always repeated in the same part of the cage, and it generally involved an orientation toward some feature of the cage. The effect of the reinforcement was to condition the bird to respond to some aspect of the environment rather than merely to execute a series of movements. All responses came to be repeated rapidly between reinforcements—typically five or six times in 15 sec.

The effect appears to depend upon the rate of reinforcement. In general, we should expect that the shorter the intervening interval, the speedier and more marked the conditioning. One reason is that the pigeon's behavior becomes more diverse as time passes after reinforcement. A hundred photographs, each taken two sec. after withdrawal of the hopper, would show fairly uniform behavior. The bird would be in the same part of the cage, near the hopper, and probably oriented toward the wall where the hopper has disappeared or turning to one side or the other. A hundred photographs taken after 10 sec., on the other hand, would find the bird in various parts of the cage responding to many different aspects of the environment. The sooner a second reinforcement appears, therefore, the more likely it is that the second reinforced response will be similar to the first, and also that they will both have one of a few standard forms. In the limiting case of a very brief interval the behavior to be expected would be holding the head toward the opening through which the magazine has disappeared.

Another reason for the greater effectiveness of short intervals is that the longer the interval, the greater the number of intervening responses emitted without reinforcement. The resulting extinction cancels the effect of an occasional reinforcement.

According to this interpretation the effective interval will depend upon the rate of conditioning and the rate of extinction, and will therefore vary with the drive and also presumably between species. Fifteen sec. is a very effective interval at the drive level indicated above. One min. is much less so. When a response has once been set up, however, the interval can be lengthened. In one case it was extended to two min., and a high rate of responding was maintained with no sign of weakening. In another case, many hours of responding were observed with an interval of one min. between reinforcements.

In the latter case, the response showed a noticeable drift in topography. It began as a sharp movement of the head from the middle position to the left. This movement became more energetic, and eventually the whole body of the bird turned in the same direction, and a step or two would be taken. After many hours, the stepping response became the predominant feature. The bird made a well defined hopping step from the right to the left foot, meanwhile turning its head and body to the left as before.

When the stepping response became strong, it was possible to obtain a mechanical record by putting the bird on a large tambour directly connected with a small tambour which made a delicate electric contact each time stepping took place. By watching the bird and listening to the sound of the recorder it was possible to confirm the fact that a fairly authentic record was being made. It was possible for the bird to hear the recorder at each step, but this was, of course, in no way correlated with feeding. The record obtained when the magazine was presented once every min. resembles in every respect the characteristic curve for the pigeon under periodic reinforcement of a standard selected response. A well marked temporal discrimination develops. The bird does not respond immediately after eating, but when 10 or 15 or even 20 sec. have elapsed it begins to respond rapidly and continues until the reinforcement is received.

In this case it was possible to record the 'extinction' of the response when the clock was turned off and the magazine was no longer presented at any time. The bird continued to respond with its characteristic side to

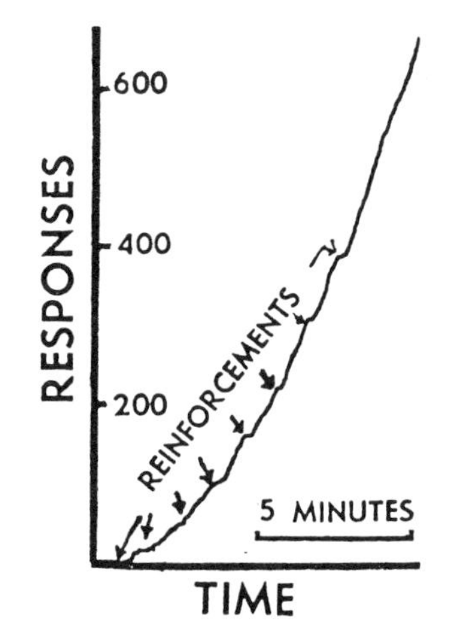

Fig. 1. 'Reconditioning' of a superstitious response after extinction. The response of hopping from right to left had been thoroughly extinguished just before the record was taken. The arrows indicate the automatic presentation of food at one-min. intervals without reference to the pigeon's behavior.

side hop. More than 10,000 responses were recorded before 'extinction' had reached the point at which few if any responses were made during a 10 or 15 min. interval. When the clock was again started, the periodic presentation of the magazine (still without any connection whatsoever with the bird's behavior) brought out a typical curve for reconditioning after periodic reinforcement, shown in Fig. 1. The record had been essentially horizontal for 20 min. prior to the beginning of this curve. The first reinforcement had some slight effect and the second a greater effect. There is a smooth positive acceleration in rate as the bird returns to the rate of responding which prevailed when it was reinforced every min.

When the response was again extinguished and the periodic presentation of food then resumed, a different response was picked up. This consisted of a progressive walking response in which the bird moved about the cage.

The response of hopping from side to side never reappeared and could not, of course, be obtained deliberately without making the reinforcement contingent upon the behavior.

The experiment might be said to demonstrate a sort of superstition. The bird behaves as if there were a causal relation between its behavior and the presentation of food, although such a relation is lacking. There are many analogies in human behavior. Rituals for changing one's luck at cards are good examples. A few accidental connections between a ritual and favorable consequences suffice to set up and maintain the behavior in spite of many unreinforced instances. The bowler who has released a ball down the alley but continues to behave as if he were controlling it by twisting and turning his arm and shoulder is another case in point. These behaviors have, of course, no real effect upon one's luck or upon a ball half way down an alley, just as in the present case the food would appear as often if the pigeon did nothing—or, more strictly speaking, did something else.

It is perhaps not quite correct to say that conditioned behavior has been set up without any previously determined contingency whatsoever. We have appealed to a uniform sequence of responses in the behavior of the pigeon to obtain an over-all net contingency. When we arrange a clock to present food every 15 sec., we are in effect basing our reinforcement upon a limited set of responses which frequently occur 15 sec. after reinforcement. When a response has been strengthened (and this may result from one reinforcement), the setting of the clock implies an even more restricted contingency. Something of the same sort is true of the bowler. It is not quite correct to say that there is no connection between his twisting and turning and the course taken by the ball at the far end of the alley. The connection was established before the ball left the bowler's hand, but since both the path of the ball and the behavior of the bowler are determined, some relation survives. The subsequent behavior of the bowler may have no effect upon the ball, but the behavior of the ball has an effect upon the bowler. The contingency, though not perfect, is enough to maintain the behavior in strength. The particular form of the behavior adopted by the bowler is due to induction from responses in which there is actual contact with the ball. It is clearly a movement appropriate to changing the ball's direction. But this does not invalidate the comparison, since we are not concerned with what response is selected but with why it persists in strength. In rituals for changing luck the inductive strengthening of a particular form of behavior is generally absent. The behavior of the pigeon in this experiment is of the latter sort, as the variety of responses obtained from different pigeons indicates. Whether there is any unconditioned behavior in the pigeon appropriate to a given effect upon the environment is under investigation.

The results throws some light on incidental behavior observed in experiments in which a discriminative stimulus is frequently presented. Such a stimulus has reinforcing value and can set up superstitious behavior. A pigeon will often develop some response such as turning, twisting, pecking near the locus of the discriminative stimulus, flapping its wings, etc. In much of the work to date in this field the interval between presentations of the discriminative stimulus has been one min. and many of these superstitious responses are short-lived. Their appearance as the result of accidental correlations with the presentation of the stimulus is unmistakable.

(Manuscript received June 5, 1947)

Editor's Comments on Papers 11 Through 14

Once Pavlov, Thorndike, Small, Yerkes, and other early investigators had demonstrated the feasibility of studying learning in animals with controlled experiments, the field grew with rapidly accelerating speed. Within the field of psychology, the study of animal learning had become something of a major industry by the 1930s and 1940s. Building on the concepts and methodologies introduced by the early researchers, students of animal learning constructed theoretical and empirical edifices that were broad in scope and vision; much of the development of psychology in this century has been influenced by studies of animal learning. It is impossible in a volume of this sort to include any but a few of the important papers that arose from this enterprise. The papers reprinted in this section were chosen because they reflect some of the important directions taken by work that developed out the general approach instituted by Thorndike and Pavlov.

Paper 11 is a report by two Polish scientists, Miller and Konorski, arguing that there are two different kinds of conditioned reflex. The question of whether there are different kinds of learning continues to be debated. With different procedures, different species, and different stimuli, different behavioral results are obtained. To what extent do these differences represent variations in a common, core learning process, and to what extent do they represent qualitatively different kinds of plasticity? For Pavlov, the conditioned reflex was the only unit of plasticity required to account for learning, although he used the reflex concept in a very broad way, including, for example, a "reflex of freedom" in his behavior typology. In Paper 11 (which was translated by B. F. Skinner), Miller and Konorski present an argument for distinguishing two different kinds of conditioned reflex, based on the amount and the role of response-produced feedback involved in the response. This early paper paved the way for the subsequent development of the distinction between Pavlovian conditioning and instrumental learning. As it was developed by Skinner (1935), Schlosberg (1937), Mowrer (1947), and others (see Rescorla and Solomon, 1967, for a review), the two-process view of learning has proved to be conceptually powerful and heuristically useful. Konorski made many other contributions in a distinguished and prolific career, continually shifting and changing his theories to conform with newly collected data. Summaries of important parts of his work can be found in his books (Konorski, 1948, 1967).

Paper 12 reports on one of the few experiments ever performed by the prominent learning theorist E. R. Guthrie. Using a variant of Thorndike's puzzle box and ingeniously employing photography to record behavior, Guthrie and Horton demonstrated remarkable stereotypy in their cats' learned behavior. The mechanical appearance of the behavior of the cats makes a strong impression, and this demonstration helped render plausible a deterministic interpretation of learned, goal-directed behavior. The precision provided by the photographic technique in isolating a single part of the behavior sequence permitted constancies in behavior to be discovered despite the complexity of the behavior sequence as a whole. However, Guthrie and Horton failed to notice that they were recording a common, species-typical, headrubbing response. Moore and Studdard (1979) have shown that the cats were probably giving a "greeting" response to the experimenters, who were present during the sessions. Guthrie's very inflential theoretical interpretation of the behavior therefore appears to be incorrect. This illustrates vividly the degree to which psychologists of the time were insensitive to natural behaviors shown by the animals they studied.

The study of learning based on rewards is intricately linked to the study of motivation and valuation. Paper 13 by Wolfe is an important demonstration of acquired values in chimpanzees. By training the chimpanzees to trade poker chips for food, Wolfe was able to instill a rudimentary sort of monetary system in his animals. The chimps performed in a variety of discrimination problems with token chips, rather than primary reward, as their payment. Such secondary reinforcement is believed by many to be based on the sorts of anticipations developed in Pavlovian conditioning; cues that signal the occurrence of reward become reinforcing themselves. Wolfe's experiments with chimps clearly showed the power of secondary reward.

The potential mediating role of secondary reinforcement posed serious problems for investigators interested in the applicability of the principle of association by recency to animal learning. An number of experiments demonstrated that animals could learn new response patterns even when a substantial delay was interposed between the occurrence of the behavior and the delivery of the reward. Did this mean that recency was unimportant? To determine this, steps had to be taken to remove as many sources of secondary reinforcement as possible to ensure that the delay interval was bridged through the presence of cues functioning as secondary reinforcers. The Grice experiment (Paper 14) used a discrimination task to avoid possible sources of secondary reinforcement. The steepness of the delay-of-reinforcement gradient obtained with the rats in this experiment was an influential discovery, affecting the subsequent development of theories that emphasized the role of secondary reinforcement in controlling behavior (for example, Spence, 1956; Mowrer, 1960). Although in other kinds of learning situations the degree of recency required for learning in Grice's experiment is not always necessary (for example, Garcia et al., 1966), the huge effect seen in the Grice experiment establishes recency as an important determinant of some kinds of learning.

REFERENCES

Garcia, J., F. R. Ervin, and R. A. Koelling, 1966, Learning with prolonged delay of reinforcement, *Psychon. Sci.* **5**:121–122.

Konorski, J., 1948, *Conditioned Reflexes and Neuron Organization*, Cambridge University Press, Cambridge, England.

Konorski, J., 1967, *Integrative Activity of the Brain*, University of Chicago Press, Chicago.

Moore, B. R., and S. Studdard, 1979, Dr. Guthrie and *Fels domesticus* Or: Tripping over the cat, *Science* **205**:1031–1032.

Mowrer, O. H., 1947, On the dual nature of learning—a reinterpretation of "conditioning" and "problem-solving," *Harvard Ed. Rev.* **17**:101–148.

Mowrer, O. H., 1960, *Learning Theory and Behavior*, Wiley, New York.

Rescorla, R. A., and R. L. Solomon, 1967, Two-process learning theory: Relationships between Pavlovian conditioning and instrumental learning, *Psychol. Rev.* **74**:151–182.

Schlosberg, H., 1937, The relationship between success and the laws of conditioning, *Psychol. Rev.* **44**:379–394.

Skinner, B. F., 1935, Two types of conditioned reflex and a pseudo type. *J. Gen. Psychol.* **12**:66–77.

Spence, K. W., 1956, *Behavior Theory and Conditioning*, Yale University Press, New Haven, Conn.

11

Reprinted from *J. Exp. Anal. Behav.* **12**:187–189 (1969)

ON A PARTICULAR FORM OF CONDITIONED REFLEX[1]

S. MILLER AND J. KONORSKI

[*Translator's Note: This important paper appeared under the title "Sur une forme particulière des reflexes conditionnels" in* Les comptes rendus des seances de la société de biologie. Société polonaise de biologie. *Volume XCIX, page 1155, June 1928. When my paper "Two types of conditioned reflex and a pseudo-type" was published in* The Journal of General Psychology *(1935,* 12, *66-67), Konorski and Miller replied in a paper called "On two types of conditioned reflex", which appeared in the same journal (1937,* 16, *264-272). They sent me a copy and I was therefore able to answer in the same issue. The paper was called "Two types of conditioned reflex: a reply to Konorski and Miller". (1937,* 16, *272-279.)*

The present translation was sent to Professor Konorski and some changes suggested by him have been made. The word particulière *has a much richer meaning in French than in English. In addition to* personal *or* private, *it suggests something* special *or* unusual. *A key phrase appears in French as follows: The dog flexes its leg* pour former ainsi le complexe conditionnel total.

Professor Konorski has supplied the Postscript, giving his present views.

B. F. Skinner]

The work presented below is based on Pavlov's theory of conditioned reflexes. As a starting point, we have taken the experiments of Zeleny, Manuilov, and Krylov on the synthetic activity of the cerebral cortex, who have established the following fact: when a compound consisting of two stimuli, A and B, is accompanied by an unconditioned stimulus R, which reinforces only the whole of the compound without reinforcing either of its components presented separately, a conditioned reflex is established only to the compound AB, whereas the conditioned response is not evoked by either of the two stimuli presented separately. Let us take now: for A any stimulus whatsoever, for example, a tone produced on the piano; for B all the sensations generated by a particular movement such as lifting the leg—that is, a set of muscular, tactile, and other sensations. We can generate this set either by causing the dog to lift its leg as a reflex response (active movement) or by lifting its leg (passive movement). Finally, for R, let us take first (Case I) the presentation of food. According to Pavlov's theory, after some time the compound AB will function alone as a conditioned stimulus, for it alone has been reinforced by the presentation of food. But at the same time a new phenomenon will appear which is not predicted by Pavlov's theory: after some time the lifting of the leg, whether reflex or passive, becomes superfluous, because the sound of the piano itself will elicit the movement. The stimulus A will then be capable of provoking by itself the appearance of the stimulus B—in order to complete the conditioned compound. On the other hand, if we take for R beating the dog or puffing air into its ear—that is to say, presenting any stimulus which elicits a defense reaction (Case II)—then the stimulus A will act as an inhibitor of movement B: this movement will not appear in spite of the presentation of the stimulus which normally elicits it, unless that stimulus is stronger than the stimulus R.

We see then that there are two kinds of stimuli: some, such as food, when they always follow the compound composed of an external stimulus plus the sensations generated by a movement (and only that compound), will cause the external stimulus to elicit this movement; others, like a blow, presented after the compound, will cause the stimulus to inhibit the movement. We call the first stimuli positive, the other stimuli negative. All stimuli may be placed in one of these two categories. It is obvious that positive stimuli

[1]Reprints may be obtained from B. F. Skinner, Psychology Dept., Harvard University, William James Hall, Cambridge, Mass. 02138.

correspond in psychology to pleasant stimuli and the negative to unpleasant. Thus, the experiments described provide an objective physiological definition of pleasant and unpleasant stimuli. We shall not use the latter terms, since they are not purely physiological. The phenomena that we have just described have the same general properties as conditioned reflexes: they originate without doubt in the cortex, and they are not innate but are formed and disappear during the life of the individual. It is for this reason that we regard them as conditioned reflexes, but their mechanism is different from that of the conditioned reflexes of Pavlov, hence we call them conditioned reflexes of the second type.

The principal characteristics of the second type of conditioned reflex are as follows: the conditioned stimulus A (conditioned stimulus of the second type) can be either any transient agent or, as we will show in our next communications, a long-lasting condition. The response B appears in the form of any movement of the animal or of the inhibition of movement. The reinforcing stimulus R can be a positive or negative stimulus which evokes a specific reaction, in opposition to the conditioned reflexes of Pavlov where the response to a reinforcing stimulus must be the same as that of the conditioned stimulus.

In addition to these two varieties of conditioned reflexes of the second type, there are others. Case III: if the stimulus A is followed by the negative stimulus R and if after the compound consisting of stimulus A plus movement B, the stimulus R does not appear, then, after some time, the stimulus A begins to evoke the movement B. Case IV: if the stimulus A is followed by a positive stimulus R, and if, after the compound AB, the stimulus R does not follow, then, after some time, the stimulus A begins to inhibit the movement B (it should be noted that our experiments on these two cases are not yet completed).

It is important to see whether conditioned reflexes of the second type can be reduced to the conditioned reflexes of Pavlov. We have demonstrated that: 1) the conditioned reflex of the second type is not the ordinary first type of conditioned reflex $A \rightarrow B$, because: (a) A is not always followed by B; (b) even if the conditioned reflex $A \rightarrow B$ were established, this reflex no longer reinforced by the unconditioned stimulus eliciting the movement B would have disappeared; 2) the conditioned reflex of the second type is not the second-order conditioned reflex of Pavlov; 3) the conditioned reflex of the second type (Case II) is not Pavlovian external inhibition [we do not present demonstrations of points 2 and 3]. Hence, unable to reduce the reflexes of the second type to the conditioned reflexes of Pavlov, we should consider them consequently as the second fundamental mechanism of the function of the cerebral cortex.

The laws which apply to conditioned reflexes of the second type (laws of inhibition, irradiation, concentration, and so on) are probably the same as those of conditioned reflexes of the first type. We have verified the following laws: (1) the law of generalization holds for conditioned reflexes of the second type; (2) in conditioned reflexes of the second type, one can produce differential inhibition; (3) extinction of the reflex is identical with that of the conditioned reflex of the first type; (4) in conditioned reflexes of the second type, we have applied conditioned inhibition, and it has been demonstrated that the inhibiting stimulus inhibits not only the primary reflex but also the other reflexes.

Differences between conditioned reflexes of the first and second types are considerable. (1) In the ordinary conditioned reflex, the conditioned stimulus always elicits the same reaction as the reinforcing stimulus, while in conditioned reflexes of the second type these reactions are different. Hence we can form "families" of reflexes, all the reflexes of a given family being reinforced by the same unconditioned stimulus but differing among themselves by their centripetal and centrifugal links. (2) The role of Pavlov's conditioned reflex is limited solely to signalization, while the role of conditioned reflexes of the second type is entirely different: whether they serve to complete a given stimulus in order to compose the conditioned compound, if this compound is followed by a positive agent, or whether they prevent the appearance of the conditioned compound, if this compound is followed by a negative agent. (3) In conditioned reflexes of the first type, the reaction is effected by organs innervated through the central or autonomic nervous system, while, in conditioned reflexes of the second type, the effector can probably be only a striate muscle.

POSTSCRIPT

This paper, published 40 years ago, was the first of a series of papers written by Miller and myself in the prewar period; it was concerned with a new experimental procedure which led us to the concept of "type II conditioned reflex". The experimental results obtained by us in that period are now available, since they have been presented systematically in Chapter VIII of my recent book "Integrative Activity of the Brain", published by the University of Chicago Press. Here I would like to make some comments on this first paper.

Our original thesis claiming that for the formation of type II conditioned reflexes the compound composed of an exteroceptive stimulus and proprioceptive stimulus must be differentiated from its components is now of only historical interest. In fact, it was shown by special experiments that proprioception of a trained movement does not play any essential role in type II conditioning (Konorski, op. cit., Chapter XI).

The notion that the source of the trained motor act may derive either from active (reflex) movement, or from purely passive movement is also not correct, since it has been shown that passive movement as a rule cannot be instrumentalized, unless it includes reflex elements (Konorski, op. cit., Chapter XI).

The thesis that the effector of the type II conditioned reflex is "probably only the striate muscle" seems now to be seriously undermined by a recent series of most important experimental research guided by Professor Neal Miller.

I cannot understand why we claimed that all *stimuli can be classified either to the category of "positive" (or attractive) or "negative" (or aversive) stimuli. In fact, at that time we were well aware of the existence of "neutral" stimuli, which, after habituation, are neither positive nor negative. I don't think that we repeated that thesis in our later papers.*

Finally, the sharp distinction between, not only the procedural side of type I and type II conditioned reflexes, but also between their physiological mechanisms seems to me now largely exaggerated. In fact, further investigation shows with increasing clarity that both types can be explained on the basis of the same general principles of connectionistic processes.

To sum up, we may come to the conclusion that almost every single thesis of the above paper is more or less erroneous. I consider this fact very fortunate, because it shows that further experimentation has led to an increasing clarification of our ideas concerning one of the most important problems in brain physiology: the intimate nature of type II, alias operant, alias instrumental, alias voluntary activities of the organism.

12A

Reprinted from pp. 15–26 of *Cats in a Puzzle Box*, Rinehart, New York, 1946, 67 p., by permission of Holt, Rinehart and Winston and Edwin R. Guthrie and George P. Horton

BEHAVIOR IN THE PUZZLE BOX

E. R. Guthrie and G. P. Horton

The whole purpose and method of this experiment are different from the purpose and method of earlier experiments with the puzzle box or the purpose and methods of the maze. Other experiments have been concerned with goal-achievement and have defined learning in terms of accomplishment, improvement, or other end results of the action, not in terms of the nature of the change in the action itself. In other experiments records are kept of the reduction in errors, reduction in time, the effects on time and errors of different forms of incitement and degrees of incitement, the degrees of complexity or difficulty of the goal, or the effects of different spacing of trials. None of these experimental aims is concerned with the problem of how the animal learns the solution or with the possible application of associative learning in bringing about improvement in performance. *Success is a quality extrinsic to the process of learning and depends on the animal's environment and the accidents of that environment.*

Our concern was with several problems: (1) What is the nature of the behavioral changes that occur in the course of the experiment? (2) How does success affect the process? (3) Does the principle of association apply to all changes in response to situation? This made paramount the need to observe in detail just what the animals did in terms of movement rather than in terms of accomplishment. We were interested in the means of achievement, not merely in the end fact of achievement.

It was for this purpose of observing the specific movements of the animal that the release mechanism was arranged so as to allow a wide variety of possible escape movements, and, at the same time, to make possible escape by the repetition of the same movement if that occurred. Thorndike's box in one form did not allow a repetition of a former movement to be effective even if it occurred. A loop hung from a string confronted the cat. The movement that had succeeded when this was hanging edge-on would not be effective when it was hanging face-on. Thorndike's cats had to acquire a complicated skill in dealing with a changing instrument.

We hoped that the automatic picture taken by the successful movement itself would secure a good record of that movement. To supplement the still pictures, occasional use was made of a continuous or an interrupted motion-picture record. As another precaution both experimenters watched each trial (except for one experiment done at Princeton by Horton alone) and a record was kept of outstanding features.

The automatic picture of the animal at the moment the door began to open represents a cross section of the behavior series. In some of the records this cross section was unsatisfactory because it came at slightly different points of what appeared to be two instances of almost identical movement series. If, in approaching the release post, the cat on the second occasion was going through a series of movements substantially identical with the movements of the first occasion, but on the second occasion happened to be one half inch farther removed from the post, the instant of impact would be slightly delayed and the picture would show the cat in a posture which appeared substantially different. In a number of such instances we were convinced that the second series had duplicated the first, but the second picture represented a cross section of the movement series at a slightly later point. This was particularly true when escape was achieved by interrupting a beam of light not visible to the animal.

The total time elapsed was of minor interest to us. Adams, like Hobhouse, is correct in pointing out that elapsed time is often misleading because the animal is grooming or not occupied in escape.

Approximately eight hundred escapes were observed and photographed. A number of these do not enter into the formal record because they were incidental to adjusting and perfecting the mechanisms of the box.

A number of broad generalizations resulting from our observations may be stated here. In the first place it is obvious that what is seen depends on what the observers are looking for. Lewin's descriptive categories could be easily applied to the animal confronted with the barrier. The cats spent most of their time at the barrier at the front where the barrier was transparent. Only after a considerable time spent at pawing, clawing, sniffling along this glass partition did the cat turn to the rear of the box. Here also time was spent at the wall rather than in the neighborhood of the post toward the center. The post might be unnoticed for many minutes, and most of the escapes were achieved not by attention to the post but by inadvertent contact with it while moving from one part of the barrier to another.

Certain features of the cats' behavior could readily be described in Tolman's phrases of "means-end-readiness" and "sign-gestalt." But neither Tolman nor Lewin is interested in the animal's movements; their attention is on achievement or accomplishment. Our interest was not in accomplishment but in the changing pattern of movement that results from experience.

Early in our observation we found ourselves distinguishing between "advertent" and "inadvertent" movements. Observers could readily agree on these descriptions. The "advertent" movement is guided by attention, normally visual. The cat, for instance, reaches out at the pole at which it is looking and continues the movement until the paw is in contact with the pole. "Inadvertent" movements consisted in brushing the pole with neck, side, flank or tail, or in backing into the pole. "Advertent" movements usually result in an act. "Inadvertent" movements remain mere movements. Reaching out and touching or biting seen objects are acts in which all cats have had an indefinite amount of practice, and the end result (pawing the pole, biting the pole, etc.) is attainable from a variety of stances by a variety of movements adjusted to stance and to distance.

An advertent response to the pole tempts one to describe it in Tolman's language. But the highly mentalistic nature of Tolman's terms and their misleading associations with the use of language (e.g., hypotheses, etc.) make such description dangerous. When we say: "The cat thinks that . . . ," or "The cat feels that . . . ," we forget that the word "that" is followed by a proposition, and we should hesitate to attribute propositions to cats. We need more objective phrases, phrases which do not read into the cat's behavior any possible use of inner speech.

To Tolman and others there appears to be a wide gap between an animal's "seeing a connection" between moving the pole and escaping and a simple inadvertent contact with the pole which is followed by escape. *But in our opinion the difference is one of timing and degree rather than one of kind.*

Let us illustrate: An animal has for many trials stepped on the base of the pole and walked without hesitation out of the opened door. Now the pole is moved to a new position. This time the cat repeats in detail its movements, raises its paw and sets it down where the pole had been and then bumps its nose on the glass door. Our impression that the cat had "seen what the pole was for" must be mistaken because there is an exact repetition of the former movements and these are now meaningless.

Tolman would probably reply to this that the connection really "seen" by the cat was the connection between movement of stepping and escape. We judged it to be better practice to limit our description to what we saw, namely that the movements occurred in a certain order. We have no confidence in the "intervening variable." We feel insecure in holding that the cat "thought" that the movement would let it out.

What the cats were observed to do and what the photographic record of posture at the moment of success indicates are described in the following account of performances of individual cats.

Cat A, the first animal used, represents one of the most variable records. There were certain errors in the operation of the apparatus which may account for variability. The mechanisms opening the door and activating the still camera were not so finely adjusted as in the runs of the following cats. The displacement of the pole necessary to open the door was not the same on each trial, and this proved an essential condition of some of the behavior which will be described. The speed with which the door opened was variable and the entrance door did not operate smoothly. All these factors were corrected in later experiments.

Cat A was given thirty-four runs in Box A on October 21, 1936. A motion-picture record was also made, and this has been issued by the Psychological Film Exchange with the title, "Cats in a Puzzle Box." Time records are lacking because we depended on the cinema, and this had to be interrupted occasionally when film ran out.

Each cat was first placed in the enclosed starting box at the rear of the apparatus; the doors were then opened, and the cat was allowed to exit by the open door. Three such "free runs" were usually allowed in order to speed up the trials by establishing the location of the exit. All trials numbered from the trial following these "free runs."

Cat A, stopped by the closed glass door on its third trial (the camera failed on the first two trials), spent many minutes exploring the box and finally brushed the pole with its flank while moving back and forth across the front of the box. On the fourth trial the escape movement was entirely different. The cat was headed in the opposite direction and was pawing the pole. The fifth trial finds it in almost the position of the first escape except for the angle at which the tail is held. Escapes seven and eight were accomplished by what we judged to be the "same" movement, but this resembled no others in the series.

Trials 14, 18, 19, and 20 have a like substantial identity of posture at the moment of escape. The cat has approached the door with the pole on its right and

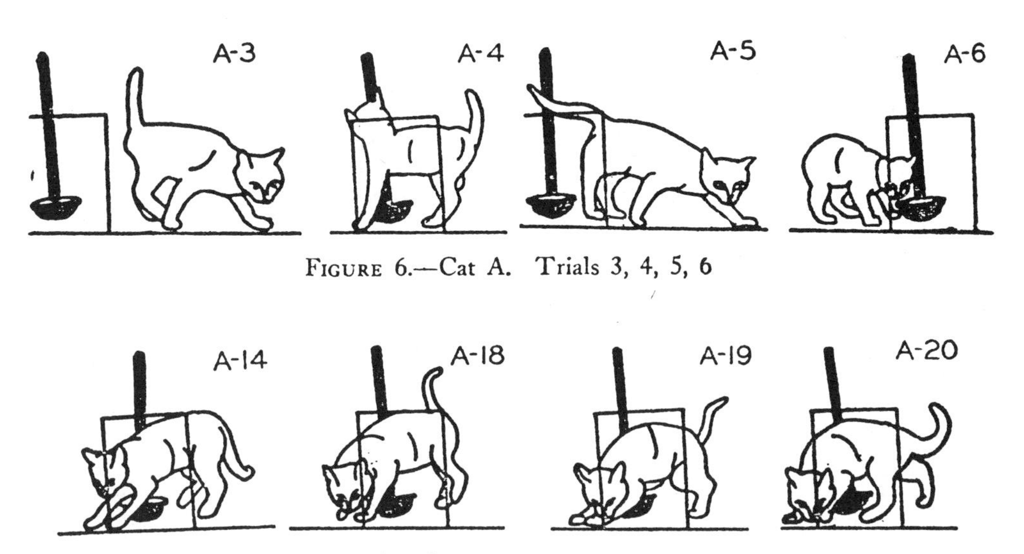

FIGURE 6.—Cat A. Trials 3, 4, 5, 6

FIGURE 7.—Cat A. Trials 14, 18, 19, 20

with head down toward the corner of the door has brushed against the pole.

In numbers 15, 16, 17, 21, 22, 23, and 24 the cat has passed the pole on its way to the door and has struck the pole with its left side. There are minor differences of stance.

In numbers 26, 28, 29 are three new and radically different positions. (The camera had failed on 1, 2, 11, and 25.)

On March 12, 1937, almost five months later, this cat was given sixteen trials in the improved box with an electric timer. The pole was in a new position, approximately eighteen inches from the door instead of six inches. The first escape (trial 35, counting from the first) required only forty-five seconds as compared with approximately fifteen minutes for the first of the previous series. The cat had

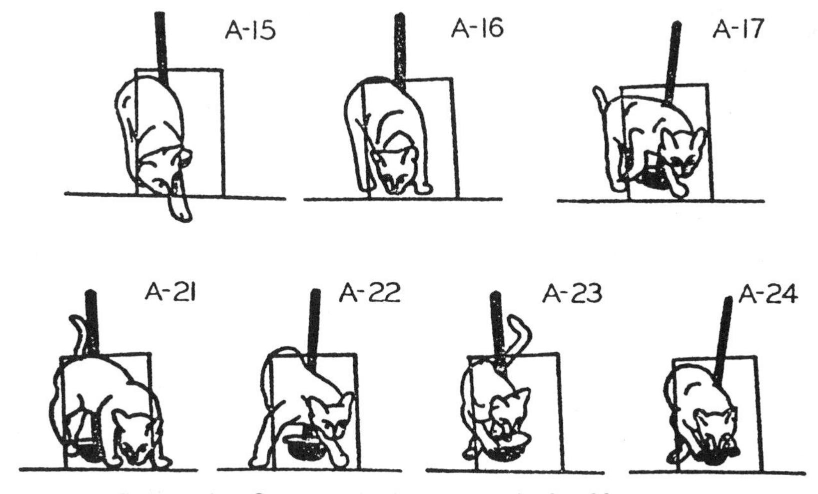

FIGURE 8.—Cat A. Trials 15, 16, 17, 21, 22, 23, 24

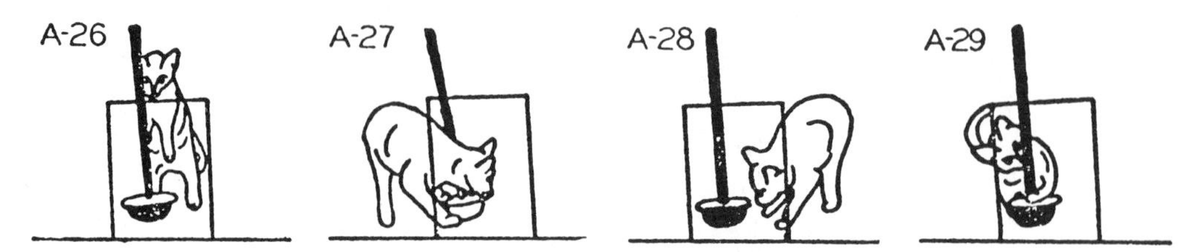

FIGURE 9.—Trials 26, 27, 28, 29

gone rather promptly to the door and had struck the pole with its rear while examining the door. Trials 38, 42, 44, 45, 46, 47, 48, and 49 were accomplished by almost identical movements (pawing), times being twenty, nine, fifteen, ten, and nine seconds. In trial 43, which failed to conform to this series, the time is longer, thirty-six seconds, and the action was a use of the left paw as in trial 37 but at the top of the pole rather than at the base.

It is perhaps significant that this cat had twice in the first series escaped by "advertent" use of its left forepaw. Tolman's description in terms of "means-end-expectancy" appears to fit the case very well in view of the added fact, not shown in the pictures, that the cat turns promptly from the pole to the door after pawing the pole. This action must have been prepared while the cat was pawing the pole.

But this description is highly superficial. Why should the actual movement series and the postures in which the picture catches the cat be so remarkably alike? Surely this cannot be explained in terms of the cat's knowledge of the necessary

FIGURE 10.—Cat A. Trials 38, 42, 43, 44, 45, 46, 47, 48, 49

muscular contractions any more than a human being's movements can be explained in such terms. The cat does not use names for its movements and choose a movement by name. The cat has no insight into or hypotheses concerning these movements. The theories of Tolman and Lewin have no place for the movements by which goals are reached, and this occasional highly stereotyped repetition of actual movements has no place in their theories. We may take occasion here to mention that there were many more of these stereotyped repetitions than appear on the record or in the pictures because it many times happened that this repetition was not successful in operating the door. Movement series were in some instances repeated as often as forty times, but a slight change in the position of the cat now rendered the movement useless.

A movement-by-movement account of eight hundred escapes is out of the question because, among other reasons, it would not be read. This account must be limited to certain outstanding features noticed by the experimenters and in most cases based on the picture record.

Cat B was given twenty-four trials on October 26 and nine trials on October 27, 1936. The box used was the first, referred to as *A*. As in all cases the cat was allowed three exits from the box through the open door. Cat B showed great hesitation at the exit on these preliminary runs.

The first two escapes were by a direct attack on the door which, contrary to plan, opened. This defect in the door was repaired. On the third run the cat attacked the door vigorously but later stepped on the base of the pole with its right rear foot, effecting release. With few exceptions (7, 12, 28) this was its method of release for the rest of the runs. It is noteworthy that in the sixth and fourteenth runs the cat looked back at the pole and used its right forepaw in addition to the right rear. For a number of runs after this the cat gave the impression of *using* its right rear with or without the right front. On trial 24 it stood with the right rear on the base for several seconds before pushing so as to operate the release. The use of the rear foot now resembled those acts that we have called "advertent" even though the cat no longer turned to look at the action.

After twenty-four hours the cat, in trial 25, repeated without any hesitation its

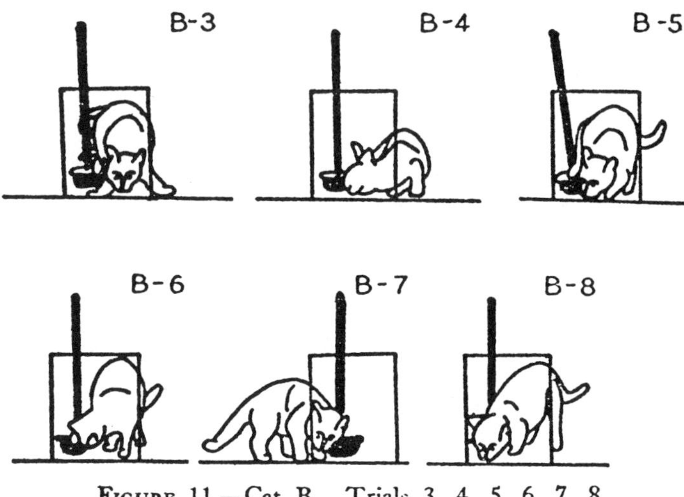

FIGURE 11.—Cat B. Trials 3, 4, 5, 6, 7, 8

manner of escape of the day before. On the twenty-sixth trial the door did not open at the first use of the right rear foot *and the cat repeated the action,* using the right front foot in addition.

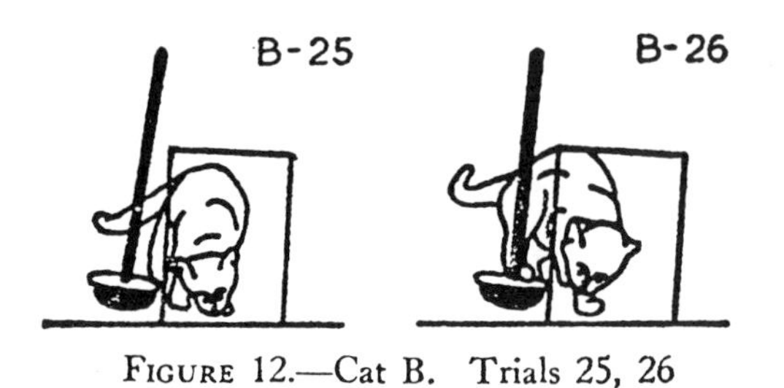

FIGURE 12.—Cat B. Trials 25, 26

After the eighth trial this cat had paid no attention to the "reward" of a bit of canned salmon. On the twenty-eighth trial the second day the cat made no effort to escape for some minutes; it then walked by the pole and stepped on this with its left rear foot (inadvertently?) and sat in the open exit and then returned to the box. When it finally emerged it was allowed to wander about the large animal room for some time. On being placed in the box for the twenty-ninth run it walked directly to the door and operated the release with its right hind foot. This was repeated for the succeeding three trials.

Twenty-three days later this cat was used for another series of trials (numbers 40-53) with the pole in a new position. It advanced as before and raised its right rear foot over the base of the release pole but did not put it down. After eleven minutes of wandering about the box it lay down before the door. After seven minutes more it rose and stepped on the base as in the early series but did not immediately leave by the opened door. On the third trial (number 42) of the new series

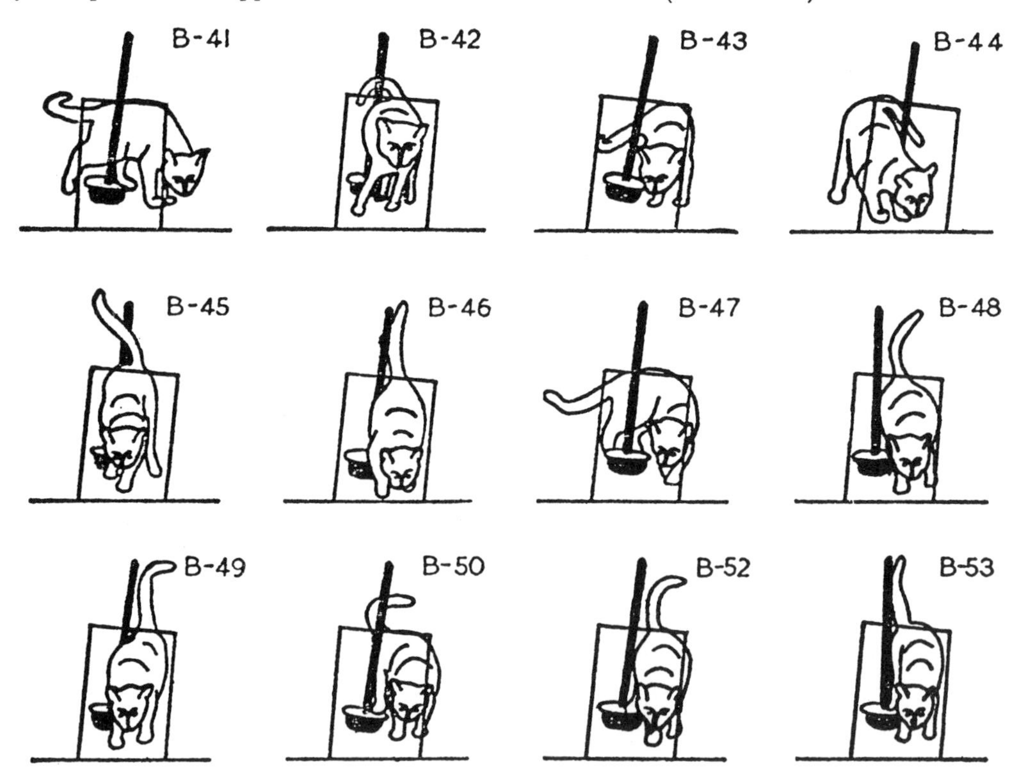

FIGURE 13.—Cat B. Trials 41, 42, 43, 44, 45, 46, 47, 48, 49, 50, 52, 53

the cat struck the pole with its tail as that was lashed. On the next two trials this was repeated after grooming or other behavior. On the sixth trial of the new series (number 45) the cat stood before the door and swung its tail as in the previous successful escape, but it was too far from the pole and the release mechanism did not operate. This was repeated seven or eight times. We might note in passing that the cat's tail is not ordinarily regarded as a member used for "manipulation" or "voluntary action." After this "use" of the tail the cat then twice (in the course of other behavior) raised its right rear foot over the base, but did not put it down as in the first trial of the new series. It then stood before the door and operated the release with the tail as before. On the next trial (46) the right rear paw was again used and likewise for the next seven trials.

This release by use of the right rear paw was normally effected by advancing to the door and then retreating, which brought the right rear paw back to the release mechanism. On the next to the last occasion the cat went for the first time directly to the pole and applied the right rear paw but not with sufficient force to operate the release. On the last trial there was no hesitation; the cat walked to the door and then retreated to step on the base of the pole.

During this series a little milk in a saucer was used as the reward. In the last three trials the cat did not drink on being released.

Cat C, the third cat, was given only six trials but was so inactive that its use was discontinued. The fourth cat (D), a full-grown male, had been brought to the laboratory by a student.[1]

On its first trial after the "free runs" this cat made a slight movement of its body to the right while pawing at the door, and this operated the release. This was repeated for the rest of twenty-three runs with two exceptions. On runs 11 and 12 the cat walked toward the front with the pole on its left, and the pole was struck with the left side.

Just three weeks later this cat was used for another series of runs with the pole in a new position, eighteen inches from the door. The cat had in the meantime been used for demonstrating the puzzle box to numerous classes. The cat advanced toward the door just as in the last trials of the first series, made the slight move toward the right that had resulted in release, clawed at the door, repeated the now ineffective movement (the pole was now on the cat's left) and after three minutes and five seconds in the box struck the pole with his tail in moving back and forth before the door, and the release mechanism operated. In the second run this same movement brought the flank against the pole, and in the next trials release was effected by repetition of this movement, striking the tail against the pole.

Eleven days later this cat was placed in the apparatus to check the mechanism, and the movement it had been using failed to operate the release. The cat finally struck the pole with his flank in passing it. Two more trials exhibited this new movement. On the next day three trials showed a resumption of the earlier release through striking the post with the tail.

On May 7, over five months later, Cat D for three trials repeated without delay his established pattern of escape, striking the door with the tail in passing.

Cat E's behavior was notable for its variety. This was a very active half-grown kitten. Escapes 7, 8, 10, and 13 were accomplished in much the same manner. In trials 9 and 12 the escape pos-

[1] Note: The puzzle box A was used with the pole six inches from the escape door.

tures are almost identical but like no others in the series.

second trial the cat took up rather promptly the same stance before the door,

FIGURE 14.—Cat E. Trials 7, 8, 9, 10, 12, 13

but a direct nosing of the door opened it. The door mechanism was adjusted. On the third trial the cat approached the door with the pole on its left. In the two previous trials the pole had been on the right. The pole was moved by the cat's side as it nosed at the door. In the fourth

In trial 15 the cat had attended to the pole for the first time, and the escape was radically different from what had gone before. It consisted in biting the pole with the head inclined to the left. This was repeated for trials 16, 17 (pictures lacking because of camera failure), 18, and 20.

FIGURE 15.—Cat E. Trials 15, 18, 20

trial the cat approached the door as in trials 1 and 2, and release was effected as in trial 1, by striking the base of the pole with the "elbow" while clawing at the door. Trial 5 was a close repetition of trial 4, both taking twenty-five seconds. In trials 6 and 7 the behavior sequence of trial 3 was repeated. In trial 8 the cat delayed two minutes in the entrance and then repeated the pattern of trial 1. With

Cat G was, like the others, allowed to enter the box and escape through the open door three times. Its first run with the door closed took three minutes. The final movement of release was pawing at the base of the exit door, the "elbow" striking the base of the pole in the course of this pawing. The cat was, of course, immediately before the door when it opened and went promptly out. In the minor variations this characteristic pattern of movement persisted for the rest of twenty trials on November 9, 1936, and for twenty more trials on November 10. On the thirty-second trial the cat approached the door with the pole on its left and then used the right paw, which was now away from the pole. This did not operate the mechanism, and the cat turned from the exit and reapproached with the pole on its right, performing successfully. the cat as it stood in the exit. More than fifteen times it went through the routine by which it had escaped in the original series. It then desisted and wandered about the cage. After three minutes it returned and resumed its early pattern of escape movements. Escape was finally by a direct push on the door, which, contrary to our design, opened.

Two days later the cat was given runs 42 to 60 (camera failure on 51). In trial

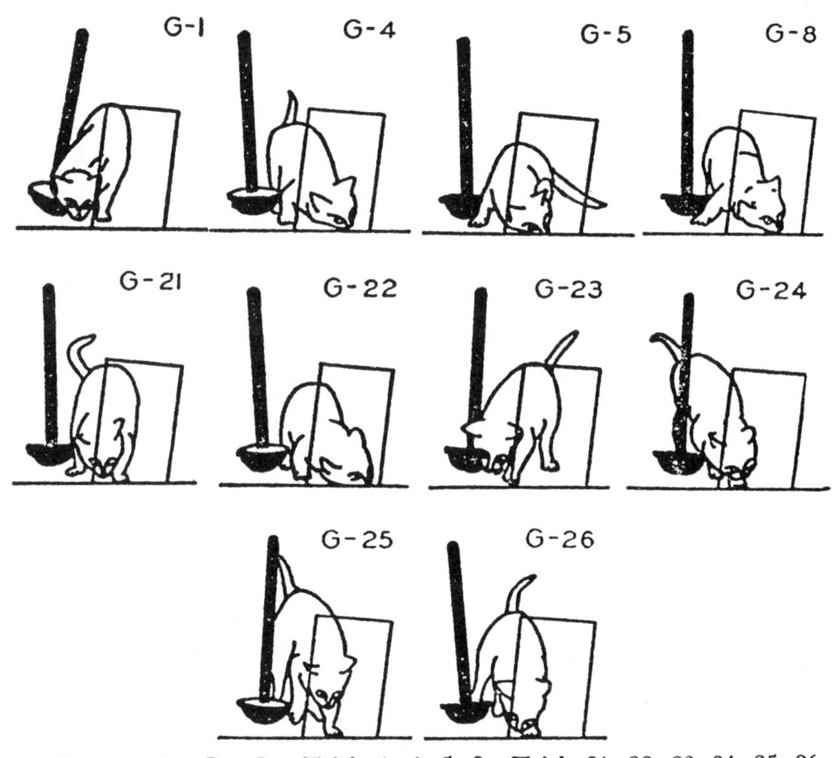

FIGURE 16.—Cat G. Trials 1, 4, 5, 8 Trials 21, 22, 23, 24, 25, 26

Cat G had now escaped from the box on forty different occasions. Thirty-seven of the pictures show the cat in substantially the same position, nose down toward the edge of the exit door and the right paw clawing at the edge. On November 19 it was given one trial only with the pole in a new position eighteen inches from the door and now to the rear and left of 42, immediately on entering the box, it resumed the position and movement that had released it in the first series. This movement was repeated and returned to several times. Eventual escape after seven minutes, thirty-five seconds, was secured by turning and biting the pole. On the next trial this successful movement occurred after two minutes, thirty seconds. Every picture after this shows substantially the same position. The cat is biting the pole at the same point with head canted toward the right.

63, 64); the first bite did not succeed and two or three repetitions occurred until the door opened. With one or two minor differences the pictures of the releasing

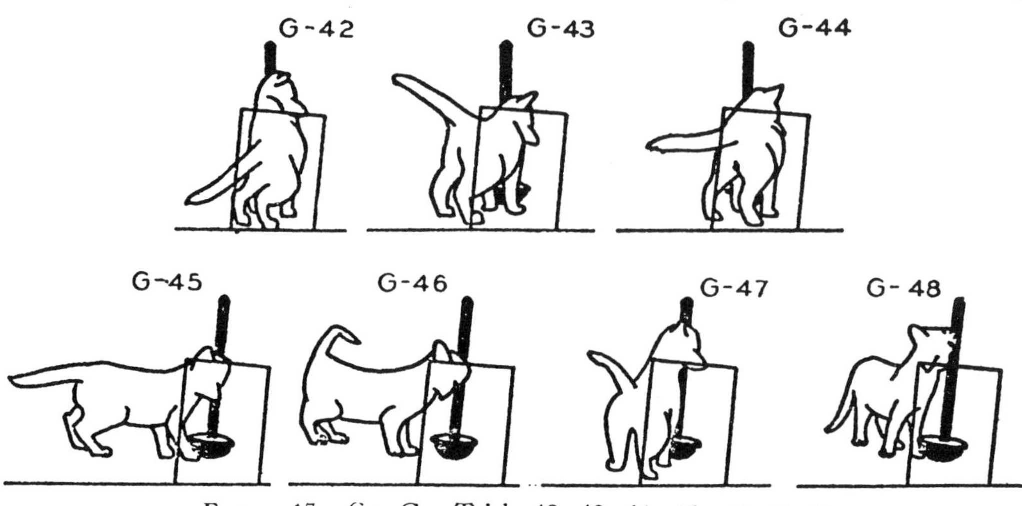

FIGURE 17.—Cat G. Trials 42, 43, 44, 45, 46, 47, 48

From trials 47 to 53 the movement consisted of a trip to the door, a turn toward the pole, and the biting which effected release. On trial 53 the exit door stuck partly open and the cat was not able to leave the box until this had been adjusted. On the rest of the trials of the series until trial 57, the preliminary trip to the door was omitted and the cat approached the pole immediately on entering the box except for a slight hesitation between door and pole on trial 57.

On November 23, two days later, the cat was given fourteen more trials. The first (58) required only ten seconds and was a repetition of the biting movement. On the second (59) the cat bit the pole but so gently that the mechanism did not operate. It went to the closed door immediately; then it returned to the pole and used a foot as well as its jaws. On the next trial (60) the first three bites failed, and the fourth was successful. This was repeated for the next four trials (61, 62, 63, 64); the first bite did not succeed and two or three repetitions occurred until the door opened. With one or two minor differences the pictures of the releasing position are strikingly the same from trials 41 to 71.

To the photographic record can be added the observer's comment that in most of these trials not only is the final position highly stereotyped, but the sequence of movements from entrance to exit is just as stereotyped. In some cats this is true of long series of movements about the cage, series that occupy a minute or more. And the camera record fails to note that the stereotyped movement occurred as usual but was, on account of a slight divergence in path, not successful in operating the pole.

Two days later Cat G was given ten more runs. The first required only five seconds and was of the usual pattern. On the third the cat advanced to the door and returned to the pole. On the fourth it moved to bite but did not finish the movement and looked toward the door. It then bit and operated the release. On three later trials it advanced toward the door, turned, and bit the pole.

Cat I (for which tracings are lacking) was of particular interest to the observers because its behavior showed a great deal of variety. This cat was suffering from an eye infection, and as the experiment progressed this condition appeared to affect its behavior. On the first run it was very active. At the end of four minutes and fifty seconds of exploration and clawing, it was clawing at the base of the exit door and suddenly backed into the pole, released the door, left the box, and ate the bit of fish placed outside. On the second run this backing movement away from the door was repeated several times but the pole was not struck until after five minutes. The right rear paw was placed on the base of the pole. On the third trial the backing movement was repeated six times before the forepaw struck the base of the pole on the sixth backing. On the fourth trial the time was over eight minutes before the cat released itself by backing into the pole after a number of unsuccessful backings. On the fifth trial thirteen backings occurred before success. On the next trial the cat backed once but ceased activity, and the door was eventually opened by the experimenters.

One day later there was prompt release (one minute, fifty seconds) by the third backing. On the eighth and ninth trials release was accomplished by the sixth and twenty-seventh backings, some of which carried the cat in a complete circuit of the box. On the tenth trial the thirty-third backing effected a collision with the pole and release; on the eleventh trial, the twenty-fifth backing; on the twelfth trial, the thirty-first backing. The thirteenth trial was discontinued because the cat lay down after one backing movement. Thirteen days later the cat interspersed thirty-nine backing movements with grooming, mewing, and clawing at the door. It then, being in the front left corner of the box, jumped toward the wire netting of the top. This was repeated four times. After one minute elsewhere the cat returned to this corner and jumped and clung to the wire mesh for ten seconds, fell, struck the pole, and made its exit. Time, nineteen minutes, twenty-eight seconds.

On the next, the fourteenth trial, the cat backed three times, then jumped for the top in the usual corner and on the second jump fell on the pole and was released. The fifteenth trial the following day was a mixture of backing and jumping. Whenever the cat reached the door in its wanderings about the box it backed; whenever it reached the corner from which it had jumped it looked up and crouched for a jump. This was also true of trials 16, 17, and 18.

Three months later this cat was again placed in the box. Its behavior was similar to that observed in December. The first of the new trials exhibited both the backing and the jumping movements. The release was by backing but was accomplished only after much other activity or occasional refusals to be active. On May 7 the same cat was placed in the box for five trials, and escape was by backing. Pictures of the release of this cat show little uniformity because the peculiar backing movement, which at first followed clawing at the door, eventually was initiated at some distance from the door and toward the end of the trials from various positions in the box. Its cue became obscure. But the occasion for jumping remained an approach to the left front corner of the box. Whenever the cat approached this corner from the center of the box, it either jumped, or, on days when its general behavior indicated bad general effects of the eye infection, merely crouched for a spring, hesitated, and then moved away.

[*Editor's Note:* In the original, material follows this excerpt.]

12B

Reprinted from pp. 39–42 of *Cats in a Puzzle Box*, Rinehart, New York, 1946, 67 p., by permission of Holt, Rinehart and Winston and Edwin R. Guthrie and George P. Horton.

INTERPRETATION OF RESULTS

E. R. Guthrie and G. P. Horton

[*Editor's Note:* In the original, material precedes this excerpt.]

To be embodied in our final description, then, are a number of features: (1) a tendency to repetition in remarkable detail of movement series, sometimes lasting a minute or more; (2) new behavior sometimes including a complete trial which from beginning to end appears new; (3) a tendency on the part of the final series of movements, the action of the last few seconds but occasionally an involved and longer series, to be more stable than the preliminary behavior.

In our opinion these generalizations are consistent with the principles of associative learning. On the cat's first trip through the box, its behavior is subject to accidental determiners, the incidental stimuli encountered, the general condition of the cat, the past experience of the cat. But a series of movements once repeated in this general puzzle-box situation tend to become independent of minor distractions. We interpret this as due to the establishment of one movement as a cue for the next. On the first occasion it is merely followed by the next. On the second it has its effect in occasioning the next, rendering this following movement comparatively independent of distraction. We would not limit this to the movement-produced stimuli from proprioceptors but would include all stimulation consequent on a movement, exteroceptive as well as interoceptive.

In our opinion the second occasion tends to repeat the behavior of the first, errors and all, except in so far as remaining in the puzzle box for a long time tends to establish new responses to the puzzle-box situation. *The reason for the remarkable preservation of the end action leading to escape is that this action removes the cat from the situation and hence allows no new responses to become attached to the puzzle-box situation.* Escape protects the end action from relearning. This is made more plausible by our observation that early success predicts a stable method of escape, while a long trial makes the use of the same method less likely. In the long trial, early responses are displaced by later responses, and the cat is compelled to do something else while remaining among the same external cues. All that is necessary for learning is activity. Our "rewards," consisting of food or escape from the box, were of dubious effect on the action. Most of the cats became indifferent to the food in the course of the experiment; and all the cats had escape followed rather promptly by being replaced in the dark confinement of the starting box except for the last trial of each series. *The cat learns to do in the circumstances what it has been caused to do in the circumstances.*

This may operate to change the behavior of the cat in the box. For instance, if the usual movement of escape fails to operate the release mechanism, when the cat turns as before and approaches the door, it finds the door closed. This is a new situation. Occasionally the cat collides with the closed door. The cat now remains in the box but is acting from a new initial position and may or may not pick up at some point the thread of its old behavior series. In some cases much behavior intervenes. In other cases there is a prompt repetition of the final action. This is illustrated in the case of cat K, which escaped by pausing beside the pole and moving slightly to that side. The door did not open. The movement was repeated, with success. From that time on for several trials the cat paused and made two distinct movements against the pole, though the first was succeessful.

When the movement of escape is an act oriented toward the pole, usually with visual attention, and the opening of the door is then noticed and exit is prompt, the act has this appearance of voluntary action and is repeatable on failure. The cat has paused at the pole. Approach to the door is inhibited by the sight of the closed door. With the cat remaining in place, its last movement in these circumstances was pawing or biting the pole. If the cat is not oriented toward the door, the movement of escape (this time unsuccessful) is followed by turning and approaching the closed door. The cat is not now in the position for its successful escape action and must wander about until it recaptures the situation that led to attack on the pole. In other words, the difference between this prompt repetition on failure and a long delay before repetition is not that one instance is an action somehow "voluntary" and of a higher order, and the other "mere" association, or that one movement is "senseless" and the other "meaningful." The difference between senseless and meaningful actions is an interpretation put on the event by an observer whose real data can be analyzed in terms of situation and response.

This question of how new behavior originates or why the cat does not invariably and endlessly repeat each behavior so long as it remains in the situation deserves more attention.

There are many answers, not just one. We have mentioned the interference of accidental stimuli, the buzzing of a fly, the voice of another cat in a near-by cage, a sudden movement of the observer. Any of these may interrupt a series in the course of repetition and so change the cat's relation to its surroundings that it does not again pick up the thread of its routine.

We have also mentioned that failure of a movement series previously leading to escape may bring the cat up short against the glass door and thus start a new train.

We do not enter on endless repetitions of routines because most actions are self-terminating. When we have descended the stairs, it would require energetic and rapid work by a steam shovel and builders to enable us to repeat immediately. A man who has just put on his hat might add another, but certainly not a third. It is characteristic of most of the cats' actions in the puzzle box that they end as well as begin, and their ending changes the situation so that there is no immediate repetition. Even so simple an act as clawing at the door alters the cat's relation to the door. It is now a cat that has clawed rather than a cat that is about to claw. What its behavior was prepared for by previous experience did not happen. Acts which bring the animal to a situation and posture from which the action began are subject to repetition. The pacing of a caged animal is very different from the behavior of the cat in a puzzle box. The cage-broken animal has through thousands of abandonments of clawing and biting and exploring substituted the only actions which do not bring pain. These actions exhibit the same repetitiousness evident in series leading to escape.

On those occasions in which the second run does not start the cat on a series of movements like the first there appear to be two possible alternative causes for this divergence. If the first trial has occupied many minutes and the cat has remained in the box and varied its behavior, the new associations with the situation may have overlaid the first. Even so, in a few cases of protracted trials, the initial behavior of the next trial is a startling reproduction of the initial behavior of the first. The other cause is in some instances obvious as in the case of Cat K. This cat shows four postures, numbers 8, 16, 26, and 45, which are strikingly alike and which bear no resemblance to the rest. The cat stands with forelegs crossed, head turned toward the door, rear brushing the pole as it turns. Our notes indicate that on these trials the cat had managed to turn about in the narrow starting box and, when the entrance doors were opened by a lanyard, had backed into the puzzle box. This maneuver started it on a completely different series of movements which is preserved intact through the intervening trials. The times of these four trials were five, six, five, and two seconds. When escape is effected promptly, there appears to be no opportunity for relearning in the box to change the method of escape.

We may summarize our generalizations of the puzzle-box behavior as follows: The behavior of the cat on one occasion tends to be repeated on the next, even to occasional prolonged series of movements about the cage. Exceptions to this are either the result of a different entrance, which initiates a different line of action, or the result of accidental distractions, which may deflect the behavior from its former sequence; moreover, when the cat has been in the box for a long time, responses made later in the trial may supplant those made earlier.

The most stable response is the one which ends in release from the box. In some instances this is a comparatively long series of movements ending in exit. The reason for its stability appears to be that escape removes the animal from the puzzle-box situation so that no reassociation may occur. All response series tend to recur, but some of them are lost when the animal remains in the box and is compelled to new action. *The cat learns to escape in one trial* and will repeat the specific movements of its first escape except in so far as new trials by accidental variations of situation cause new associative connections to be established. A protracted trial breaks up the sequence of movements by superimposing new behavior in the same situation. Time, therefore, tends to be reduced with repetition.

In so far as the puzzle-box situation varies (as in the case of the cat that backed in occasionally), the cat will learn not one but two or more specific movement series leading to escape. We believe that when the puzzle-box situation varies indefinitely, as it did in the Thorndike box with the hanging loop, it is necessary for the cat to establish a large repertoire of specific escape movements adjusted to the specific differences in the situation. In other words, the cat establishes a skill, rather than a stereotyped habit. But the skill is made up of many specific habits. The gradual reduction of time reported by Thorndike is a consequence of the varied situation confronting the cat.

We have been interested in comparing our present interpretation with the description of the behavior of a cat in a different form of box, a description written by Stevenson Smith and E. R. Guthrie in 1921 and published locally in a volume entitled *Chapters in General Psychology*. The interpretation read as follows:

"The cat when in confinement is instinctively organized to respond to the sight of bars, cracks, corners, and even solid walls by approaching and pulling and pushing with paws, claws, nose and teeth. . . . He is also instinctively organized to turn away from these same objects when they offer more than a certain amount of resistance to his attack.

"The situations which the cat faces in

the puzzle box are for the most part composed of visual stimuli which attract him, followed by stimuli to his proprioceptors and sense organs of touch which repel him, and these two classes of stimuli *are given by the same object*. . . .

"With repetition the sight of the bars becomes the conditioning stimulus for retreat so that the conditioned response inhibits the original response. One by one the movements of approach to the various confining surfaces of the box are inhibited by the conditioned responses of retreating, until at last the animal is attracted by the door-opening device. The reason that this last movement is not inhibited is that the device itself never serves as the source of a stimulus which is instinctively avoided. Although the cat turns away from the button in response to the open door, he does so not because the button repels him but because the open door attracts him. Approaching the button and approaching the open door are the only approach responses which are uninhibited by conditioned avoidance responses, and while the door is closed the button alone calls forth an uninhibited response.

"When the cat is but partially trained he makes useless responses to various parts of the box. He always ends, however, by making the successful response and by escaping. As some useless responses are given on some occasions and others on other occasions, depending upon the cat's chance position in the box, the successful response, always occurring, is likely to be the one most practiced. This may be another factor in lowering the threshold of response to the door-opening device." [3]

It will be noticed that this differs from the present account in its assumptions that the probability of an associative response is a function of practice. The present account explains the stability of the final movement of escape through the protection of that response from unlearning, a protection furnished by the fact that escape separates animal and situation and gives no opportunity for learning new responses to the situation.

It has been our conclusion from our observation of this series of experiments that the prediction of what any animal would do at any moment is most securely based on a record of what the animal was observed to do *in that situation on its last occurrence*. This is obviously prediction in terms of association.

Plans to extend the observations to human subjects remain only tentative. It is our conviction that human behavior shows the same repetitiousness of manner and means as well as of ends. The chief difference lies in the greater repertoire of movements of manipulation possessed by men and in the human capacity for maintaining subvocal word series somewhat independent of posture and action. Thought, which is language in use, is itself as routinized as the movements of the cat. But the fact that thought routines are not completely dependent on action enables them to exert their direction and interference at new points, and men in a quandary may discover the exit while sitting still or during irrelevant action. Man can name and count. Learning in men is more intricate a process than in the cat. But it is possible that human learning and feline learning both are equally in essence the acquisition of new signals for action through the association of signal and act; and that this association alone and of itself, without dependence on reward or punishment or effect, is adequate for the establishment of the new signal.

[3] Stevenson Smith and E. R. Guthrie, *Chapters in General Psychology;* D. Appleton Co., New York, 1919.

13

Reprinted from pp. 2–13, 18–28, 70–72, of *Comp. Psychol. Monogr.* **12**:1–72 (1936)

EFFECTIVENESS OF TOKEN-REWARDS FOR CHIMPANZEES

J. B. Wolfe

I. INTRODUCTION

Extensive investigations have been made since the turn of the century on the relative effectiveness of various rewards as learning-producing and as work-eliciting agencies. Notable among the rewards used have been food, water, sex objects, and freedom from noxious stimuli. In the main the rewards have been objects or situations for which animals exhibit, from the observer's point of view, unconditioned seeking or avoiding tendencies. Hence, they may be called primary rewards.

One important aspect of rewards or incentives, however, has received little experimental attention, namely, secondary rewards. By secondary rewards are meant objects, external to the organism, which through use have become associated with a primary reward such as food, which evoke striving behavior, and which can serve as agents in establishing or in eliciting a distinctive sequence of acts leading to the secondary rewards. It is the writer's contention that under certain conditions the elicitation of a previously established sequence of acts by means of an object associated with a primary reward is evidence that the object is a secondary reward. If an object, for example, which has previously been associated with food elicits a series of acts which have never before led to the object but have led to food, then the food-associated object has been established as a secondary reward.

The problems investigated in the present study fall into the category of secondary or surrogate rewards. Attempts were made to determine in an exploratory fashion the effectiveness of such rewards for chimpanzees after it was evident that the objects used were secondary rewards.

The particular objects used in this study were poker chips. It was felt that manipulatable and easily transportable objects such as poker chips would have certain advantages as surrogate rewards: they are especially advantageous in that they are small, compact, available in standard sizes and colors, and relatively indestructible; they can be used in a variety of ways in investigating complex behavior; they are objectively separable from the primary rewards; and they may be used conveniently in work with organisms of different species including the human, particularly the young and the feebleminded.

The experiments herein reported concerning the effectiveness of chips (hereafter designated as tokens or token-rewards) for chimpanzees may be grouped around four central problems. The first problem reported is that of determining the relative effectiveness of tokens and food in eliciting a work-task. Certain performances rewarded by tokens were compared with similar performances rewarded immediately by food similar in kind and amount to that for which the tokens might be exchanged. A work technique was used inasmuch as it permitted the systematic variation of rewards while the task remained constant.

The second group of experiments was designed to determine the influence of token-rewards upon the effectiveness of delayed food rewards. In learning experiments with rats, chickens and human beings, it has been shown that a reward loses much of its effectiveness if the giving of it be postponed for an interval of time following the execution of the rewarded act. On the assumption that the same relationship would hold for chimpanzees in a work situation, i.e., that a delayed food reward would be less effective in eliciting a work-task than would an immediate food reward, token-rewards were introduced in certain delayed primary reward situations and were not introduced in others. The effectiveness of the secondary reward would be displayed if the work were enhanced by the introduction of the object which could later be exchanged for food.

The third group of experiments was undertaken to investigate the chimpanzees' ability to discriminate between tokens having different exchange values. Tokens of certain colors and sizes were assigned particular reward values including food, water, and activity-privileges. After the subjects had received training in associating the tokens with their particular reward values, conditions were arranged to provide the subjects with specific drives or motives and then with opportunities to select and to use the tokens appropriate to those motives.

The final and fourth group of observations in this study was made upon the capacity of token-rewards to elicit competitive behavior in social situations. The animals when paired were presented with tokens which could be worked for, kept, or collected from the floor. These observations were made to determine whether or not tokens would elicit behavior similar to that elicited by food when given to groups of chimpanzees.

II. APPARATUS AND SUBJECTS

The apparatus used in these experiments consisted of two restraining cages, four automatic food-delivering machines, a water-vending device, two subsidiary cabinets, a work machine, and an assortment of tokens.

The main restraining cage was circular in form and all-metal in material. The arrangement of the framework which was covered with heavy wire netting may be seen in figure 3. The cage was 148 cm. in diameter and 103 cm. in height. Two grilles, through which the animals worked, were 90° apart. They were 30 cm. wide and 72 cm. high. Each was barred with two 1.25 cm. steel rods vertically placed. Wooden shutters were hinged outside the grilles and could be closed when the procedure so demanded. The smaller restraining cage, which was used in some of the experiments, was built of material similar to that of the main cage and was 94 cm. high, 81 cm. wide and 120 cm. long. One end of the cage was a grille made of 1.25 cm. steel rods horizontally placed.

The automatic food-delivering machines, or venders, were constructed so that the insertion of a token (poker chip) effected the release of a food-reward. The mechanisms were encased in white wooden cabinets built of paneled sides on a solid base and covered with two hinged lids. The only two features which were visible from the front of a vender were the metal food cup and the black slot into which the tokens might be inserted. The dimensions of the cabinets were as follows: height, 78 cm.; width, 38 cm.; length, 58 cm. Figures 1 and 2 show the internal design of a vender. The large drum, which carried the food, was 48 cm. in diameter and 8 cm. in thickness. On its circumference were twenty food compartments, each one of which was 8 cm. deep. The drum was wood covered with 22 gauge galvanized iron and was mounted on a shaft which ran in ball bearings. The drum itself turned in a trough, lined with galvanized iron, which terminated in a chute leading to the food cup. Motion of the drum was effected by a weight fastened to a wire wound about a smaller drum on the main shaft. A twenty-tooth steel ratchet wheel fastened to the small drum controlled the movement. Since the

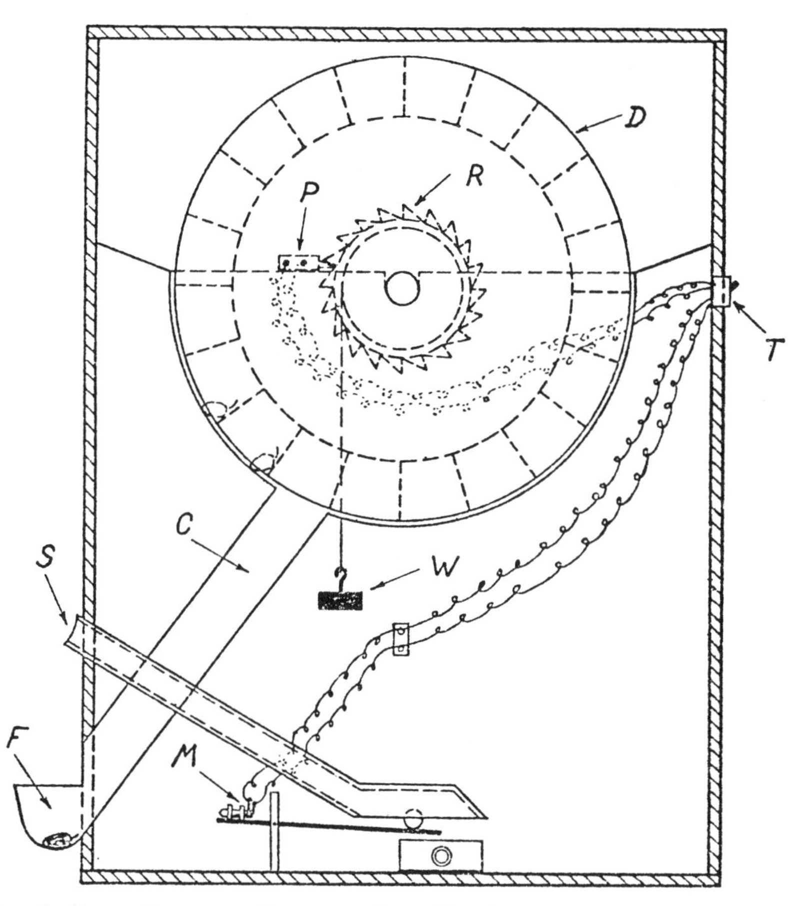

FIG. 1. CROSS SECTIONAL DIAGRAM (SIDE VIEW) OF FOOD VENDER SHOWING INTERNAL ARRANGEMENT OF THE MECHANISM

D, food drum; *C*, food chute; *F*, food cup; *S*, token slot; *M*, trigger switch; *T*, two-way toggle switch; *R*, ratchet wheel; *P*, electrically operated pawl; *W*, weight to effect movement of the food drum.

weight attached to the wire prevented backward motion, only one pawl was needed to regulate forward movement. A brass door-catch served as the pawl. This pawl was governed by a 110 volt, direct current solenoid.

The mechanism for operating the large drum consisted of a wooden slot which conducted the tokens to a delicately balanced trigger. Mounted on the short end of this trigger was a glass-inclosed mercury switch which made contact only when the token

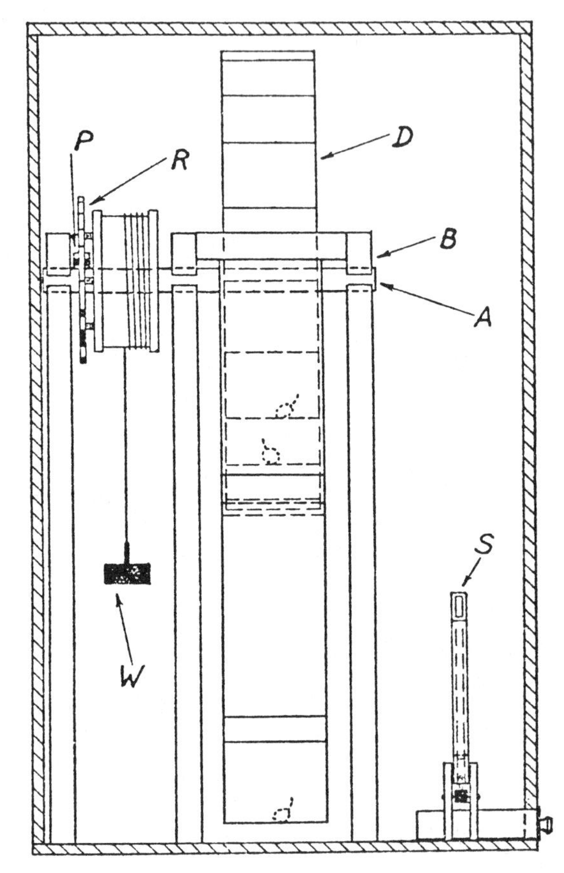

FIG. 2. CROSS SECTIONAL DIAGRAM (END VIEW) OF VENDER
D, food drum; *R*, ratchet on small drum; *W*, weight fastened to small drum to provide motion; *A*, main shaft to which small drum and food drum are attached; *B*, bearings in which main shaft turns; *P*, pawl; *S*, token slot.

was on the long end of the trigger. This contact was momentary, closing the circuit through the solenoid just long enough to draw the pawl and to allow one tooth on the ratchet wheel to pass. Thus the drum was permitted to turn through an angle of 18°, a distance sufficient to discharge the contents of one food compartment into the chute. A small drawer just under the switch served to catch the tokens as they rolled off the trigger. The drawer was flush with the side of the cabinet and could be removed with a key for emptying.

A two-way toggle switch mounted on the back of the cabinet permitted an activation of the solenoid optional to the one involving tokens. By reversing the toggle switch the solenoid, and hence the vender, could be operated by a push button of the ordinary bell-ringing type from any place in the room.

The entire movable assembly could easily be taken out of the cabinet for cleaning. The venders were of sufficient strength and weight to withstand without damage the occasional pushes or pulls by the subjects.

The water-vending device consisted of a small white box, 10 cm. by 20 cm. by 25 cm., covered with an aluminum lid which opened automatically upon the closing of a push button switch by the experimenter. A water cup inside the box could be reached only when the lid was open.

The two subsidiary cabinets were used as receptacles for tokens which granted activity privileges. They were plain white boxes 78 cm. high, 19.5 cm. wide and 25 cm. long. Black slots similar to the ones on the venders were on the front, and drawers for receiving the tokens were in the bottom of each.

A weight-lifting device (fig. 3) was used as the work apparatus. It could be operated by the subject's extending its arm horizontally forward and moving a handle up through an arc of 90°. The handle was attached to two brass tubes which in turn were fastened by plain bearings to shafts on the two sides of an open wooden trough. Hence the tubes were free to move through an angle of 90°, or from a horizontal to a vertical position. The trough was 50 cm. long and 11 cm. wide. Each of the two tubes was composed of two parts, one of which fitted inside the other so that by means of a sliding collar and set screw it was possible to adjust the length to fit animals with varying reaches. The tubes were 1.5 cm. in diameter and were adjustable in length from 20 cm. to 35 cm. The difficulty of the task of lifting the handle was controlled by weights placed on a stem fastened by a bifurcated cord to the ends of the handle. The cord, below the point of bifurcation, passed across a pulley so that the weights swung free of the supporting framework.

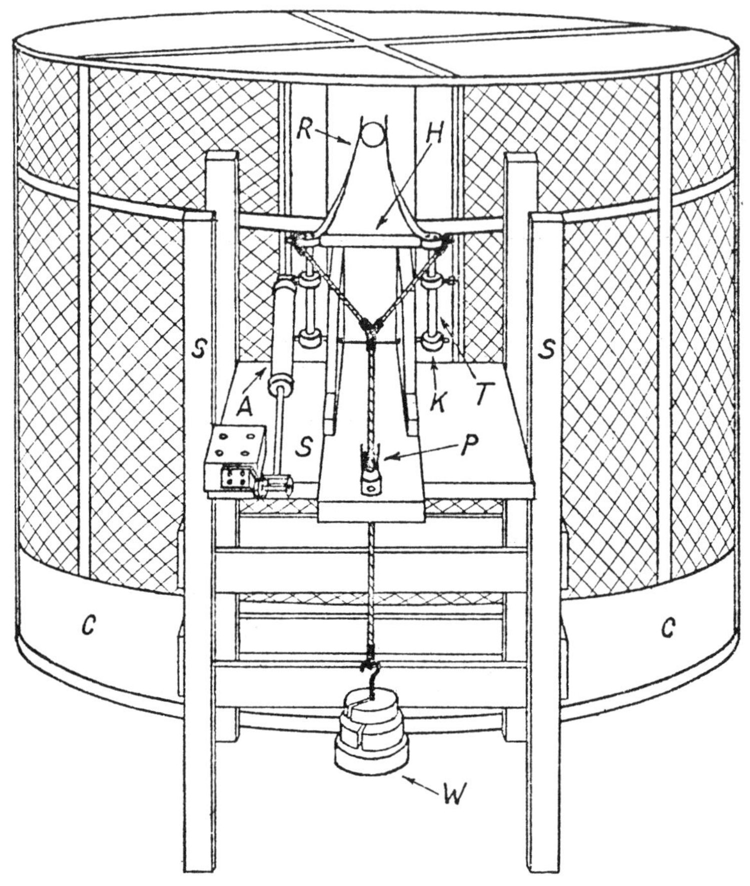

FIG. 3. DIAGRAM OF WORK APPARATUS AND RESTRAINING CAGE
R, reward-holder; *H*, handle; *A*, air brake; *W*, weights; *K*, bearing; *P*, pulley; *T*, adjustable tubes; *S*, supporting frame work; *C*, restraining cage.

A spring-brass reward-holder was attached to the handle. The reward could not be reached by the animal until the handle had been lifted to a vertical position. A one-way air brake was connected to one of the brass tubes and served to let the handle fall back gradually to a horizontal position after the subject had released it. The supporting framework for the apparatus was adjustable in height to accommodate animals of different sizes. All moving parts of the apparatus were kept lubricated.

Composition poker chips of standard size, 3.75 cm. in diameter, and of one-half standard size were used for tokens throughout the experiments. The colors of the chips were red, green, yellow, dark blue, light blue, white, and black. Brass slugs, which were the same size as the large composition tokens, were also used.

Six immature chimpanzees, three males and three females, served as subjects. All were members of the New Haven experimental group of the Yale Laboratories of Primate Biology and are described in the laboratory records as follows:

TABLE 1

NUMBER	NAME	SEX	BIRTH DATE	AGE AT BEGINNING OF WORK
28	Alpha	Female	September 11, 1930	2 years 2 months
48	Bula	Female	April 27, 1930	2 years 5 months
26	Bimba	Female	Estimated, 1929	3 years, estimated
21	Bon	Male	Estimated, 1927	5 years, estimated
31	Velt	Male	Estimated, 1929	4 years, estimated
11	Moos	Male	Estimated, 1927	6 years, estimated

Alpha was born at the Anthropoid Experiment Station, Orange Park, Florida. Bula was born on the Abreu estate, Quinta Palatino, Havana, Cuba. The other subjects were from Africa. Alpha, Bula, Bimba, and Moos had been in the Yale Laboratories for over two years prior to the experimentation herein reported. Bon and Velt had been in the laboratories only a few months before the work bégan. With the exception of Moos, who had been used almost continuously for over two years in various studies, none of the subjects was well accustomed to experimentation.

III. GENERAL PREPARATORY TRAINING OF SUBJECTS

The general preparatory training of the subjects consisted of three parts: operation of the venders, discrimination between tokens with and without food value, and operation of the work apparatus. These will be considered in turn. Subsequent to or during the training, observations were made to determine whether or not the tokens functioned as secondary rewards.

A. Operation of venders

Before the problem of the effectiveness of token-rewards for chimpanzees could be attacked, it was necessary to establish in the animals an association between using tokens and getting food. To achieve that end white tokens and grapes were used. In order to get the grapes the tokens had to be inserted into a vender.

Procedure. The procedure which was used with success in training the subject to use the tokens to get food included three steps: (1) a period of orientation during which the animal was brought to the experimental room as many times as were necessary for it to gain confidence in the experimenter and to become thoroughly accustomed to the working quarters; (2) a period during which the experimenter continued to pick tokens up from the floor and insert them into a vender until the animal would reach into the food cup as soon as the experimenter started to place a token in the slot; (3) a period during which the subject came to insert the tokens. The manner in which the act of inserting the tokens was achieved varied from subject to subject.

Results. In table 2 a summary is presented as an introduction to the more complete description of the manner in which the individual subjects mastered the technique of getting food with the tokens. The individual descriptions follow the table. Since the procedures had to be adapted to the characteristics of the animals and consequently were not entirely uniform, the figures reported cannot be used for comparing the animals with each other.

Bon was allowed twenty sessions of approximately thirty minutes duration each for adjustment to the experimenter and to the experimental room before the tokens were introduced. When the tokens were presented, the subject would not take them either from the hand of the experimenter or from the floor. After forty tokens had been picked up from the floor and inserted into the vender by the experimenter, the subject began anticipating the appearance of the food by reaching into the food cup as soon as a token had been released in the slot and before the mechanism had had time to vend the food. Attempts to induce the subject to take the tokens in hand or to push them into the slot when they were balanced at the mouth of it failed, however, for nine experimental periods, or while 180 tokens were being inserted by the experimenter. At that point Bon began to put the tokens into the food cup and to hold his hand in anticipation of food. He very carefully took in succession seventeen tokens, which had been placed in the mouth of the slot after opportunity had been given him to pick them up from the floor, and dropped them into the food cup. Then the subject picked up a token from the

TABLE 2

Showing for each subject the number of orientation sessions given (Step 1), the number of tokens inserted for demonstration into the vender before the food was anticipated (Step 2), and the total number of tokens inserted before the subject inserted a token independently (Step 3)

SUBJECT	STEP 1	STEP 2	STEP 3
Bon	20	40	237
Alpha	22	10	120
Bula	18	17	93
Bimba	1	4	10
Moos	0	1	1
Velt	3	25	35

floor and inserted it into the slot. A total of 237 tokens had been deposited by the experimenter before he executed this act.

Alpha at the end of twenty-two half-hour sessions in the experimental room seemed adjusted to the situation and ready for experimentation. When the tokens were then introduced, the subject after examining them with hand, mouth, and nose dropped them. At that point the experimenter began inserting them into the vender and the subject secured the vended food. The connection between the insertion of a token and the appearance of food became apparent to the animal after ten tokens had been used. Extensive verbal, gestural, and manual attempts in ensuing sessions to get the subject to insert the tokens proved ineffective. Finally the experimenter balanced a token in the mouth of the slot and Alpha pushed it in. This she did on twenty successive trials before she came finally to insert the tokens herself. Always before a token was balanced in the slot, the animal was given an opportunity to pick it up from the floor and insert it herself. Including the twenty tokens the subject had pushed in from the mouth of the slot, a total of 120 tokens had been used before the subject learned to insert them independently.

Bula required eighteen experimental sessions before adaptation to the experimental room was achieved. When tokens were placed on the floor near her following the adaptation sessions, she did not examine them in any way. After the experimenter had inserted seventeen tokens, the subject began to anticipate the appearance of the food as soon as a token was released in the slot. After four more tokens had been used, she began to push them in when they were balanced at the mouth of the slot by the experimenter; this she continued for eleven successive trials. She then made six unsuccessful attempts to insert a token which she had picked up from the floor. The failures were due in part to the hasty manner, almost a throwing movement, in which she tried to put the tokens into the slot. Her unsuccessful efforts were continued with decreasing frequency in the following ten experimental sessions during which time a total of twenty-two attempts were made. Then Alpha, who had previously learned to use the tokens, was brought into the experimental room so that Bula might see her use the tokens. After Alpha had used twenty tokens, Bula tried to insert one but was pushed away from the vender by Alpha. Alpha was then restrained by the experimenter, and Bula persisted in her attempts until she succeeded in inserting a token into the slot. Thereafter she had little difficulty with the task.

Velt was introduced to the tokens after three sessions in the experimental room. He initiated no attempts to inspect or to handle the tokens. Twenty-five were inserted into the vender by the experimenter before Velt began to anticipate the food by reaching into the cup as soon as a token had been released in the slot. When the experimenter began to hand him the tokens to insert, he dropped them on the floor immediately. However, he would push them into the slot if they were balanced at its mouth. This act he repeated ten times before he independently performed the complete series of acts leading from the picking up of the token to the securing of the food.

Bimba was given only one session for orientation. From the start she was accompanied by Alpha, a subject experienced in using the tokens. When the tokens were placed on the floor near the vender, Alpha began to insert them at once. Coincident with Alpha's insertion of the fifth token Bimba reached into the food cup ahead of Alpha but was crowded out by the latter before the food appeared. After Alpha had used a total of ten tokens, Bimba picked up one and attempted to insert it, but she was pushed away by Alpha who took the token. At that point Alpha was restrained, whereupon Bimba succeeded in inserting a token after she had dropped it four times. She subsequently had little difficulty in inserting tokens.

Moos was introduced to the experiment without his being given any sessions for orientation, since he had been used extensively in laboratory work and was acquainted with the experimenter. He examined the tokens which were on the floor but lost interest in them very quickly. Then the experimenter picked up a token, directed the subject's attention toward it, and inserted it into the vender. The subject took the reward from the food cup, ate it, and immediately picked up a token which with difficulty he inserted into the slot. He showed marked excitement when he obtained the food. After he had used ten tokens, he could insert them in one motion.

[*Editor's Note:* Material has been omitted at this point.]

IV. EFFECTIVENESS OF TOKENS AND FOOD IN ELICITING A WORK-TASK

As one approach to the determination of the effectiveness of food and token-rewards for chimpanzees a comparison was made of the extent to which the subjects would execute a work-task for a token-reward, on the one hand, and for a food-reward, on the other. Two measures of work performance were obtained. One was in terms of the amount of time required for the animals to do a fixed amount of work, a constant weight technique; the other was the amount of work, in pounds lifted, which the animals would do, a variable weight technique. In the former the unit of measurement was the time in seconds required for the animals to execute the work-task ten times. In the latter the unit of measurement was the maximum weight which the subjects would lift as determined by regularly increasing the weight until a point was reached beyond which they would not work. With both techniques the tokens could be used in obtaining food as soon as they were secured.

As apparatus, the circular restraining cage, the work machine, and a vender were used. The work machine and the vender were placed before the grilles of the restraining cage. White tokens and grapes were used as rewards. Bon, Moos, Bula, and Bimba served as subjects.

Each subject was given the same amount of practice in working for grapes and for tokens before regular experimentation began. During this practice, measurement was made of the time required by the animals to remove the two kinds of reward from the work apparatus. This was done because one subject in learning to operate the apparatus had, at first, exercised extreme caution in removing grapes. For this subject (Bon) practice was continued until the one reward was removed as quickly as the other. None of the other subjects showed consistent difference in speed of removing the two forms of reward from the holder.

A. Constant weight technique

Procedure. This experiment, which utilized the constant weight technique, consisted of eighty sessions for each subject. During forty of the sessions food was the incentive; during the other forty, tokens were used. In each session a subject was allowed to work for only ten units of reward in order to prevent fatigue and waning motivation.

The weights used in the work-task could be lifted by the animals without undue strain. For the three smaller subjects, Bon, Bula, and Bimba, the constant weight was 12 pounds. For Moos, the largest subject, it was 18 pounds.

The order of presentation of rewards was balanced. On any one day both kinds of reward were used. If grapes were the reward in the afternoon, they would be used again the following morning; then tokens would be used for two consecutive sessions, and so on to the end of the experiment. The reward-holder of the work apparatus was always loaded with the reward before the grille shutter was opened for the first trial of any experimental session. The next reward was placed in the holder the instant the handle fell back to a horizontal position, and so on to the end of the period.

Measurements of performance were in terms of time. When grapes were used as incentive in the work apparatus, the unit of measurement was the time required to execute the work-task ten times. When tokens were used as incentive, two time intervals were recorded. One was the total time spent per session in working for and in using ten tokens. The other was the total time required in exchanging the ten tokens for grapes. The ten intervals which composed the total time spent in using the tokens began at the instant the subject removed a token from the reward-holder and ended when it got the grape from the food cup of the vender. The difference between the total time spent in securing and in using the tokens and the time spent only in using the tokens was the measure of the animal's performance when tokens were the incentive. The data collected, then, consist primarily of the times, in seconds, required by each subject to secure comparable amounts of the two kinds of reward. The difference between these times indicates the relative effectiveness of the grapes and the tokens in eliciting the work-response. Stop watches served as timing instruments.

Results. Tables 4, 5, and 6 present a summary of the results for the four subjects. Interruptions due to temporary illnesses occurred for two of the subjects at the middle of the experiment. Bon was treated for intestinal parasites, and Bimba was allowed to rest for two weeks in order to recover from a cold. For these reasons the data from all subjects are divided into two parts which, for convenience of discussion, are called test data and retest data. For Bon and Bimba the test data consist of the records made before their illnesses. For the other two subjects, the test data were simply the records of the first half of the experiment. The retest data were in all cases the records of the second half of the experiment. Since the performances of Moos and Bula were not interrupted between the test and retest periods, it was possible to combine for each of them the data from the two periods in order to provide a sufficient number of measures to justify determinations of the significance of the differences between the token and the food data. The sigma scores which are based upon ten measures are presented merely to indicate trends. It should be emphasized, however, that each of the ten measures consisted of ten trials.

TABLE 4

Showing for each subject the average, variability, and difference scores of the time, in seconds, spent in working for grapes in the constant weight situation

Test period = 10 experimental sessions of 10 trials each
Retest period = 10 experimental sessions of 10 trials each
Combined = both the test and retest periods combined

SUBJECTS	PERIODS						
	Test		Retest		Difference of M's	Combined	
	M	σ	M	σ		M	σ
Bon	89.9	4.23	67.8	3.46	22.1	78.9	
Moos	58.2	4.87	55.0	5.15	3.2	56.6	5.23
Bula	69.9	7.16	68.7	5.37	1.2	69.3	6.36
Bimba	104.3	5.97	90.2	4.84	14.1	97.2	

TABLE 5

Showing for each subject the average, variability, and difference scores of the time, in seconds, spent in working for tokens in the constant weight situation

Test period = 10 experimental sessions of 10 trials each
Retest period = 10 experimental sessions of 10 trials each
Combined = both the test and retest periods combined

SUBJECTS	PERIODS						
	Test		Retest		Difference of M's	Combined	
	M	σ	M	σ		M	σ
Bon	90.8	4.38	69.1	4.95	21.7	80.0	
Moos	57.6	3.96	57.0	3.29	0.6	57.3	3.65
Bula	71.3	7.61	69.9	7.23	1.4	70.6	7.41
Bimba	114.6	11.63	106.2	8.30	8.4	110.4	

Tables 4 and 5 show the speed of working for grapes and for tokens, respectively. A rough indication of behavioral consistency is yielded by a comparison of the test and retest data. Table 4 shows that the average performances of Bon and Bimba in the test and the restest periods differed by 22.1 and 14.1 seconds, respectively, in the speed with which the animals worked for ten grapes. They both worked more rapidly in the second half of the experiment. Moos and Bula worked only a little faster in the last half of the experiment, the differences in the averages of the first and second half being for the two animals only 3.2 and 1.2 seconds, respectively. The variabilities of the subjects' performances as shown by the sigma scores were similar in the test and

TABLE 6

Showing for each subject the average, variability, difference, and significance of difference scores of the time, in seconds, spent in working for both grapes and tokens in the constant weight situation

Test period = 10 experimental sessions of 10 trials each
Retest period = 10 experimental sessions of 10 trials each
Combined = both the test and retest periods combined

SUBJECTS	PERIODS	REWARDS						D	D/σd
		Grapes			Tokens				
		M	σ	σ_m	M	σ	σ_m		
Bon	Test	89.9	4.23		90.8	4.38		0.9	
	Retest	67.8	3.46		69.1	4.95		1.3	
	Combined	78.9			80.0			1.1	
Moos	Test	58.2	4.87		57.6	3.96		−0.6	
	Retest	55.0	5.15		57.0	3.29		2.0	
	Combined	56.6	5.26	1.21	57.3	3.65	0.84	0.7	0.48
Bula	Test	69.9	7.16		71.3	7.61		1.4	
	Retest	68.7	5.37		69.9	7.23		1.2	
	Combined	69.3	6.36	1.46	70.6	7.41	1.70	1.3	0.58
Bimba	Test	104.3	5.97		114.6	11.63		10.3	
	Retest	90.2	4.84		106.2	8.30		16.0	
	Combined	97.2			110.4			13.2	

retest periods. When the test and retest data from Moos and Bula are combined, the variabilities do not increase materially.

The results obtained when the subjects were working for tokens (table 5) show for Bon and Bimba a marked increase in speed of working in the retest period over that of the test period. Moos and Bula worked at about the same rate in both of the periods. The differences, in seconds, in average speed of working in the test and retest periods were for Bon 21. 7, for Bimba 8.4, for Moos 0.6, and for Bula 1.4. The variability or sigma scores of the test and retest periods were remarkably constant for three of the subjects. For Bimba the variability was lower in the retest than in the test period, but it was relatively high in both periods. Little change in the variability scores for Moos and Bula occurred when the data from the test and retest periods were combined.

Consider, now, the differences between the grapes and the tokens as work-producing agencies. In table 6, parts of the data shown in tables 4 and 5 are presented and treated for purposes of such a comparison. The means of the combined measures show for three of the subjects negligible differences (1.1, 0.7, and 1.3 seconds) in speed of working for grapes and for tokens. The significance of the differences computed for two of the subjects, Moos and Bula, were 0.48 and 0.58 critical ratio units. For the fourth subject, Bimba, considerable difference was shown. She worked faster for grapes than she did for tokens. The average difference in the speed with which she worked for the two rewards was 13.2 seconds.

The differences between the average speed of working for grapes and that of working for tokens are all but one in the same direction when the periods of work are treated either separately or combined. The differences are in the direction of a greater time expended in securing the token-rewards. In addition, the extent of the differences between the speed of working for grapes and the speed of working for tokens which were found in the test period persisted in the retest period, except in the case of Moos. In the test period Moos worked faster for tokens by 0.6 second per ten trials while in the retest period he worked faster for grapes by 2.0 seconds per ten trials.

The variability of the performances, as expressed in standard deviations, was greater for three subjects when they were working for tokens than it was when they were working for grapes. Moos, to the contrary, showed greater variability when working for grapes during the test period, the retest period, and the two periods combined. This may have been due to the unusual stimulation which he appeared to derive from using the tokens to get food from the venders. When he first learned to use the tokens, he jumped about in the cage very excitedly, had an erection of the penis, and vocalized considerably. Thereafter, when using the tokens, he was noticeably more alert and anxious than were any of the other subjects.

B. Variable weight technique

Procedure. With the variable weight technique, in which the relative effectiveness of tokens and grapes in eliciting the work-task was measured in terms of the difference between the greatest number of pounds which the animals would lift for each form of reward, ten experimental sessions were devoted to each of the four subjects. Since considerable food was consumed per period and since the work done was strenuous, only one session was given per day. A session terminated when a subject would no longer work for either tokens or grapes. It was assumed that if an animal did not begin work within three minutes after a reward had been placed in the holder it was no longer interested in it.

The weights which the subjects were required to lift in getting the rewards were increased by steps of 2 pounds each from the starting point of 12 pounds. At any given weight, opportunities to secure both a token and a grape were given. Then the weight was increased, and again both rewards were introduced but in an order opposite to that used with the preceding weight, and so on until the animal quit working.

In order to establish whether or not an animal would lift more for one kind of reward than for the other, the following test was made. If an animal had secured one kind of reward at a given weight but had not got the other in the allotted time, then the task was increased in difficulty and both forms of reward were again used as lure. Now if the subject secured only the previously preferred reward again, it was considered to have worked harder for it than for the other form of reward. This criterion was adopted since the order of presenting the rewards might have made it appear that one form of reward was more effective in eliciting the work-task than the other. Suppose, for example, that the first reward presented with the 20-pound weight be a token and that the subject after securing it will work no further. It might be that the subject was fatigued or near satiation at that point and would have quit working no matter what the reward next to be presented was.

Results. The quantitative results are presented in table 7. Three of the subjects, Bon, Moos, and Bula, showed little difference in the weight which they would lift whether the reward was grapes or tokens. Bimba, on the other hand, rather consistently lifted more to get grapes than to get tokens. These findings are

TABLE 7

Showing the greatest weights which the subjects lifted for each kind of reward (grapes and tokens) on ten different days and the difference between these weights

G and T represent grapes and tokens, respectively, and the numbers in the columns which they head represent in pounds the greatest weights lifted. The numbers under D are the differences between the two maximum weights lifted each day. The numbers preceded by the minus sign represent differences in favor of the tokens.

DAY	BON			MOOS			BULA			BIMBA		
	G	T	D	G	T	D	G	T	D	G	T	D
1	20	18	2	16	16	0	20	20	0	18	16	2
2	24	26	−2	20	20	0	22	20	2	26	22	4
3	26	26	0	20	20	0	24	20	4	28	24	4
4	26	26	0	22	22	0	20	22	−2	30	24	6
5	30	30	0	22	24	−2	26	26	0	30	26	4
6	34	32	2	24	24	0	32	30	2	30	30	0
7	36	36	0	22	22	0	28	26	2	26	24	2
8	36	34	2	22	22	0	30	30	0	32	30	2
9	34	34	0	22	20	2	24	26	−2	30	26	4
10	36	36	0	22	22	0	30	30	0	30	30	0

consistent with results obtained with the constant weight technique. It will be remembered that Bimba, there, worked faster for grapes than for tokens.

A consideration of the performances of the individual animals reveals that Bon on three occasions lifted a greater weight to get a grape than he did to get a token and that on one occasion the reverse was true. The differences in each case were only 2 pounds and may have been due to the order in which the rewards were presented as he never in any case continued to work until he met the criterion of harder-work which had been adopted. On each of these occasions he seemed to quit work because the weight was heavy rather than because a particular reward was unattractive.

Moos' general behavior in the form of attempts to lift the heavier than 22-pound weights did not suggest that he tried harder to get the one rather than the other kind of reward. That on one day he lifted 2 pounds more to get a token than he did to get a grape and that he did the opposite on a later day are probably insignificant since he did not continue to work for the given rewards until the criterion was met. It will be noted that the points, in pound weights, beyond which Moos did not work were not as great as they were in the cases of the other subjects. This was no reflection upon either his strength or his motivation. He was the largest subject in the experiment and no doubt could have lifted the heaviest weights, but the weight-stem of the apparatus was not sufficiently long to accommodate enough weights to test the maximum that he might have lifted. In view of this fact, to the apparatus was attached a restraining device which restricted considerably the use of the shoulder muscles in lifting the weights. This device reduced the size of the weight which he could lift because it made the act primarily one of arm movement.

On only one occasion did Bula meet the criterion of working harder to get grapes than to get tokens. That was in the third period when she secured two grapes by lifting heavier weights than she would for tokens. In order to test whether or not she was working to her maximum for grapes and for food-tokens, a light blue token, which she had previously learned to use in getting out of the experimental room and returning to the living quarters, was introduced on three occasions after she had stopped working for food and for food-tokens. These tests were made in the sixth, eighth and tenth sessions of the experiment. How Bula had learned to use the light blue token and under what conditions she used it will be told in a later section. Suffice it here to say that she became proficient at its use to the extent that it had to be out of her sight or she would not work for any other kind of reward. It should be emphasized that in these tests the blue token was placed in the reward-holder of the work apparatus only after the subject had quit working for the two regular rewards and also after the weight had been increased by 6 pounds. On all three occasions she pulled in the light blue token, stood before the door of the restraining cage until she was released, straightway inserted the tokens into the proper receptacle, and climbed upon the experimenter preparatory to leaving the room. She had not been lifting her maximum by at least 6 pounds to get the grapes and food-tokens.

Bimba met the criterion of working harder for one kind of reward than for the other in five of the ten periods. She lifted greater weights for grapes. In three of the remaining periods she lifted 2 pounds more for grapes than for tokens.

Additional experimental work was done with Bimba since she had demonstrated in the regular experimental periods that she would work harder, in general, for grapes than for tokens. Her work was extended until on each of five days she stopped working for the tokens before she did for the grapes. Then, following the usual three minute period in which she would not work for a token, more tokens were placed in the reward-holder. After four or five tokens had been added, she lifted the weight and secured the tokens. This suggests that the reward value of the tokens may be cumulative.

Summary of results with work techniques. Three of the subjects showed little difference in speed of working at a constant work-task whether the reward in the apparatus was food or tokens. The fourth subject worked faster for food. The variability of the performances of the subject which worked faster for food was greater when tokens were the incentive than when food was the incentive in the apparatus. The variability of performance of the other three subjects was not markedly different under the two conditions of reward; for two of these subjects the differences which did occur were in the direction of more variability for the tokens than for the food reward.

With the variable weight technique, it was found that three of the four subjects lifted on any given day about as much to get tokens as they did to get food. The fourth subject, on the other hand, rather consistently lifted greater weights for the food. These findings are consistent with those obtained with the constant weight technique. The subject who had worked faster for the food than for the tokens at the constant work-task also worked harder for the food than for the tokens under the variable weight procedure.

[*Editor's Note:* Material has been omitted at this point.]

There are a number of ways of characterizing a token (poker chip) which has through use become associated with a primary reward. It may be considered as: (1) a sign or symbol of something to come or happen, e.g., food, water, or an activity-privilege; (2) an instrument, a tool, a means to an end, a manipulandum; (3) something which, because of certain past associations, now has intrinsic value. One is unable to choose between these possibilities from the present experiments. To regard a token as a sign or symbol (1) seems justifiable provided a distinction is made between a physical object and a language process or a pure stimulus act. But symbolism has so many different connotations that its use is confusing.[12] The second and third characterizations seem preferable.

IX. SUMMARY

1. This experimental investigation was concerned with (*a*) a determination of whether or not poker chips, or tokens, could serve as surrogate or secondary rewards for chimpanzees and (*b*) a determination of the relative effectiveness of such tokens, once they were established as secondary rewards, in comparison with that of a primary reward, food.

2. The observations were grouped around four central problems: (*a*) the effectiveness of tokens as work-producing agencies; (*b*) their influence upon the effectiveness of delayed food-rewards; (*c*) their effectiveness as tools in attaining ends in harmony with specific drives or needs; (*d*) their effectiveness in eliciting competitive behavior in social situations.

3. The experiments made use of restraining cages, food-vending machines, a water-vending device, subsidiary cabinets as

[12] Morris, C. W. The concept of the symbol. J. Phil., 1927, *24*, (I) 253-262, (II) 281-291.

token receptacles, a work machine, and an assortment of poker chips. Six chimpanzees, three males and three females, ranging in age from two years and two months to approximately six years, served as subjects in the study. In the several experiments varying numbers of these subjects were used.

4. The subjects readily learned to manipulate the tokens in obtaining food from a vender, to discriminate between tokens with and without food value, and to operate the work apparatus. That the tokens, by virtue of their use in the preliminary training, came to function as secondary or surrogate rewards was demonstrated.

5. For three out of four subjects the tokens were about as effective as was food in eliciting a work-task when measured either in terms of speed of doing a fixed task or in terms of amount of work done within a stated time interval. The fourth subject did more and faster work for food than for tokens. For three of the four subjects the variability in speed of working was slightly greater when the incentives were tokens than when they were food.

6. The tokens were effective as incentives in the delayed primary reward situations provided there was no long delay between the insertion of the tokens into a vender and the appearance of food. The condition most favorable to the elicitation of a single work act in the delayed reward situations was the one in which the subject secured a token, kept it during an interval of delay and finally exchanged it for food. Little difference was shown among the other three conditions in which the subjects either executed the work-task and waited until the end of a period for food, executed the work-task and waited with valueless tokens in their presence until the end of a delay, or worked for a token, inserted it into a vender and then waited for the food-reward.

7. Of the three situations in which the subjects could work for a period of time and at the end of that time receive from a vender food-reward in proportion to the amount of work done, the two situations most and about equally favorable to work were those in which tokens and grapes were used as incentive in the work apparatus. The third situation which offered no incentive in the work apparatus was not very conducive to work.

8. The number of tokens a subject had in its possession was one of the factors determining how many tokens would be worked for during a fixed period of time when none of the tokens could be exchanged for food until the end of the period. The more tokens possessed, the fewer additional tokens would the subjects work to secure.

9. In the discrimination situations chimpanzees (*a*) learned to select of two tokens the one which had twice the reward value of the other, (*b*) selected tokens in harmony with a hunger or a thirst drive provided the drives were fairly intense, and (*c*) associated tokens with certain activity-privileges.

10. In social situations the tokens elicited competitive behavior similar to that evoked by food. From one subject they elicited begging behavior.

11. In short, this study shows that tokens came to function as secondary or surrogate rewards for chimpanzees, that in certain situations they were about as effective as food in eliciting certain kinds of behavior, and that chimpanzees can make a variety of discriminations between tokens having different reward values and can use them in harmony with their drives or motives.

14

Reprinted by permission from *J. Exp. Psychol.* **38**:1–16 (1948)

THE RELATION OF SECONDARY REINFORCEMENT TO DELAYED REWARD IN VISUAL DISCRIMINATION LEARNING[1]

BY G. ROBERT GRICE[2]

Psychological Laboratory, State University of Iowa

INTRODUCTION

Delayed reward is a problem of central importance for reinforcement theories of learning. In his goal gradient hypothesis, Hull (4) postulated that the strength of the association formed between a response and its accompanying stimuli is inversely related to the length of the delay by which the reward follows the response. The gradient was assumed to be logarithmic in form. This hypothesis was supported by the delayed reward experiments of Hamilton (3) and Wolfe (11), both of which showed learning to be a decreasing function of the delay of reward. These studies indicated also that the gradient extended for a considerable period of time following the response. Subsequently, Hull (5) pointed out the possibility that the learning after long delays in the Wolfe and Hamilton experiments might have been the result of immediate secondary reinforcement. In both of these maze learning experiments the *S*s were detained in delay compartments for a period of time prior to entering the goal box. It was suggested that the delay boxes, being followed by reward, might have become secondary reinforcing agents.

In an attempt to minimize secondary reinforcement immediately following the response to be learned, Perin (6, 7) employed a Skinner-type box in which the lever pressing response, the delay, and the reward all occurred in the same compartment. The results of these experiments indicated that learning under such conditions was im-

[1] A dissertation submitted in partial fulfillment of the requirements for the degree of Doctor of Philosophy in the Department of Psychology in the Graduate College of the State University of Iowa. The writer is indebted to Professor Kenneth W. Spence who directed the investigation.

[2] Now at the University of Illinois.

possible with delays of about 30 sec. or longer. On the basis of these findings, Hull (**5**) postulated a short primary delay of reinforcement gradient of about 30 sec. The longer goal gradient was then derived from this by assuming the development of secondary reinforcement within the 30-sec. period, and the gradual moving forward of such secondary reinforcing property in the stimulus response sequence.

Further evidence of the importance of secondary reinforcement in delayed reward learning is provided in an experiment by Perkins (**8**). Perkins employed a covered T-maze of which both the top and bottom were opal-flashed glass, thus eliminating all extra-maze cues. The possibility of secondary reinforcement by the two different delay boxes was eliminated by interchanging the two delay compartments so that each was followed by reward half of the time. When a group for which the delay compartments were interchanged in this manner was trained with 45 sec. delay, it was found to learn significantly more slowly than another 45 sec. delay group for which the same delay box was always followed by food. This difference in rate of learning was attributed by Perkins to the action of secondary reinforcement in the case in which the delay compartments were not interchanged, and to the elimination of differential secondary reinforcement by the shifting procedure. Perkins then went on to study learning in the T-maze as a function of delay with the delay compartments interchanged. He obtained a function which dropped more sharply than that obtained by Wolfe, and which showed only a barely significant amount of learning with 120 sec. delay.

Spence (**10**), in a recent analysis of this problem, has suggested that even the shortened gradients obtained by Perin and Perkins may be the result of immediate secondary reinforcement. His suggestion is that while these experiments may have succeeded in eliminating differential secondary reinforcing stimuli from the external environment, such stimuli may have existed *within* the animal. Thus the particular pattern of proprioceptive stimulation following the correct response would presumably persist within the organism for a short period, and might still be effective at the time of reward. In such an event, the proprioceptive pattern of stimulation coincident with the reward would acquire secondary reinforcing properties through its association with the immediate food reinforcement. On subsequent occurrences of the response, the proprioceptive stimuli resulting from the act, being similar to the proprioceptive traces persisting until the moment of reinforcement, could, through generalization, provide immediate secondary reinforcement. The length of delay during which learning could occur, would depend then upon the length of time that the changing proprioceptive stimulus trace remained sufficiently similar to that at the time of the response to permit generalization.

Under such an interpretation there is no need for the assumption of a primary delay of reinforcement gradient, as all instances of learning under delayed reward conditions would be accounted for by immediate secondary reinforcement. This formulation relieves the learning theorist of the embarrassing problem of explaining how a reward can work backwards to strengthen a stimulus-response association, when the response was made some time earlier.

One possible experimental test of this interpretation is to introduce delay of reward in a non-spatial type of discrimination learning problem, e.g., visual discrimination learning. In spatial discrimination problems such as the T-mazes used by Wolfe and Perkins, the *S* is forced to make different spatial responses, turning right or left, which provide very different proprioceptive stimulation. Since one of these responses is consistently followed by reward, and the other is never reinforced, differential proprioceptive secondary reinforcement may be built up so long as the proprioceptive traces of the different acts are discriminably different at the moment of food reward. However, in the non-spatial visual discrimination situation, it is possible to eliminate such immediate secondary reinforcement based on differential proprioceptive stimulation. Since the positions of the positive and negative stimuli are shifted irregularly from left to right, each motor response of turning left or right is correct half of the time and incorrect half of the time. This condition means that neither pattern of proprioceptive stimulation acquires greater secondary reinforcing strength than the other. The remaining possibility of immediate secondary reinforcement in this type of learning problem is that the stimulus traces of the visual stimuli provide the basis of the differential stimulation. So long as a trace of the positive stimulus is effective at the time of reward, this trace may become a secondary reinforcing agent, and the positive stimulus itself might then come to provide immediate secondary reinforcement. The limit within which delayed reward learning could take place in a visual discrimination problem would depend then on the time that the after-effects (stimulus traces) of the positive and negative stimuli remain discriminably different.

A further implication of the above analysis for visual discrimination learning is that it should be possible to improve learning with delayed reward by experimentally introducing immediate secondary reinforcement. For example, if the positive stimulus itself, or stimuli within the range of its generalization gradient, were present at the time of reward, this stimulus should acquire secondary reinforcing properties, and would then provide immediate reinforcement, thus increasing the speed of learning, and lengthening the delay with which learning might occur. Furthermore, if the *S*s were *forced* to respond with chatacteristically different motor patterns to the different choice stimuli, the proprioceptive traces of the response to the correct stimulus should, within limits, acquire secondary reinforcement properties, and extend the delay of reinforcement gradient.

Several investigators have employed delayed reward in visual discrimination situations. Wood (12) using chicks in a brightness discrimination found learning to be a decreasing function of delayed reward up to five min. However, the chicks were delayed only following correct choices. Following errors they received immediate electric shock. Thus, the avoidance of shock on correct trials may have provided immediate reinforcement, and the experiment is not an uncomplicated delayed reward situation. Wolfe (11), in a black-white discrimination experiment with rats, found no learning with delayed reward when the delay followed both correct and incorrect responses. However, when the door was blocked providing immediate frustration of wrong responses, results were obtained similar to those of his T-maze experiment. Like the Wood experiment, this one is also probably not a genuine delayed reward situation. The delay compartment, which after the delay always leads immediatley to food, would acquire secondary reinforcing properties. Under Wolfe's procedure, only correct responses are followed by entrance into the delay compartment. Thus, immediate secondary reinforcement is provided for correct responses but not for incorrect responses. Wolfe's incidental finding, that learning is greatly retarded when delays follow choices of both stimuli, has been verified by two other investigators. Riesen (9), using a red-green discrimination with chimpanzees, found greatly retarded learning with one- and two-sec. delays, and one of two *S*s failed to learn with four sec. delay. There was no evidence of learning with eight sec. delay. However, several specially trained *S*s, trained to respond differentially to the stimuli, were able to learn with eight sec. delay. Gulde (2), in a study with rats in a black-white discrimination, found no evidence of learning in 200 trials with five sec. delay.

In the present experiment, the learning of a black-white discrimination problem was studied as a function of the time of the delay of reward, in order to ascertain the limit and form of this relationship. The function presumably depends on the stimulus traces of the black and white stimuli. Second, the effect of introducing immediate secondary reinforcement was studied. This was accomplished first by the black and white goal boxes, so that following a choice of either black or white, the animal, after the delay, always entered a goal box of the same color as the stimulus chosen. The final experiment was to force the animal to make characteristically different motor adjustments to the black and white stimuli. Learning in this situation was then compared with that in a problem of equal delay of reward in which no such characteristically different motor reponses were made.

Subjects and Apparatus

The *Ss* were 75 experimentally naive, female, albino rats from the colony maintained by the department of psychology of the State University of Iowa. Their ages ranged from 80 to 110 days at the beginning of the experiment. They were assigned at random to the experimental groups.

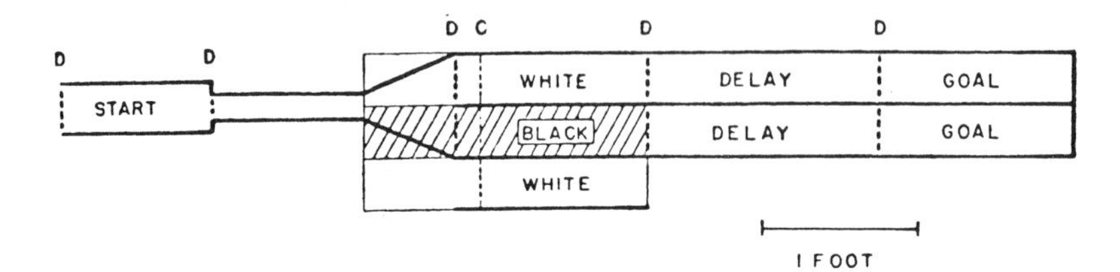

Fig. 1. Ground plan of the experimental apparatus. Doors are represented by heavy dotted lines at the points D. Curtains are represented by the light dotted line at C.

The apparatus consisted of a black-white discrimination box. The ground plan is shown in Fig. 1. The rat was placed in the starting box from which it passed through a two-in.-wide alley into the choice chamber. From this point it could enter either a black or a white painted alley. The floor of the half of the choice chamber leading to the black alley was painted black and the floor of the half leading to the white alley was white. In each alley two in. from the entrance there were black or white curtains the same color as the alley. The section of the apparatus which made up the black and white alleys and the floor of the choice chamber consisted of three identical alleys with the two outer ones white and the middle one black. By sliding this section back and forth, the black and white alleys could be shifted from right to left. After passing through either the black or white stimulus alley, the animal entered a neutral gray alley which was used as a delay compartment, and could be varied in length from 18 to 72 in. The goal boxes were continuations of these alleys and were 15 in. long. All alleys except the narrow starting alley were four in. wide and four in. high. With the exception of the black and white alleys, the entire apparatus was painted a neutral gray. The choice chamber and the stimulus alleys were covered with clear glass and all others were covered with hardwarecloth. Vertical sliding doors, which prevented retracing, were located at the entrances to the stimulus alleys, at the beginning of the delay compartments and at the entrances to the goal boxes. The doors were operated by *E* from behind a one-way vision screen at the starting end of the apparatus.

The lighting was indirect from two shaded 200-watt bulbs. The brightness of the floor of the white alley just in front of the curtain was 1.515 apparent foot candles. The brightness of the floor of the black alley at the same point was 0.071 apparent foot candles.

In order to force the animals to make characteristically different motor responses to the black and white alleys, different obstacles could be placed in them. One of these sets of obstacles consisted of a 15-degree incline, nine in. long, which began one in. beyond the curtain. The other consisted of two blocks $2\frac{1}{4}$ by five in. and the same height as the alley. One block was placed on the left side of the alley one in. beyond the curtain. The second was placed on the right side of the alley two in. beyond the first. This forced the animal, after passing the curtain, to pass through a five-in. section of alley $1\frac{3}{4}$ in. wide, make a sharp jog to the left, and continue through another narrow five-in. section before entering the delay compartment. Both the blocks and the inclines were available in black and white, so that the blocks could be in the white alley and the incline in the black, or the reverse. Removable black and white goal boxes were made of quarter-in. plywood. These boxes fitted inside of the gray goal boxes and could be shifted from the left to the right side.

Preliminary Training

All animals were adapted to a 24-hour feeding schedule for at least a week prior to preliminary training. They were fed in individual cages and received eight gm. of Purina Laboratory Chow daily. This diet and feeding procedure were the same throughout the experiment. During the experiment they were fed immediately after the daily runs.

For the preliminary training, neutral gray alleys were substituted for the black and white stimulus alleys. On the first day the rats were placed in the goal box and allowed to eat ten 0.3 gm. pellets of Purina Chow. They remained in the goal box for 15 min. The animals which were assigned to the zero delay group were fed in the removable gray alleys, and all others were fed in the regular gray goal boxes at the end of the 18-in. delay compartment. Next, the animals were placed in the gray alley between the curtain and the door at the entrance to the alley and were allowed to run through the alley and the delay compartment to the goal box where they received one .15 gm. pellet of food, which was the standard reward throughout the experiment. There were four such runs, two on the left side and two on the right. The zero-delay animals received food in the removable alley and did not run through the delay compartment. On the second day there were four more such runs. On the third and fourth days the animals were placed in the starting box and allowed to run through the gray alleys to food. They received 10 runs each day, forced half to the right and half to the left in a random order. The forcing was accomplished by closing one of the doors at the choice point. The purpose of these 20 forced trials was to help equalize position habits. The animals in the differential response group received their forced runs with gray blocks and a gray incline, otherwise identical to the black and white ones, placed in the gray alleys. Runs through the blocks and over the incline were divided equally between left and right.

Experimental Procedure

Gradient Experiment:

In the experiment proper, all animals were trained to go to the white alley. All animals had either no initial color preference or a preference for the black. Three animals with initial white preferences were eliminated. There were 10 free choice trials per day for the first 200 trials and 20 trials per day after that. The white alley was alternated from left to right in the order RLRRLLRLLRLRLLRRLRRL. All trials were separated by at least two min. Animals were run until they reached the criterion of learning, which was 18 out of 20 trials correct, with the last 10 perfect. One animal was discontinued after failing to learn in 700 trials and four after failing to learn in 1440 trials.

Groups of animals were run under six different delay of reward conditions. The zero delay group received food immediately in the white alley. The 0.5 sec. delay group was allowed to run through the 18-in. delay compartment with the doors open and was rewarded in the goal box following choices of the white alley. The 1.2 sec. delay group ran through a 36-in. delay alley with the doors open and the 2 sec. delay group ran through a 72-in. alley. The times 0.5, 1.2 and 2 sec. were determined by timing the animals on each run. The time measured was that from the leaving of the black or white alley to the entrance into the goal box. These times are the means for each group, of the median times for each animal in the group. The median times ranged from 0.4 to 0.6 sec. for the 1.5 sec. group, from 1.1 to 1.3 sec. for the 1.2 sec. group, and from 1.5 to 2.4 sec. for the 2.0 sec. group. The distributions of delay times for individual animals were all positively skewed in the manner typical of such time data. The mean semi-interquartile ranges of these distributions were 0.1 sec. for the 0.5 sec. group, 0.2 sec. for the 1.2 sec. group and 0.2 sec. for the 2.0 sec. group. The five-sec. delay group was delayed for five sec. in the 18-in. delay compartment by leaving the goal box door closed for five sec. after the animal entered the delay compartment. The delay for the 10 sec. group was accomplished in a similar manner. In all groups the delay was the same following correct and incorrect choices. Following all choices of the white, the rat received the pellet of food in a glass cup placed at the end of the goal box and was allowed to eat in the goal box. Following choices of the black there was no food or cup in the goal box, and the animal was allowed to remain in the box approximately the same amount of time as in the case of correct choices.[3] There were 10 animals in each group except in the 10 sec. delay group in which there were only five.

[3] The use of no food cup, rather than an empty one, in the goal box following incorrect responses reduces the possibility of secondary reinforcement following wrong responses, and probably, in part, accounts for the unusually rapid learning of the zero delay group.

Secondary Reinforcement Groups:

One group of 10 rats was run under the same conditions as the regular five sec. group except that the goal box following choices of the white alley was white with a white food cup, and the box following choices of the black was black with no reward. The delay compartments were gray as in the regular five sec. group. The black and white goal boxes were placed inside the regular gray ones and were shifted from left to right to correspond with the stimulus alleys.

Another group of 10 rats was run under the five sec. delay conditions, with the blocks in one color alley and the incline in the other. Half of the animals had the incline in the white alley with the blocks in the black, and half had the reverse arrangement. White was correct in both cases, and both the delay compartments and the goal boxes were gray as in the original five sec. delay group.

Results

Gradient Experiment:

The number of trials required by each *S* to reach the criterion of learning is shown in Table I. The median number of trials for the

TABLE I

Number of Trials Required by Each Animal to Reach the Criterion, and the Median Number for Each Experimental Group

Sec. of delay	0	0.5	1.2	2	5	10	5 sec. Black and White Goal Boxes	5 sec. Differential Responses
	20	100	220	300	320	840	80	320
	20	110	160	150	350	850	180	380
	10	60	160	190	250	1440+	160	600
	20	140	280	440	560	1440+	130	270
	20	90	230	360	700+	1440+	130	210
	10	40	140	370	350		140	390
	10	100	250	260	600		150	350
	10	90	230	440	650		200	170
	30	130	180	240	1440+		270	140
	20	80	170	280	1260		180	270
Median	20	95	200	290	580	(1440+)	155	295

zero delay group was 20; for 0.5 sec., 95 trials; for 1.2 sec., 200 trials; for 2 sec., 290 trials, and for 5 sec., 580 trials. In the five sec. group one animal was discontinued after failure to learn in 700 trials and another after 1440 trials. No median is available for the 10 sec. delay group since three of the five *S*s failed to learn in 1440 trials. It was not possible to apply the *t*-test of statistical significance to the differences between the groups because the assumptions of normality and homogeneous variance were not fulfilled, and because there were some indeterminate values in cases where animals failed to learn. However, a test proposed by Festinger (1) which makes no asumptions as to the form of the distributions was applied. All differences between adjacent groups were significant at the five percent level of confidence or higher. The results of the test for the comparisons made with this test are shown in Table II.[4]

TABLE II

RESULTS OF THE FESTINGER TEST OF THE SIGNIFICANCE OF THE DIFFERENCES BETWEEN THE GROUPS IN THE NUMBER OF TRIALS REQUIRED TO REACH THE CRITERION OF LEARNING

The P-values are the levels of confidence at which the hypothesis may be rejected that the two groups compared were drawn from the same population.

Groups Compared	*d*	P
0 vs. .5 sec.	5.00	.01
0 vs. 1.2 sec.	5.00	.01
.5 vs. 1.2 sec.	4.05	.01
.5 vs. 2.0 sec.	5.00	.01
1.2 vs. 2.0 sec.	3.25	.05
1.2 vs. 5.0 sec.	4.20	.01
2.0 vs. 5.0 sec.	3.20	.05
5 sec. vs. B & W Boxes	4.90	.01
5 sec. vs. different responses	3.00	.05

Learning curves for the six groups for 700 trials are shown in Fig. 2. These curves represent the percent of correct choices for successive blocks of 20 trials. Each *S* is assumed to continue at the 100

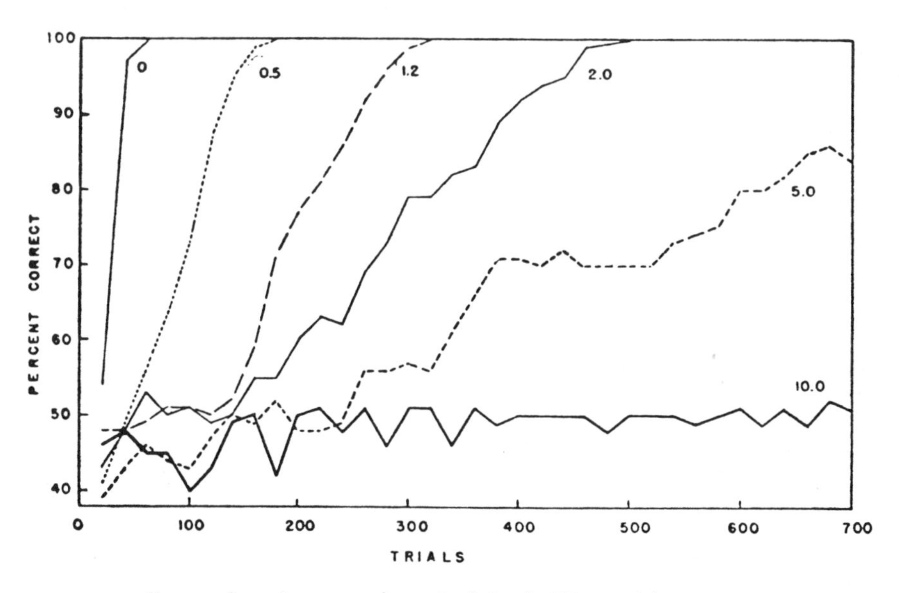

FIG. 2. Learning curves for each of the six different delay groups

[4] The test of statistical significance proposed by Festinger (1) is based on a pooled ranking of the measures in the two groups to be compared. It is then possible to make statements of probability about the sum of ranks of one group. The test yields a statistic, '*d*,' which may be referred to tables giving the values of '*d*' for various *N*'s required for significance at the one and five percent levels of confidence. Strictly speaking, the hypothesis tested is that the two groups of measures are samples drawn from the same population, but the test is believed by Festinger to be most sensitive to differences between means. The only assumptions involved are that the two samples are independent and drawn at random.

percent level after reaching the criterion. Tests with several animals showed this to be the case with only very slight deviation.

Gradient curves to show learning as a function of delay are shown in Figs. 3 and 4. In Fig. 3 the reciprocal of the number of trials required by each group to reach the level of 75 percent correct choices is plotted against the time of the delay of reward.[5] Fig. 4 shows a

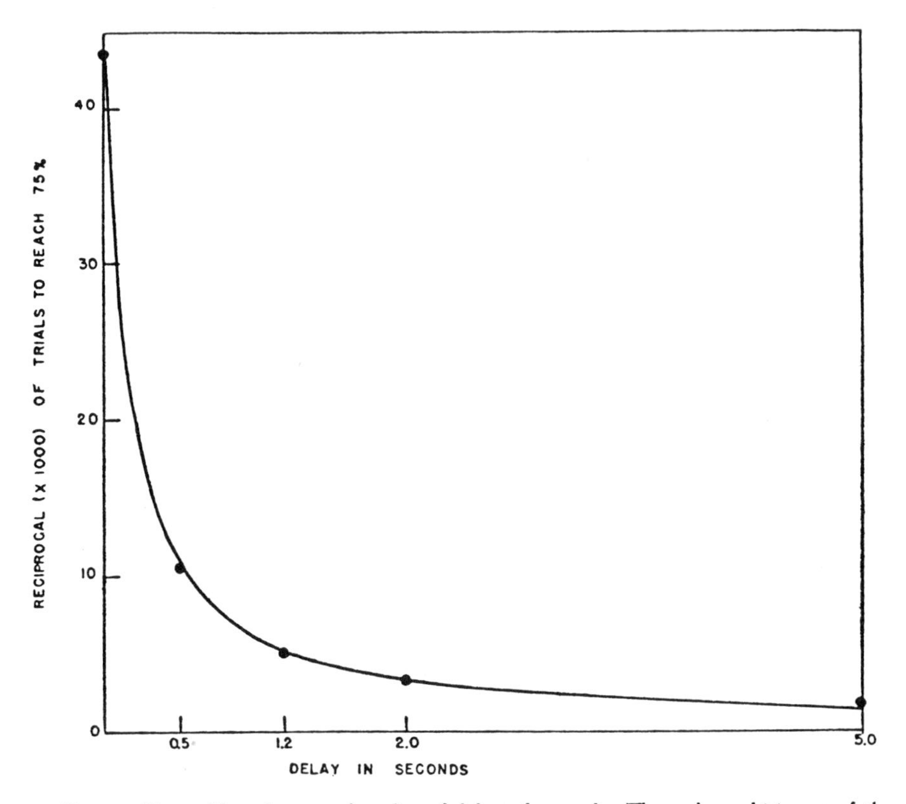

Fig. 3. Rate of learning as a function of delay of reward. The reciprocal × 1000 of the number of trials to reach the level of 75 percent correct choices is plotted against the time of delay. Experimental values are represented by black dots and the smooth curve is fitted to these data.

gradient based on the percent of correct choices for each group during trials 141–180. The sigma value, based on the percent correct, is assumed to be a measure of the difference in habit strength between

[5] Mathematically this function may be represented by a hyperbola of the reciprocal type. The equation for the curve fitted to the data of Fig. 3 is:

$$R = \frac{1}{.023 + .14T},$$

where R is the reciprocal × 1000 of the number of trials to reach 75 percent correct, and T is the time in sec. of the delay of the reward.

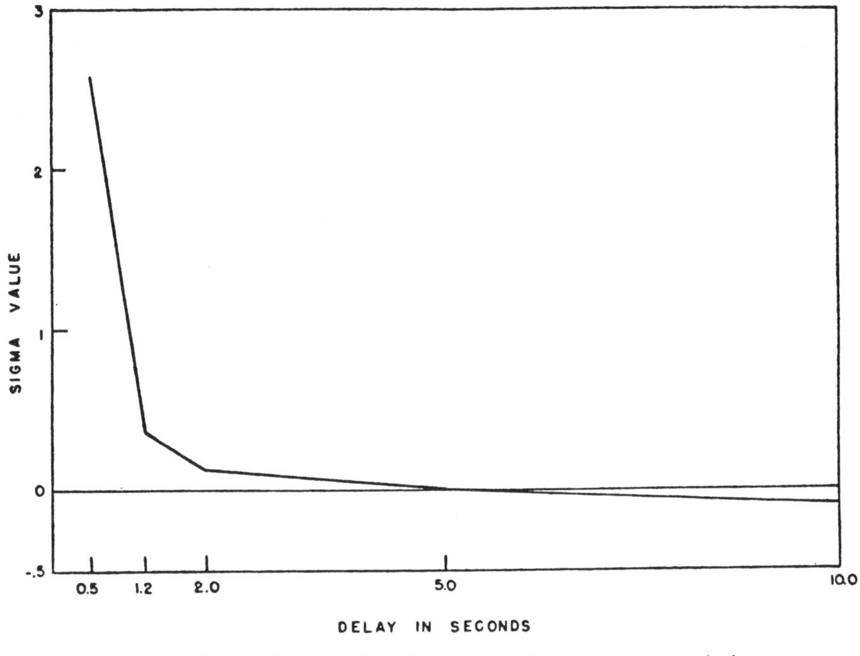

FIG. 4. Sigma values based on the percent of correct responses during trials 141–180 for each delay group

the correct and incorrect responses.[6] The fact that the 10 sec. group is below zero reflects the fact that this group had, at this stage of learning, not yet overcome a slight initial preference for the black alley.

Secondary Reinforcement Experiments:

The two groups of concern in this portion of the experiment are the five sec. delay group with the black and white goal boxes, and the five sec. delay group in which the animals were forced to make different responses to the two stimulus alleys. These groups may be compared to the five sec. group in the gradient experiment, where no differential secondary reinforcement was present. The numbers of trials required by each animal to reach the criterion are shown in Table I. The medians were 155 for the black-white goal box group and 295 for the differential response group as contrasted with 580

[6] This measure, suggested by Hull (5), is based on the assumption that the excitatory strengths of the two competing responses oscillate from moment to moment according to the normal probability function. Consequently, their difference would also oscillate in this manner. The result is that, when the tendencies are equal, there is a 50 percent choice of each response, and as the difference between them increases, the percent of choice of the stronger increases until the ranges of oscillation no longer overlap, and one response is chosen 100 percent of the time. Any percent of occurrence of one response may be converted into an amount of difference value by means of the normal integral table. This gives a standard score representing the difference between the excitatory (habit) strengths of the competing responses.

trials for the five sec. control group. Again the t-test of statistical significance could not be applied. According to the test used above, the difference between the black-white goal box group and the five sec. control group is significant at the one percent level of confidence, and the difference between the differential response group and the control group is significant at the five percent level.

The learning curves for the three groups are shown in Fig. 5. They are plotted as percent correct for blocks of 20 trials. The rate of learning for the black-white goal box group is clearly much more rapid than the other two. The difference between the differential

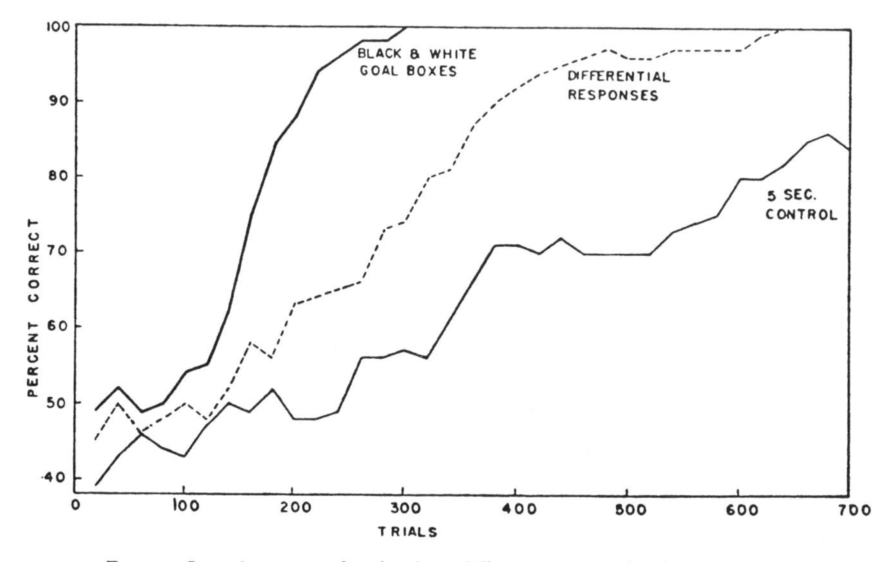

Fig. 5. Learning curves for the three different groups with five sec. delay

response group and the control group was tested by applying the t-test for related measures to the differences between blocks of 20 trials for the 300 trials from 141 to 440. The mean number correct in each block of 20 trials was obtained for each group. Since the differences between these means approximated a normal distribution the t-test was appropriate. A t of 3.61 was obtained, which for 14 degrees of freedom means that the hypothesis that the mean difference is zero may be rejected at the one percent level of confidence. This means that the two curves differ significantly, and that the differential response group did learn at a significantly faster rate than the rats in the control condition.

Discussion of Results

One fact of primary interest in the above data is the steepness and short duration of the obtained delay of reinforcement gradient.

It should be pointed out that the discrimination problem itself was a very easy one, as shown by the unusually low median number of 20 trials required to learn the problem with no delay. It is significant that a delay of even one-half sec. required almost five times that number of trials, and that at five sec., the median trials to learn increased to 580 and two animals failed to learn in 700 and 1440 trials respectively. Ten sec. delay was apparently beyond the limit under which learning was possible for three of the five *S*s, and the other two were able to learn only after about 850 trials. The discrepancy between

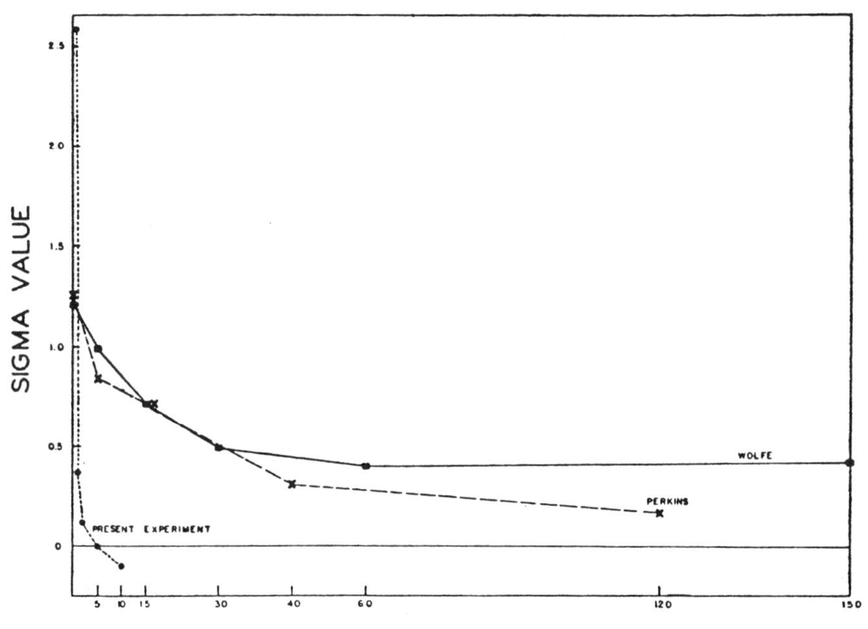

FIG. 6. Sigma values from blocks of trials for the various delay groups in the Wolfe, Perkins and present experiments. The blocks of trials included for each of the three experiments are as follows: Wolfe—trials 7, 8, 9, and 10; Perkins—first 36 trials; present experiment—trials 141–180.

the results of this study and those of Wolfe (11) and Perkins (8) is shown strikingly in Fig. 6, in which sigma values similar to those of Fig. 4 are plotted against length of delay. Perkins' experiment differed from Wolfe's in that differential secondary reinforcement in the delay boxes was eliminated by alternating the two delay boxes in a random order and by rotating the maze in the room. However, there was the possibility of secondary reinforcement based on the reinforcement of the proprioceptive trace of the consistently correct turning response. However, this possibility was eliminated in the present experiment, since there was no motor response pattern which was always correct. The difference between Perkins' results and those of this experiment may be accounted for by the presence of such secondary reinforcement in the one but not in the other.

The gradient obtained here is also steeper and shorter than that obtained by Perin (7). The results of the two experiments are compared in Fig. 7. Both of the gradient curves are based on the slopes

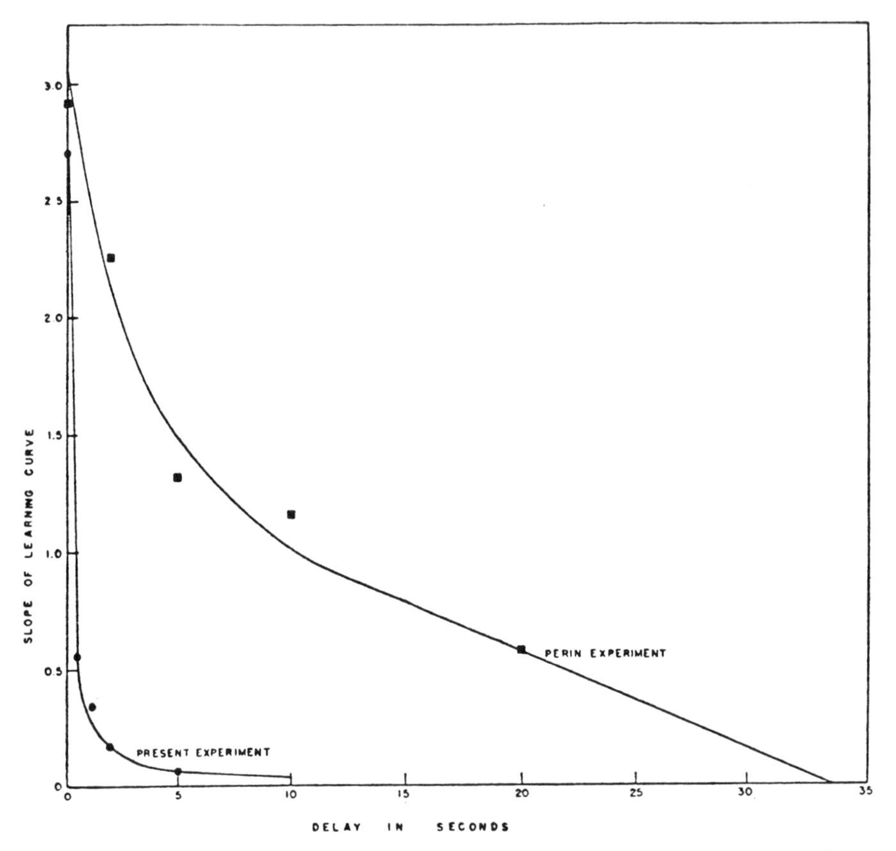

FIG. 7. Rate of learning as a function of time of delay in the Perin and present experiments. The slopes of the group learning curves are plotted against the time of delay of reward. The Perin experimental data are represented by black squares and the present data by black circles. The smooth curves are fitted to the empirical data.

of group learning curves which were plotted as percent of correct trials. Perin's learning curves ranged from zero to 100 percent correct responses. The slopes plotted for his experiment are the slopes of the tangents to the fitted learning curves at the 50 percent point. The learning curves of the present experiment range from 50 to 100 percent correct. Since the middle portions of these curves were approximately linear, the slopes plotted in the gradient curve are the slopes of straight lines fitted to the portions of the learning curves between 60 and 90 percent. These two gradients then represent the slopes of the learning curves (the rate of learning) of both experiments at the point where learning was half completed.[7] Perin's results indicate that learning may be obtained in his situation up to about 30 or 35 sec. delay, and the present data suggest that the limit is about 10 sec. or possibly less for most *S*s. Again the discrepancy may be accounted for by the possibility that Perin's rats obtained immediate secondary reinforcement from the proprioceptive cues following the act. A fact of interest is that in Perin's second experiment (7), involving a differentiation between two bar-pressing habits, no learning was obtained with delays of only two and five sec. if the bar was removed immediately following both correct and incorrect responses. It was only when the bar remained in place following the incorrect response, but was removed immediately following the correct response, that the 30-sec. gradient was obtained. The result of this difference in procedure was probably to make the difference between the correct and incorrect responses greater. The removal of the bar following response would certainly affect the postural adjustments to the bar and those immediately following the act of pressing the bar. Thus the pattern of response following removal of the bar would always be reinforced, while that following the incorrect response and the bar remaining in place would never be reinforced. Apparently the difference between the two motor patterns was not enough to provide differential secondary reinforcement unless this additional difference were introduced. Another factor is that the occurrence of massed, non-reinforced incorrect responses with the bar in place would tend to eliminate that response through the process of extinction.

If there is no true primary gradient of reinforcement, as previously suggested, the question may be raised as to why even the short gradient effect was obtained in the present experiment. One plausible answer is that the basis of the secondary reinforcement is the perseverative stimulus trace or sensory after-effect from the black or white stimulus. So long as any traces of the black and white stimuli remain discriminably different at the time of reward, there is the possibility of differential secondary reinforcement. The stimulus trace from the postitve white stimulus in the choice chamber could thus acquire secondary reinforcing potency through generalization

[7] The curve fitted to the data from the present experiment in Fig. 7 is of the same type as the one in Fig. 3. The equation is:

$$S = \frac{1}{.3704 + 2.83T},$$

where S is the slope of the learning curve and T is the time of the delay of reward. The equation fitted by Perin to his data and reproduced here is:

$$S = 1.6 \cdot 10^{-.15T} - .043T + 1.45.$$

from its perseverative trace which is contiguous with the food reward. The data suggest that the traces or the differences between them decrease rapidly during the first sec., are at a very low level at the end of five sec., and have more or less disappeared at the end of 10 sec. or are sufficiently different from those at the choice point to be beyond the range of generalization.

The black-white goal box experiment demonstrates clearly the effect of secondary reinforcement in delayed reward learning. The stimulus, being present at the time of reward, acquired secondary reinforcing properties. Thus, upon orienting toward and entering the white alley, the *S* received immediate secondary reinforcement. The result was a marked speeding up of learning, even though the food reward itself was delayed for five sec.

The fact that animals which were forced to make consistently different motor responses to the two stimuli learned at a significantly faster rate than the *S*s for which this was not the case, may be interpreted in terms of proprioceptive secondary reinforcement as described in the Perin and Perkins experiments. In this group there were not only different visual traces following the different choices but also entirely different afferent traces resulting from the different postural and motor adjustments required in the two alleys. It is reasonable to assume that in the rat such differential proprioceptive stimulation effects would continue longer than the visual after-effects. As stated above, the limits of delay with which learning may occur would depend on the length of time during which these two proprioceptive traces remain discriminably different. During the period in which this difference between the traces remains, the food reward following correct responses will produce and strengthen secondary reinforcing properties for the trace of the stimuli produced by the correct response. No such reward is associated with the traces of the incorrect response. As long as the trace of the correct response is within the range of the generalization gradient of the proprioceptive pattern stimulating the organism at the time of the response in the white alley, this proprioceptive pattern will acquire secondary reinforcing properties. It is this immediate secondary reinforcement that accounts for the superiority of the differential motor response group over the five sec. delay group with no such differential motor response.

Summary

1. Groups of white rats were run on a black-white discrimination problem with delays of reward of 0, 0.5, 1.2, 2, 5, and 10 sec.

2. A very steep delay of reinforcement function was obtained within this range, with no learning by three of five *S*s in the 10 sec. group.

3. When immediate secondary reinforcement was introduced by allowing the animal to eat in a goal box of the same color as the positive stimulus, learning with delayed reward was greatly facilitated.

4. When animals were forced to make characteristically different motor responses to the black and white stimuli, they learned at a significantly faster rate than animals which received equal delay, but made no such characteristically different motor adjustments.

5. The data are consistent with a theory which assumes no "primary" delay of reinforcement gradient, but accounts for learning under delayed reward conditions in terms of some type of immediate secondary reinforcement. Such secondary reinforcement may be based upon proprioceptive stimulation resulting from the response and continuing until the moment of the reward. The proprioceptive pattern accompanying the correct response acts as a secondary reinforcing agent by virtue of its similarity to the traces which on previous trials have lasted until the reward. In the usual visual discrimination learning situation no differential proprioceptive stimuli follow correct and incorrect choices. In such situations, learning is possible with only very short delays of reward. What learning does occur may be attributed to immediate secondary reinforcement from the visual stimuli. This secondary reinforcement is presumably based on traces of the visual stimuli which continue until the time of the primary reward.

(Manuscript received March 14, 1947)

References

1. Festinger, L. The significance of difference between means without reference to the frequency distribution function. *Psychometrika*, 1946, 11, 97–105.
2. Gulde, C. J. The effects of delayed reward on the learning of a white-black discrimination by the albino rat. Unpublished Master's thesis, Univ. Iowa, 1941.
3. Hamilton, E. L. The effect of delayed incentive on the hunger drive in the white rat. *Genet. Psychol. Monogr.*, 1929, 5, 131–207.
4. Hull, C. L. The goal gradient hypothesis and maze learning. *Psychol. Rev.*, 1932, 39, 25–43.
5. Hull, C. L. *Principles of behavior.* New York: Appleton-Century, 1943.
6. Perin, C. T. A quantitative investigation of the delay-of-reinforcement gradient. *J. exp. Psychol.*, 1943, 32, 37–51.
7. Perin, C. T. The effect of delayed reinforcement upon the differentiation of bar responses in white rats. *J. exp. Psychol.*, 1943, 32, 95–109.
8. Perkins, C. C. The relation of secondary reward to gradients of reinforcement. Unpublished Ph.D. thesis, Univ. Iowa, 1946.
9. Riesen, A. H. Delayed reward in discrimination learning by chimpanzees. *Comp. Psychol. Monogr.*, 1940, 15, 1–53.
10. Spence, K. W. The role of secondary reinforcement in delayed reward learning. *Psychol. Rev.*, 1947, 54, 1–8.
11. Wolfe, J. B. The effect of delayed reward upon learning in the white rat. *J. comp. Psychol.*, 1934, 17, 1–21.
12. Wood, A. B. A comparison of delayed reward and delayed punishment in the formation of a brightness discrimination habit in the chick. *Arch. Psychol.*, 1933, 24, No. 157, 40 pp.

Editor's Comments on Papers 15 and 16

15 ESTES and SKINNER
Some Quantitative Properties of Anxiety

16 SOLOMON, KAMIN, and WYNNE
Traumatic Avoidance Learning: The Outcomes of Several Extinction Procedures with Dogs

The capacity to escape, anticipate, and avoid noxious events is obviously an important one. It is not surprising that experience with aversive events often produces profound changes in behavior. However, there are several features of such learning that are puzzling and difficult to interpret. A considerable amount of effort has been devoted to untangling some of the perplexing features of learning produced through aversive stimulation.

That animals learn to perform acts that cancel scheduled noxious events raises many questions about the nature of such learning. How can the nonoccurrence of an event control behavior? What is the nature of an animal's anticipation of an aversive stimulus such as a loud noise or an electric shock? Can a reinforcer be found that accounts for avoidance behavior? What motivational roles are played by fear and anxiety? How do species-typical behavior patterns enter into avoidance learning?

Considerable controversy continues to surround proposed answers to questions such as these (for example, Bolles, 1970, 1972; Gray, 1975; Herrnstein, 1969; Rescorla and Solomon, 1967; Seligman and Johnston, 1973). Despite this disagreement, substantial progress has been made toward understanding learning based on aversive stimulation. The two papers reprinted in this section represent important steps in this progress.

A useful behavioral measure of the anticipation of an aversive event was discovered by Estes and Skinner and is reported in Paper 15. The suppression of appetitive, operant behavior in rats by the presentation of a signal for electric shock has proved to be a reliable and sensitive measure. Such suppression has been demonstrated in cats, dogs, monkeys, guinea pigs, pigeons, and many different strains of rat. Several reviews of this work are available (Davis, 1968; Millenson & deVilliers, 1972; Blackman, 1977). One of the most interesting uses of this measure of anxiety has been in determining the course of anxiety to cues in avoidance learning (Kamin et al., 1963).

Paper 16 by Solomon, Kamin, and Wynne reports the remarkable persistence shown by their dogs following shock-avoidance training. A relatively brief training experience was enough to establish behavior ranges that persisted for hundreds of trials. The persistence of learned avoidance behavior is one of its most interesting features. It has attracted the attention of psychologists interested in animal models of human psychopathology, and it has provided a serious challenge for reinforcement theorists because the learned behavior is maintained without an obvious source of reinforcement.

REFERENCES

Blackman, D., 1977, Conditioned suppression and the effects of classical conditioning on operant behavior, in *Handbook of Operant Behavior,* W. K. Honig and J. E. R. Staddon eds., Prentice-Hall, Englewood Cliffs, N.J..

Bolles, R. C., 1970, Species-specific defense reactions and avoidance learning, *Psychol. Rev.* **77**:32–48.

Bolles, R. C., 1972, Reinforcement, expectancy, and learning, *Psychol. Rev.* **79**:394–409.

Davis, H., 1968, Conditioned suppression: a survey of the literature. *Psychon. Sci. Monogr. Supp.* **2**:283–291.

Gray, J. A., 1975, *Elements of a Two-Process Theory of Learning,* Academic Press, New York.

Herrnstein, R. J., 1969, Method and theory in the study of avoidance. *Psychol. Rev.* **76**:49–69.

Kamin, L. J., C. J. Brimer, and A. H. Black, 1963, Conditioned suppression as a monitor of fear of the CS in the course of avoidance training, *J. Comp. Physiol. Psychol.* **56**:497–501.

Millenson, J. R., and P. A. deVilliers, 1972, Motivational properties of conditioned anxiety, in *Reinforcement: Behavioral Analyses,* R. M. Gilbert and J. R. Millenson, eds., Academic Press, New York, pp. 97–129.

Rescorla, R. A., and R. L. Solomon, 1967, Two-process learning theory: relationships between Pavlovian conditioning and instrumental learning, *Psychol. Rev.* **74**:151–182.

Seligman, M. E. P., and J. C. Johnston, 1973, A cognitive theory of avoidance learning, in *Contemporary Approaches to Conditioning and Learning,* F. J. McGuigan and D. B. Lumsden, eds., V. H. Winston & Sons, Washington, D.C., pp. 69–110.

15

Reprinted by permission from *J. Exp. Psychol.* **29**:390–400 (1941)

SOME QUANTITATIVE PROPERTIES OF ANXIETY

BY W. K. ESTES AND B. F. SKINNER

University of Minnesota

Anxiety has at least two defining characteristics: (1) it is an emotional state, somewhat resembling fear, and (2) the disturbing stimulus which is principally responsible does not precede or accompany the state but is 'anticipated' in the future.

Both characteristics need clarification, whether they are applied to the behavior of man or, as in the present study, to a lower organism. One difficulty lies in accounting for behavior which arises in 'anticipation' of a future event. Since a stimulus which has not yet occurred cannot act as a cause, we must look for a *current* variable. An analogy with the typical conditioning experiment, in which S_1, having in the past been followed by S_2, now leads to an 'anticipatory' response to S_2, puts the matter in good scientific order because it is a current stimulus S_1, not the future occurrence of S_2, which produces the reaction. Past instances of S_2 have played their part in bringing this about, but it is not S_2 which is currently responsible.

Although the temporal relationships of classical conditioning provide for an acceptable definition of anticipation, the analogy with anxiety is not complete. In anxiety, the response which is developed to S_1 need not be like the original response to S_2. In a broader sense, then, anticipation must be defined as a reaction to a current stimulus S_1 which arises from the fact that S_1 has in the past been followed by S_2, where the reaction is not necessarily that which was originally made to S_2. The magnitude of the reaction to S_1 at any moment during its presentation may depend upon the previous temporal relations of S_1 and S_2.

The concept of 'emotional state' also needs clarification in view of the experiments to be described. It has been suggested elsewhere (3) that in treating emotion purely as *reaction* (either of the autonomic effectors or of the skeletal musculature), a very important influence upon operant behavior is overlooked. In practice we are most often interested in the effect of a stimulus in altering the strength of behavior that is frequently otherwise unrelated to the emotion. A stimulus giving rise to 'fear,' for example, may lead to muscular reactions (including facial expression, startle, and so on) and a widespread autonomic reaction of the sort commonly emphasized in the study of emotion; but of greater importance in certain respects is the considerable change in the tendencies of the organism to react in various other ways. Some responses in its current repertoire will be strengthened, others weakened, in varying degrees. Our concern is most often with anxiety observed in this way, as an effect upon the normal behavior of the organism, rather than with a specific supplementary *response* in the strict sense of the term.

The experiments to be described follow this interpretation. An emotional state is set up in 'anticipation' of a disturbing stimulus, and the magnitude of the emotion is measured by its effect upon the strength of certain hunger-motivated behavior, more specifically upon the rate with which a rat makes an arbitrary response which is periodically reinforced with food. Such a rate has been shown to be a very sensitive indicator of the strength of behavior under a variety of circumstances (1), and it is adapted here to the case of emotion. Mowrer's recent summary of techniques for measuring the 'expectation' of a stimulus does not include a comparable procedure (2).

In these experiments the disturbing stimulus to be "anticipated" was an electric shock delivered from a condenser through grids in the floor of the experimental box. The stimulus which characteristically preceded the disturbing stimulus and which therefore became the occasion for anxiety was a tone, produced by phones attached to a 60 cycle A.C. transformer.

The apparatus, which provided for the simultaneous investigation of twenty-four rats, has been described in detail elsewhere (1, 3). Each rat was enclosed during the experimental period in a light-proof and nearly sound-proof box containing a lever which could be easily depressed. A curve (number of responses vs. time) for each rat and mechanically averaged curves for the group and for certain sub-groups of six or twelve rats were recorded. Under the procedure of periodic reconditioning, the control clock was set to reinforce single responses to the lever every four minutes, intervening responses going unreinforced. The rats came to respond at a relatively constant rate during the one-hour experimental period, and the summated response curves tended to approximate straight lines, except for local cyclic effects resulting from a temporal discrimination based upon the four-minute period of reinforcement. Curves *A* and *C* in Fig. 4 are for groups of twenty-four rats and represent the sort of baseline available for the observation of the effect of anxiety.

The subjects were twenty-four male albino rats under six months of age, taken from an unselected laboratory stock. Records were taken for one hour daily during the entire experiment. After preliminary conditioning of the pressing response, two sub-groups were formed; one group of twelve rats was kept at a relatively high drive, while the other twelve were held at a drive which produced a very low rate of responding. The sound and shock were first introduced after two weeks of periodic reinforcement.

Conditioning of a State of Anxiety

The averaged periodic curve for twelve rats on a high drive on the occasion of the first presentations of the tone (*T*) and shock (*S*) is shown in Fig. 1. On this first presentation the tone was allowed to sound for three minutes. Each rat was then given a shock and the tone was stopped. It will be observed that neither the tone nor the shock (at the intensity used throughout the experiment) produced any disturbance in the mean periodic rate at either presentation. This orderly base-line made it possible to follow with ease the development of the 'anticipation' of the shock during subsequent repetitions of the situation.

The tone-shock combination was presented twice during each of six consecutive hourly periods. Then, in order to clarify any changes in the behavior, the period of the tone was lengthened to five minutes and the combination was given only once during each ensuing experimental hour.

The principal result of this part of the experiment was the conditioning of a state of anxiety to the tone, where the primary index was a reduction in strength of the hunger-motivated lever-pressing behavior. The ratio of the number of responses made during the period of the tone to the average number made during the same fraction of the hour in control experiments was 1.2 : 1.0 [1] for the first experimental hour; it had dropped to 0.3 : 1.0 by the eighth.

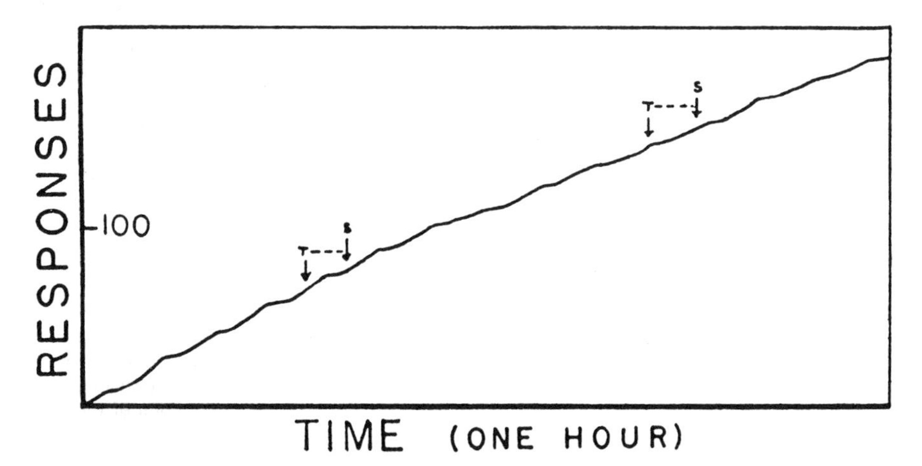

Fig. 1. First presentations of tone and shock. Mechanically averaged curves for twelve rats under periodic reinforcement. The tone was turned on at *T* and at *S* the shock was administered and the tone turned off. There is no noticeable effect of either tone or shock upon the rate of responding at this stage.

The changes in behavior accompanying anticipation of the shock are shown in Fig. 2, which gives the averaged curves for the group of six rats with the highest periodic rate during the first four days of the five-minute tone. A number of characteristics of these records should be noted. The progressively more marked reduction in periodic rate during the anticipatory period is obvious. The effect upon the rate is felt immediately after the presentation of the tone and remains at a constant value until the shock is given. (This constancy might not be maintained if the situation were repeated often enough to allow the rat to form a temporal discrimination.) Effects also appear

[1] The ratio is not expected to be exactly 1 : 1 since the number of responses made during a period of five minutes will depend upon where the period begins with respect to the four-minute interval of reinforcement.

after the shock, which were not present in Fig. 1 as the result of the shock alone. Especially in Curves *A* and *B* of Fig. 2, the shock is seen to be followed by a depression and irregularity of rate which are at least much greater than any effect in the control records. With continued repetition of the experiment, this disturbance tends to adapt out, although not completely. In Curves *C* and *D* of Fig. 2,

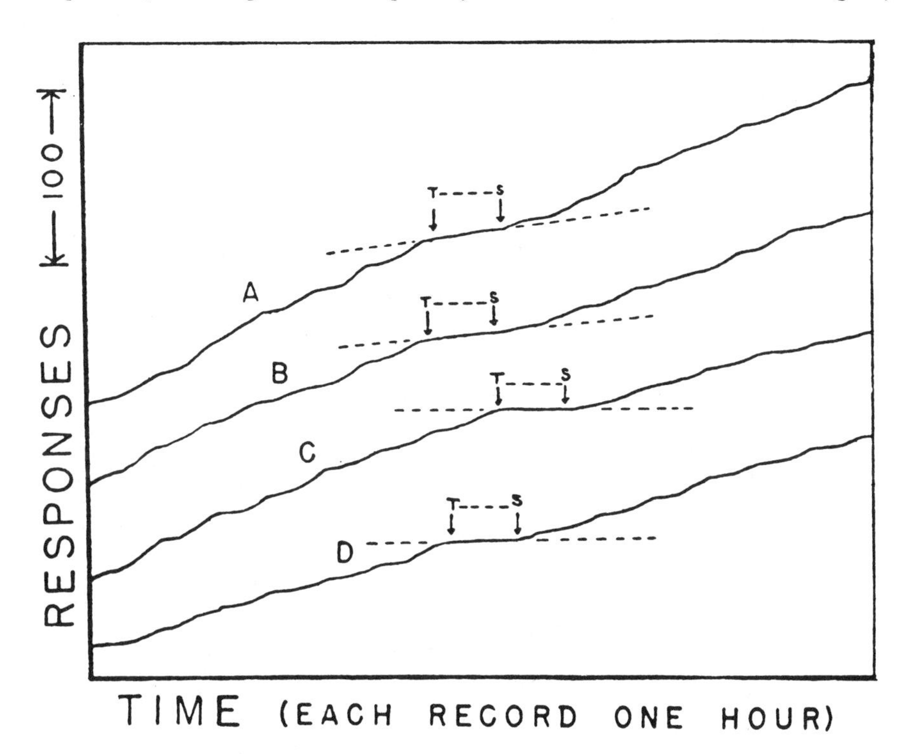

Fig. 2. Reduction in rate of responding during successive periods of anxiety. Averaged curves for six rats on four consecutive days. By the third or fourth day responding practically ceases during the presentation of the tone.

the distortion is much less marked. Curve *B* of Fig. 4 gives a similar example at a relatively late stage of conditioning.

The modification in behavior correlated with the anticipation of a disturbing stimulus cannot be attributed to a negative reinforcement of the response to the lever, since the shock was always given independently of the rat's behavior with respect to the lever. Only upon rare occasions could the shock have coincided with a response. This was especially true in the experiments upon the group at a lower drive, where a similar effect was obtained. Figure 3 shows averaged curves for a group of six rats which had been subjected to the procedure just described except that their drive was so low during conditioning that the rate of responding was virtually zero. The lower curve in Fig. 3 is for the first day on which the five-minute rather than the three-minute period was given. Up to and including this record, no effect of the anticipation of the shock could be detected, since the animals were not responding at a significant rate. The drive was then raised, and the upper curve in Fig. 3 shows the performance of the same group on the following day. By sighting along the curve, one may observe a marked depression in the rate of responding during

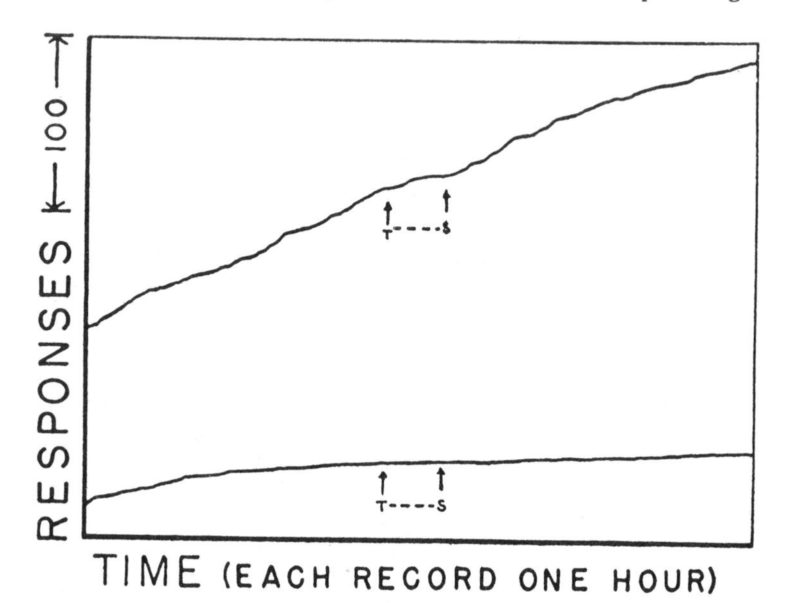

Fig. 3. Reduction in rate during anxiety following experiments at a very low drive. The lower record is a curve for six rats at a very low drive but otherwise comparable with the curves in Figs. 1 and 2. The upper curve is for the same group at a higher drive on the following day. The tone has an obvious effect, although all previous presentations have been made at a drive so low that no effect was observable.

the period of the tone. Comparison with Curve *B* in Fig. 2 shows that although the base line at the higher drive is more irregular, a depression of relatively the same magnitude is obtained. In this case, coincidental presentations of shock and response may safely be ignored, yet the tone has acquired the same depressing effect upon the behavior.

Another characteristic which deserves attention is the compensatory increase in periodic rate following the period of depression. This appears to some extent in all records obtained; but it may be seen most clearly in Curve *B* of Fig. 4, a periodic curve for all 24 rats after the emotional conditioning was quite complete. The curve was obtained about two weeks after the records in Fig. 2.

Curves *A* and *C* are controls taken (at a slightly higher drive) on adjacent days. By sighting along Curve *B*, one may observe a clear-cut increase in rate subsequent to the shock, which continues until the extrapolation of the curve preceding the break is reached. Evi-

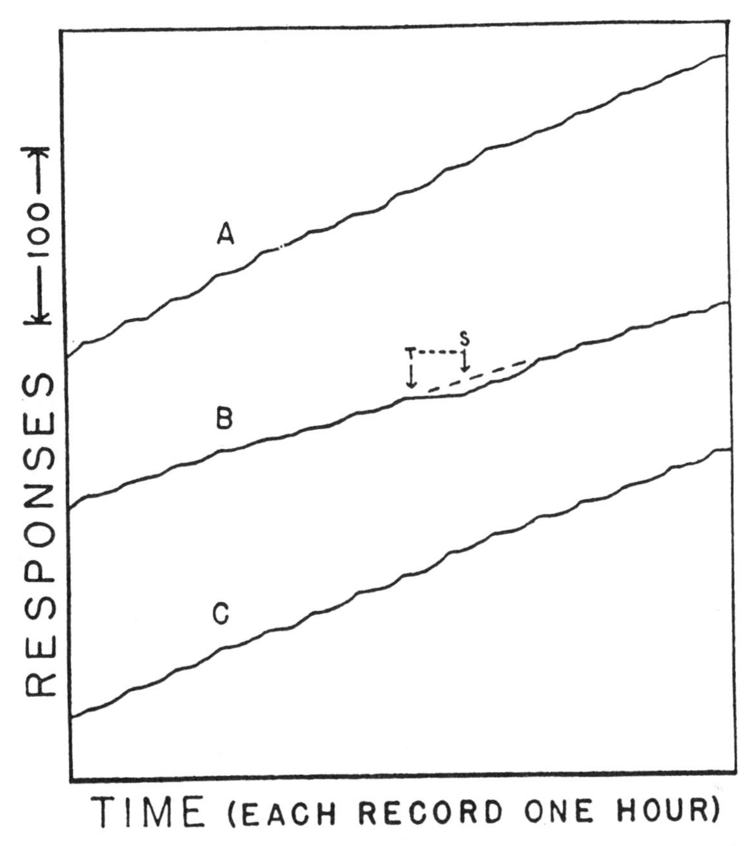

Fig. 4. Subsequent compensation for the reduction in rate during anxiety. The curves are averages for twenty-four rats taken on three consecutive days. *A* and *C* were taken under periodic reinforcement, while *B* shows the effect of the tone at a late stage in the experiment. The reduction in rate is followed by a compensatory increase, bringing the curve back to the extrapolation of the first part.

dently the effect of the emotional state is a temporary depression of the strength of the behavior, the total amount of responding during the experimental period (the 'reserve') remaining the same. Similar compensatory increases have been described under a number of circumstances, including physical restraint of the response (3).

Effects of Anxiety upon Extinction

When reinforcement with food is withheld, the rat continues to respond, but with a declining rate, and describes the typical extinction curve. The effects of anxiety upon this curve have been investigated. The first hour of a typical extinction curve, during which the combination of tone and shock was presented, is shown in the group curves of Fig. 5 and the individual curves of Fig. 6. By sighting along either curve in Fig. 5, one may observe a distinct depression in rate during the period of the tone, and (following the shock) an equally distinct compensatory increase, which appears to be maintained until an extrapolation to the first part of the curve is approximated. Figure 6 contains sample records from four rats which showed different degrees of depression during the tone.

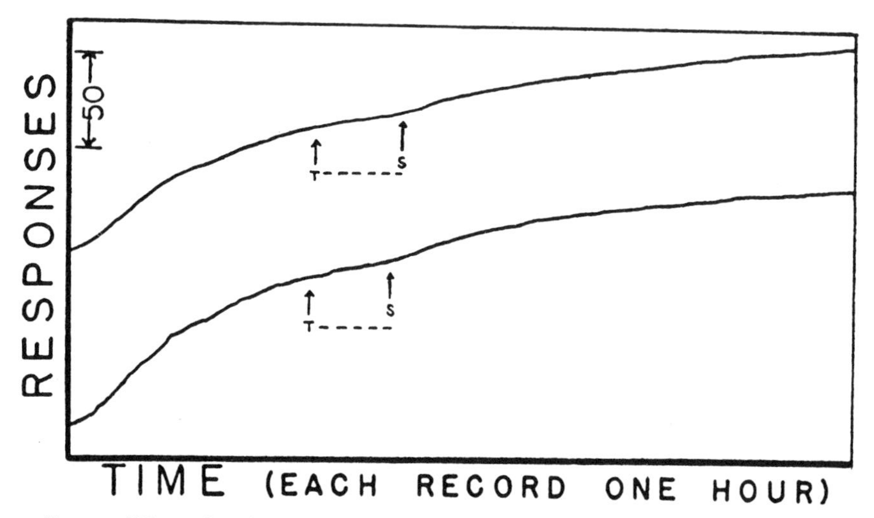

Fig. 5. Effects of anxiety upon extinction. The lower curve is an average for six rats, the upper for twelve. The tone, which had previously been followed by shock during periodic reinforcement, depresses the slope of the extinction curves, and a compensatory increase follows the administration of the shock.

During extinction, then, a state of anxiety produces a decrease in the rate of responding and the terminating stimulus is followed by such a compensatory increase in rate that the final height of the curve is probably not modified.

The Extinction of a State of Anxiety

A further property of anxiety was investigated by presenting the tone for a prolonged period without the terminal shock. In one experiment, while the rats were responding under periodic reinforcement, the tone was turned on after twenty-seven minutes of the experimental period had elapsed and allowed to sound for the remainder of the hour. The result is shown in Figs. 7 and 8. It will be observed that the recovery of a normal periodic rate is delayed

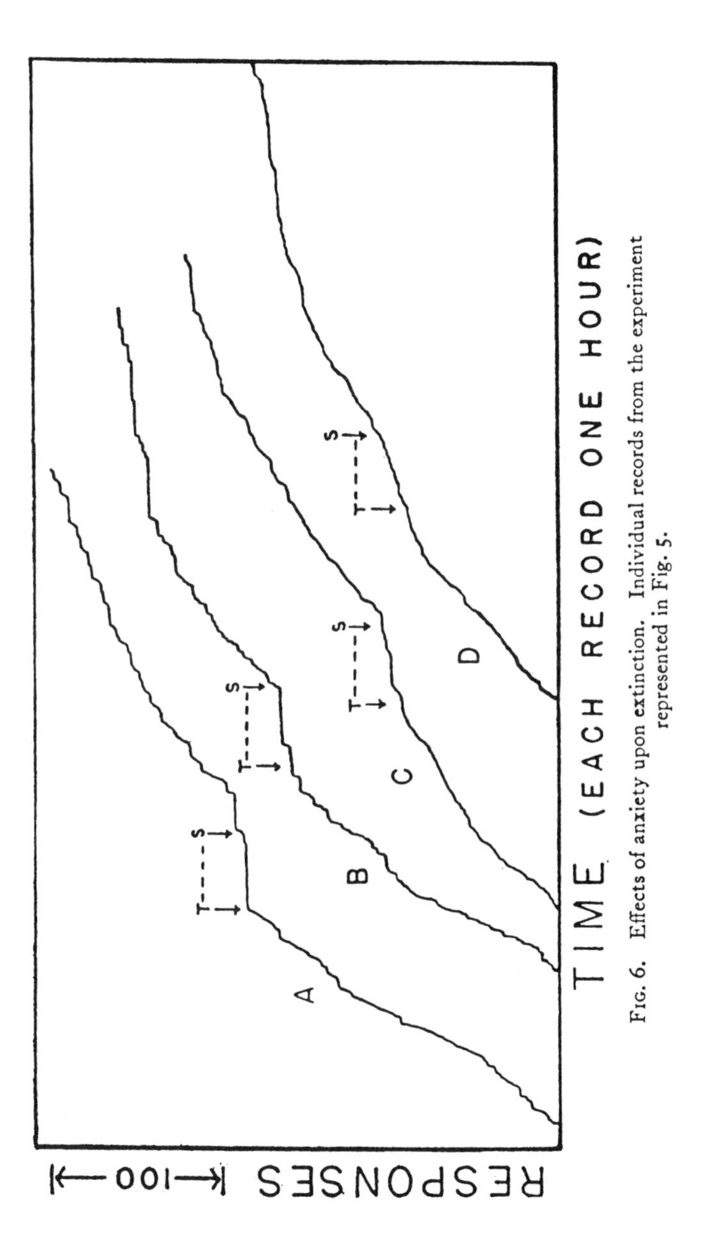

Fig. 6. Effects of anxiety upon extinction. Individual records from the experiment represented in Fig. 5.

considerably over the accustomed five-minute period of the tone. When the time is taken from the onset of the tone to the point at which the rat again reaches his previous periodic rate (measurements being made on individual curves), the mean period required for recovery is found to be 8.6 minutes. The group curve for twelve rats (the upper record in Fig. 7) shows a definite compensatory increase in rate later in the hour, although the extrapolation of the first part of the curve is not quite reached by the end of the period.

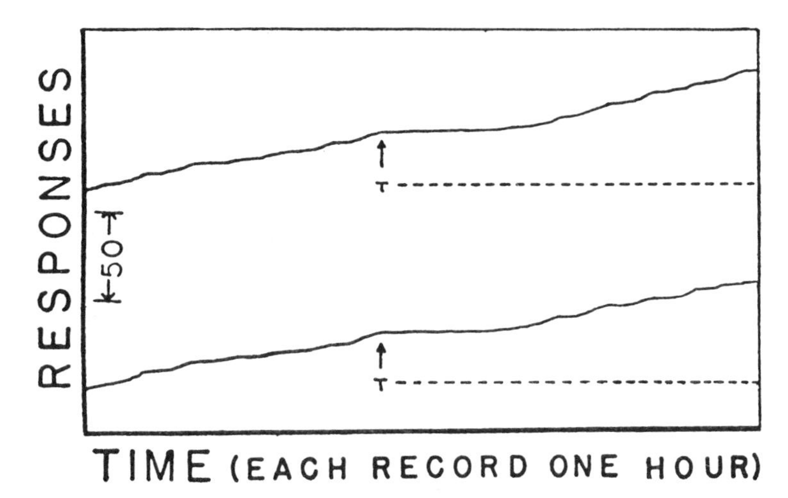

FIG. 7. Extinction of the effect of a tone when the terminating shock fails to appear. The upper record is the averaged curve for twelve rats under periodic reinforcement. The tone was turned on at *T* and continued to sound during the rest of the hour. No shock was given. The rate of responding returns to normal (and perhaps shows some compensatory increase) within ten minutes. The lower curve shows a repetition of the experiment ten days later.

The same experiment was repeated ten days later at a somewhat lower drive with the result shown in the lower curve in Fig. 7. The mean delay in recovery is here 9.1 minutes, and recovery is less complete. Except for the effects of the difference in motivation, the two records appear quite similar and exemplify the reproducibility of behavior of this sort.

Because the period of depressed activity varies among rats, individual records are needed in order to observe the course of the recovery of normal strength during the extinction of anxiety. Figure 8 shows a number of individual records with different periodic rates, the differences being attributable mainly to differences in hunger. The lag in recovery appears in nearly all records, and the compensatory increase in periodic rate in the majority. In some curves, notably *E*, *F*, and *G*, an extrapolation of the first part is reached before the end of the hour. It is not clear that this would have been

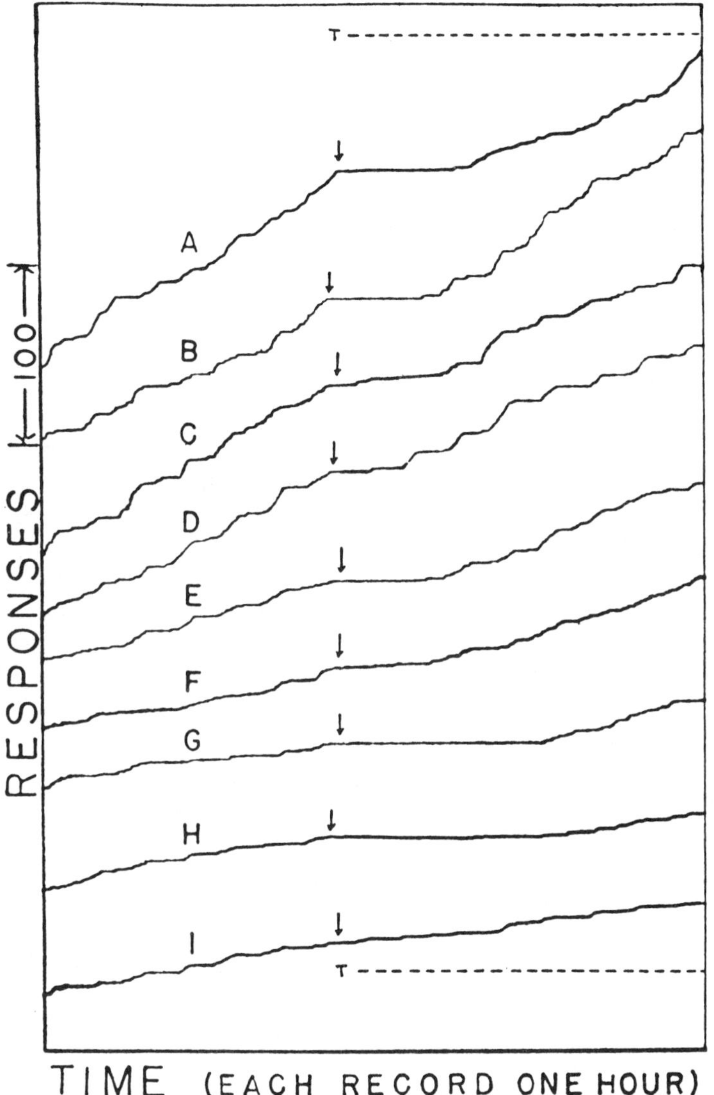

FIG. 8. Individual curves from the experiment described in Fig. 7.

the case with the other rats if the experimental period could have been prolonged, but the curves in general appear to be positively accelerated.

Spontaneous recovery from the extinction of the anxiety is fairly complete. The daily record which preceded the upper figure in Fig. 7 showed a ratio of 0.6 : 1.0 between the average periodic rate during the period of the tone and the normal rate for such an interval.

On the day following the figure, the ratio was 0.7 : 1.0 for a similar period, indicating that little or no effect of extinction survived.

Summary

Anxiety is here defined as an emotional state arising in response to some current stimulus which in the past has been followed by a disturbing stimulus. The magnitude of the state is measured by its effect upon the strength of hunger-motivated behavior, in this case the rate with which rats pressed a lever under periodic reinforcement with food. Repeated presentations of a tone terminated by an electric shock produced a state of anxiety in response to the tone, the primary index being a reduction in strength of the hunger-motivated behavior during the period of the tone. When the shock was thus preceded by a period of anxiety it produced a much more extensive disturbance in behavior than an 'unanticipated' shock. The depression of the rate of responding during anxiety was characteristically followed by a compensatory increase in rate.

During experimental extinction of the response to the lever the tone produced a decrease in the rate of responding, and the terminating shock was followed by a compensatory increase in rate which probably restored the original projected height of the extinction curve.

The conditioned anxiety state was extinguished when the tone was presented for a prolonged period without the terminating shock. Spontaneous recovery from this extinction was nearly complete on the following day.

(Manuscript received April 18, 1941)

References

1. Heron, W. T., & Skinner, B. F., An apparatus for the study of animal behavior, *Psychol. Rec.*, 1939, 3, 166–176.
2. Mowrer, O. H., Preparatory set (expectancy)—some methods of measurement, *Psychol. Monogr.*, 1940, **52**. Pp. 43.
3. Skinner, B. F., *The behavior of organisms: an experimental analysis*, New York: Appleton-Century, 1938.

16

Reprinted by permission from *J. Abnorm. Soc. Psychol.* **48**:291–302 (1953)

TRAUMATIC AVOIDANCE LEARNING: THE OUTCOMES OF SEVERAL EXTINCTION PROCEDURES WITH DOGS

RICHARD L. SOLOMON, LEON J. KAMIN, AND LYMAN C. WYNNE [1]

Harvard University

The persistence of behavior with no obvious reinforcement poses special problems for theories of learning. But numerous experimental studies confirm the clinical observation that traumatically acquired habits maintain a marked resistance to extinction despite the lack of renewed primary reinforcement (3, 4, 6, 8). The generality of the problem has been indicated by Mowrer (9), who formulates as the central problem of neurosis the paradoxical perpetuation of non-adaptive behavior.

The present research deals with the extinction of a habit acquired under the conditions of traumatic avoidance learning. The habit proved to be markedly resistant to the ordinary extinction procedure. However, extinction could be brought about through the use of special techniques. The obtained data suggested theoretical explanations of high resistance to extinction as well as explanations of the relative efficacy of the special techniques. We presume that the findings have a wide generality for theories of therapy and the nature of psychological trauma.

The findings to be reported in this paper were obtained under the following conditions. Using a modification of the Mowrer-Miller shuttlebox, learned avoidance responses were established in normal mongrel dogs. Under the impetus of very intense (just subtetanizing) electric shock, the dogs were trained to jump over a barrier from one compartment to the other in order to avoid the shock. The details of the training procedure are given in another paper (13). Ten trials comprised each day's session. After the animals had met an acquisition criterion of 10 successive responses on a given day, different dogs were subjected to different types of extinction procedures. It is the results of a series of such extinction experiments which will be the subject of this paper.

Experiment 1. Pilot Experiment

Two dogs were trained with a CS-US interval of 10 seconds, with a three-minute intertrial interval. The CS consisted of two segments: a buzzer sounded for one second, together with the raising of the gate which separated the two compartments of the shuttlebox. After these animals had met the criterion of acquisition, they were not shocked regardless of how long they remained in a given compartment without jumping.[2]

We had expected these dogs to extinguish spontaneously since the barrier was set at the height of the back of the animal, thus making the jumping response quite effortful. Instead, the experimenters found themselves running the animals day after day with no signs of extinction. Indeed, the latencies of response to the CS gradually *decreased* and the behavior of the animals became more and more stereotyped. One of the animals was continued for 190 extinction trials and the other for 490 trials. The dog which continued for 490 extinction trials had received only 11 shocks during the acquisition phase. The experimenters felt that this ordinary extinction procedure was *not efficient* in removing the jumping response. While the dogs *might* have extinguished spontaneously after several hundred, or even several thousand, trials more, the behavior and latencies of the animals gave no suggestion of this conceivable eventuality.

An attempt was then made to discourage the 490-trial dog from jumping by electrifying the opposite compartment on each trial, so that the dog jumped *into* shock. The gate was immediately lowered after each jump to prevent retracing, and the shock, at just subtetanizing level, was continued for three seconds and then terminated. The dog became more upset, and at subsequent presentations of the CS jumped more vigorously. His latencies were maintained at their already extremely fast level of 1.0–1.2 seconds. After 100 additional trials under this shock-extinction procedure, the dog was still jumping regularly into shock and gave no signs of extinguishing. As he jumped on each trial, he gave a sharp anticipatory yip which turned into a yelp when he landed on the electrified grid in the

[1] This research was supported by grants from the Medical Sciences Division of the Rockefeller Foundation. It was facilitated by the Laboratory of Social Relations, Harvard University. Dr. Wynne was a U.S. Public Health Service Postdoctoral Research Fellow in Mental Health when this work was initiated. The authors wish to thank E. S. Brush, F. R. Brush, N. Kogan, B. N. Cohn, and E. Smulekoff for research assistance and many useful suggestions.

[2] The reader is referred to another paper of this series (13) for precise details of the training procedure and the consequent behavior of the dogs during the *acquisition phase* of avoidance learning.

opposite compartment of the shuttlebox. We then increased the duration of the shock to 10 seconds and ran the dog for 50 more trials. Latencies and behavior did not change. Evidently punishment for jumping was ineffective in extinguishing the jumping response.

We tried a third extinction procedure with the same dog, one designed to prevent the dog from jumping in the presence of the CS. We placed a plate of glass in the opposite compartment, flush against the barrier and gate so that it blocked the passage between the two compartments. The plate of glass could not be seen by the dog until the gate was raised and the buzzer was sounded. The glass barrier was inserted on trials 4–7 of each extinction day. On the other six trials, the dog was shocked for jumping as before. This procedure was intended to have a "reality-testing" function for the dog. Ostensibly, the procedure let the dog "know" that the presence of the CS no longer signified shock. On the first glass-barrier trial, he jumped forward immediately at the CS and smashed his head against the glass. He drew back and was fairly quiet, and the gate was lowered after two minutes. On the subsequent trials, on which he did not strike the glass, he barked furiously, panted very rapidly, quivered, and drooled while the CS was present, but quieted down after the gate was lowered, and remained quiet during the minute before the next trial was started. On the 8th–10th trials, when the glass barrier was no longer present, he did not attempt to jump, and the gate was lowered after two minutes. This two-minute period constituted the arbitrary criterion for no response. At long last, after 647 extinction trials, the dog failed to jump in the presence of the CS alone. On the following day, the dog jumped into shock with short latencies on the first three trials, thus showing complete spontaneous recovery. He did not jump on trials 8–10. On five subsequent days, during which the glass barrier was never present, he jumped only on the second trial of the third day with a latency of 1.3 seconds. With this dog there was no gradual lengthening of latencies, but, rather, extinction was an all-or-none affair.

Two more animals were run with the same training procedures as with the first two, except that 20 trials a day, instead of 10, were carried out. The ordinary extinction procedure was used. One animal was discontinued after 310 extinction trials and the other after 280 extinction trials. During these trials, neither animal showed any signs of spontaneous extinction.

Some of the questions posed by the results of this pilot study suggested the succeeding experiments: (*a*) Did the shock-extinction procedure (punishment) fail because the animal had so much practice in jumping before the procedure was instituted? That is, *does the ordinary extinction procedure decrease the likelihood of effectiveness of the special procedures?* (*b*) Would the "reality-testing," glass-barrier procedure have been effective if it had been used directly after ordinary extinction? (*c*) Would the combined glass-barrier and shock-extinction procedure have been as effective if it had been preceded by the glass-barrier procedure rather than by shock extinction?

Experiment 2. The Effects of Ordinary Extinction Procedure

This experiment was designed to test the efficacy of the ordinary extinction procedure. The simplest hypothesis might claim that the removal of all primary reinforcement (shock) should lead to either the gradual or sudden extinction of the jumping response.

Thirteen dogs were given avoidance training in the shuttlebox situation, using the following procedures. The CS consisted of turning off a light over the compartment in which the dog happened to be and raising the gate. The light stayed off and the gate stayed up until the dog jumped into the lighted compartment, after which the gate was lowered. The light-out signal was a modification of the procedure used in the pilot experiment, and it replaced the buzzer. As in the pilot experiment, an intense shock was used. The CS-US interval was 10 seconds, and the barrier set for each dog was at the height of his back. The intertrial interval was three minutes, and 10 trials were given each day. The criterion for acquisition of the jumping response was 10 avoidances in 10 trials. A complete description of the training procedures is given in another paper (13).

After the criterion of learning was reached, no shock was administered in the presence of the CS, no matter how long a dog might delay in jumping. Such a procedure, where the shock reinforcement is no longer administered, we shall call *ordinary extinction.* Ordinary extinction was carried on for 20 days after each dog met the learning criterion. Thus each dog received 200 trials following the criterion trials, or at least 210 trials since the last shock had been received during the acquisition phase.

During the ordinary extinction procedure, not a single animal delayed jumping long enough to meet the (arbitrary) criterion of no-response, or infinite latency, which was set at *two minutes without jumping in the presence of the CS.* In other words, no extinction was obtained. In fact, there were only 11 responses with a latency greater than 10 seconds during the 2582 extinction responses.[3] Rather than showing any signs of

[3] One dog missed 10 trials because of an oversight, one dog missed eight trials because of a sore foot; this accounts for the fact that the total number of responses was less than 2600 (20 days×10 trials×13 animals).

extinction, there was a general tendency for the latencies to shorten. Figure 1 presents a graph of the mean reciprocal of response latency as a function of days after meeting the criterion of learning. Each point represents the mean for the 13 dogs for 10 trials for a given day of extinction. The arrow at the origin of the abscissa designates the mean for all dogs for the 10 criterion trials. It can be seen that the dogs were jumping with a mean latency of approximately 2.7 seconds during the criterion trials. The mean latencies gradually decreased, and after 200 ordinary extinction trials, the dogs were jumping with a mean latency of approximately 1.6 seconds.

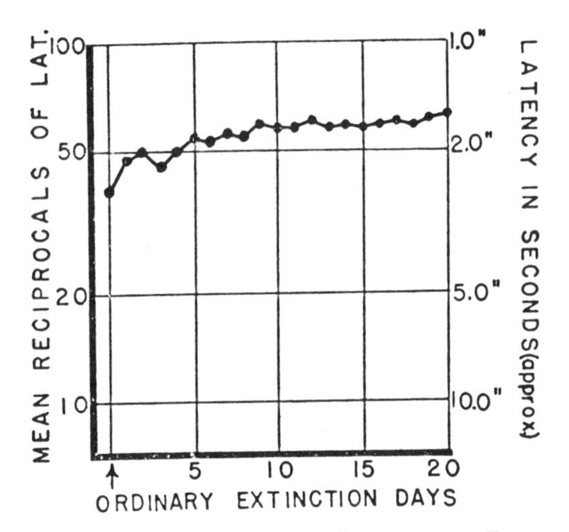

FIG. 1. MEAN RECIPROCALS OF LATENCY AS A FUNCTION OF NUMBER OF ORDINARY EXTINCTION DAYS

Each point is the mean reciprocal of response latency for thirteen dogs for the ten trials of a given day. Note that the ordinate for mean latency in seconds is approximate because the antireciprocal of the mean reciprocal of latency is not necessarily the same as the mean latency in seconds.

An inspection of the curve suggests that the latency asymptote had not yet been reached after 200 trials. This is not surprising, since it will be remembered that one of the dogs in the pilot experiment was jumping regularly at 1.0–1.2 seconds at the end of 490 extinction trials.

During the extinction trials, and accompanying the decrease in latency of the jumping response, the experimenters were impressed with certain behavioral changes. First, the behavior of the dogs, both in style of jumping and in intertrial behavior, became stereotyped. Each dog appeared to have developed his own particular "ritual." Such rituals involved actual response patterns as well as orientation in the apparatus. For example, a protocol description of one dog's behavior reads as follows: "At CS responded immediately and made clean jump with no pause. Then sat immobile in rear-door corner, with his back six inches from the rear, along wall opposite door, facing barrier gate, but nose pressed against wall. No panting or other reactions." For the next trial, the behavior was precisely the same, with the position again along the left side of the compartment with reference to the barrier, nose pressed against the wall.

Second, the frequency and intensity of overt emotional reactions, both to the CS and in the intertrial interval, decreased markedly. The types of emotional signs which usually disappeared during the course of ordinary extinction were defecation, urination, yelping and shrieking, trembling, attacking the apparatus, scrambling, jumping on the walls of the apparatus, and pupillary dilation. Whining, barking, and drooling tended to decrease in magnitude but often persisted throughout the 200 trials. Panting tended to persist relatively undiminished if it occurred during the early trials. Some dogs showed no overt emotional signs during the latter part of ordinary extinction. All dogs, early in the extinction procedure, showed some resistance—often very strenuous—to being placed into the apparatus. But most dogs, after 10 or 12 days of extinction, no longer resisted being placed into the apparatus. Many voluntarily hopped inside, displaying no visible emotional response.

Not only were there no signs of extinction in our dogs from day to day, but there was very little change in mean latency from trial to trial within experimental days. In Figure 2 are plotted the mean latencies for all dogs as a function of trials-within-days. On the first trial of each day, the mean reciprocal of latency for the 20 days was 56.1, while the mean reciprocal for the tenth trial was 55.1. This change is insignificant, representing but a small fraction of a second. There appears to be a slight warm-up effect on the first three trials, but this trend is also insignificant. Spontaneous recovery from the last trial of one day to the first trial of the next is minute, if present at all; it is indicated by the dotted line. The slight downward trend of the reciprocals between the third and tenth trials might indicate that longer experimental sessions would be more conducive to producing longer latencies, but this is doubtful.

The results of this experiment indicate that the ordinary extinction procedure is quite ineffective for eliminating the jumping response. If anything, it seems only to strengthen it. It can of course be argued that

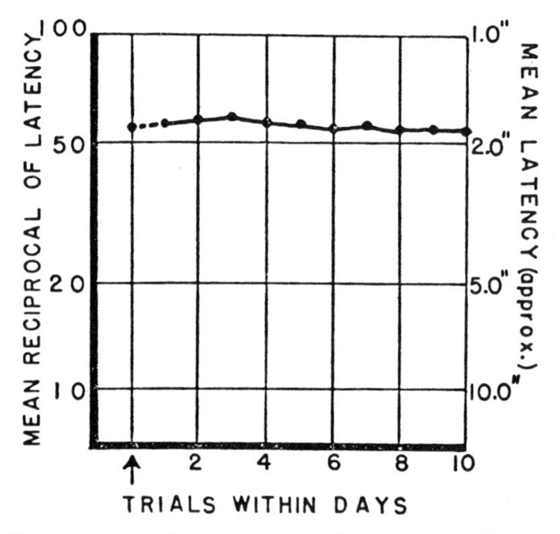

Fig. 2. Mean Reciprocals of Latency as a Function of Trials within Days

Each point is the mean reciprocal of response latency for thirteen dogs for twenty days of ordinary extinction procedure. Note that the ordinate for latency in seconds is approximate.

if the procedure had been continued for an indefinitely long time extinction *might* have occurred. But we must stress that within the time spans covered by this and the pilot experiment there were *no* signs of extinction of jumping.

Experiment 3. The Effects of the Glass-Barrier, Reality-Testing Procedure

This experiment was designed to test the efficacy of an extinction procedure which physically prevented the animal from jumping in the presence of the CS. From one point of view, this technique corresponds to "reality testing." The animal was forcibly exposed to a sequence of events in which the CS, followed by *not*-jumping, is *no longer* followed by shock. Will the dog, as a result of this new "knowledge," stop jumping?

Nine dogs were trained in the same manner as those of Experiment 2, with the exception that four had learned with a CS-US interval of 20 seconds instead of 10 seconds. Some of these dogs had been used in Experiment 2. Five of the dogs were carried through the 200 trials of ordinary extinction, just as in Experiment 2, while the other four were given only 10 ordinary extinction trials beyond the 10 criterion trials. Thus, we had a group of animals which had jumped approximately 10 times as often as the other group. Then the glass-barrier procedure was introduced as follows: On trials 4–7 of a given day, a plate of glass was placed flush against the barrier and gate on the side opposite the compartment containing the dog. The glass had three narrow vertical strips of adhesive tape placed upon it. After presentation of the usual CS, these adhesive strips enabled the dog to discriminate visually the presence or absence of the glass-barrier. On trials 1–3, and 8–10, the dog was free to jump as in ordinary extinction. If the dog failed to jump in two minutes, the CS was removed, and the latency was defined as infinite or "no-response." Of the nine dogs, seven failed to extinguish in 10 days of the glass-barrier procedure. One dog stopped jumping during the fourth day, and was run the next day again with no responses occurring. Another dog stopped jumping on the fifth day, and was run for 10 trials on the following day, giving no responses to the CS. One of the dogs which extinguished had previously been run for 200 ordinary extinction trials, while the other dog which extinguished had been run for only 20 trials of ordinary extinction. (Both of these dogs had been trained with a CS-US interval of 10 seconds.) Of the seven dogs which *did not* extinguish, four had no occurrences of an infinite latency (two minutes without jumping). Only on trials 8–10 were there any infinite latencies for the other three animals. In most cases where infinite latencies, or very long ones, occurred on the eighth trial, the animals behaved in very much the same way as they did when the glass barrier was present. The animals looked as though they "thought" the glass barrier was still there. In two instances where trial 10 produced an infinite latency, on the following day the first trial latencies were 1.3 seconds and 1.5 seconds, respectively. These were dramatic cases of spontaneous recovery. Of the seven animals which did not extinguish, only two had any latencies over 5 seconds on trials 1–3. Of the nine animals in the experiment, eight showed an all-or-none pattern whenever the latencies changed. That is, the animals either jumped quickly (from 1 to 4 seconds) at the presentation of the CS, or they did not jump for long time periods, in some cases not at all. Such long latencies were then typically followed by very short ones on the next trial. Such all-or-none changes were also noted in our study of the characteristics of acquisition (13). There was no gradual lengthening of latencies during the 10 glass-barrier days. Even the two dogs which extinguished showed a sudden transition from short latencies to infinite ones during the course of their extinction. Our protocols indicated that most dogs, when first confronted with the glass barrier, exhibited a wide variety of intense emotional responses. These usually disappeared after several repetitions of the glass-barrier procedure.

The results of this experiment indicate that the glass-barrier procedure *does* produce extinction in *some* dogs. However, *within the ten-day span covered by the experiment,* the procedure failed to extinguish the jumping response in most of the dogs. It is to be noted that there was no difference between dogs trained for 10 and for 200 ordinary extinction trials, one of each group extinguishing.

Experiment 4. The Effects of the Shock-Extinction, Punishment Procedure

This experiment was designed to test the efficacy of an extinction procedure which punished (with shock) the performance of the jumping response. The simplest expectation might be that shock-punishment would "stamp out" the jumping response.

Thirteen dogs were trained in the same manner as those in Experiment 2. Some of them had participated in Experiment 2. Seven were trained with a 10-second CS-US interval and six were trained with a 20-second CS-US interval. Seven were given 200 trials of ordinary extinction, just as in Experiment 2, while the other six dogs were given only 10 ordinary extinction trials after meeting the learning criterion. Then the shock-extinction procedure was introduced for all dogs. This was carried out as follows: The dog was shocked for three seconds in the compartment *into which* he jumped. No shock was administered if the animal did not jump in the presence of the CS. The immediacy of the shock-for-jumping was guaranteed by having the grid onto which the animal jumped electrified before the presentation of the CS. The shock level was the same as that used in training. The gate was immediately lowered after a jump in order to prevent retracing. A two-minute latency was again arbitrarily defined as infinite, or "no-response," and the CS was removed at the end of this two-minute period.

Of the 13 dogs in this experiment, 10 failed to extinguish in 100 shock-extinction trials. Three dogs extinguished; one of these animals had received 200 trials of ordinary extinction and two had received 10 trials of ordinary extinction prior to the introduction of the shock extinction procedure. All these three gave infinite latencies in the first 11 trials of punishment for jumping. They all met a criterion of no responses in 10 trials at the end of the second day of shock extinction. There was *no spontaneous recovery* once an infinite latency had occurred: if one of these three animals failed to jump he did not jump again. The transition from jumping to not-jumping was abrupt in two of the three animals. There was a considerable lengthening of latencies on the trials which preceded the onset of infinite latencies for one of these dogs.

The ten animals which did *not* extinguish showed an *entirely different course of behavior.* Nine of the 10 exhibited *no* infinite latencies. In fact, there was a tendency for the latencies to shorten on the first shock-extinction trials. These dogs jumped *faster* and *more vigorously* into the shock than they had jumped previously under the ordinary extinction procedure. (See Table 1 for a description and analysis of these data.) They often slammed into the far end of the compartment into which they were jumping. In addition, they all developed anticipatory reactions prior to jumping, which indicated that they "knew" they were to be shocked. The most common reaction was to yelp at the CS, jump vigorously, and then yelp at the shock, barking rapidly when the shock was terminated. This behavior continued for 100 trials of the shock-extinction procedure. Some long latencies did occur, but they were scattered and were usually followed by short latencies. However, one dog had six infinite latencies on the second day of shock extinction. Yet this dog failed to extinguish, and his latencies on the tenth day were short!

Several types of trend were evident in the latencies of the 10 dogs which did *not* extinguish. It is impossible, therefore, to generalize for all animals. Some dogs showed progressively shortening latencies, and some showed a slight and gradual lengthening of latencies. None showed radically lengthened latencies. One dog showed lengthening latencies for the first three days of shock extinction and thereafter showed gradually shortening latencies. Despite these different trends, trial-to-trial and day-to-day variability in latencies was very small during the 100 trials of shock extinction *for those animals which did not extinguish.* All animals were extremely upset by the procedure, exhibiting symptoms of terror.

The results of this experiment indicate that

the shock-extinction procedure *does* produce abrupt extinction in *some* dogs. However, within the ten-day span covered by the experiment, the procedure failed to extinguish the jumping response in most of the dogs. Punishment seemed to *increase* the strength of the jumping response in most dogs, as indicated by shorter latency and greater vigor of response. Again, there is no significant difference between animals trained for 10 and for 200 ordinary extinction trials.

TABLE 1

Mean Reciprocals of Response Latency for the Five Trials Preceding and Following the Onset of the Shock-Extinction Procedure

	Shock Extinction Follows 200 Ordinary Extinction Trials	Shock Extinction Follows Only 20 Ordinary Extinction Trials
Before shock extinction	50.7	50.5
After onset of shock extinction	72.9	63.1

Analysis of Variance

Source	F	p
A. Trials before and after onset of shock extinction	15.02	$>.01$
B. Number of trials preceding onset of shock-extinction procedure	1.21	—
A×B Interaction	2.50	—

if the dog jumped in the presence of the CS he jumped into shock. The shock was on for three seconds. The gate was lowered to prevent retracing. (If the dog did not jump in two minutes, the CS was withdrawn and the trial was scored as infinite in latency, or "no-response.") Then, on trials 4 through 7, the glass barrier was present, just as in Experiment 3. The CS was removed after two minutes' exposure to the glass barrier. Of course, the animal could not jump to the opposite compartment during trials 4–7. Then, on trials 8–10 the same procedure used on trials 1–3 was introduced again, and the animals were shocked if they jumped to the CS. Thus the shock-extinction procedure was used on the first and last three trials of each day, and the glass-barrier procedure was used on the middle four trials. It was decided arbitrarily to terminate this experiment at the end of 10 days if no extinction occurred.

Experiment 5. The Effects of a Combination of the Glass-Barrier Procedure and the Shock Extinction Procedure

This experiment was designed to test the efficacy of a procedure which combined in close juxtaposition the two special extinction techniques tested in the preceding experiments. Here punishment and reality-testing both occurred within the same day's experimental session.

Sixteen dogs were trained in the acquisition of avoidance in the same manner as those in Experiment 2. Of these, seven were trained with a 10-second CS-US interval and nine were trained with a 20-second CS-US interval. Ten had 200 trials of ordinary extinction, just as in Experiment 2, while the other six dogs were given only 10 trials of ordinary extinction after meeting the learning criterion. All of the dogs in this experiment had failed to extinguish in 100 trials of *either* the glass-barrier procedure *or* the shock-extinction procedure. Ten of the dogs had been subjected to the shock-extinction procedure and six had experienced the glass-barrier procedure. Then the combination of glass-barrier and shock-extinction procedure was introduced in the following manner: On the first three trials, the glass barrier was not present, but

Fourteen out of 16 dogs extinguished before 7 days of the combined procedure. The extinction criterion, as in previous experiments, was 10 infinite latencies on a given day. When a dog failed to respond on the first three trials of a day, the glass barrier was not used, and 10 trials were given in order to test for the occurrence of extinction. The remaining two animals failed to extinguish in 10 days, or 100 trials. One of them had had 200 ordinary extinction trials, and one had had only 10 ordinary extinction trials. Both had had 100 shock extinction trials prior to the combined procedures. One of the two dogs which did not extinguish in 10 days of the combined procedure showed no signs of extinction after 10 days. The other dog showed lengthening latencies and one infinite latency on the last day of the combined procedure. *All* of the six dogs which had had 10 days of *glass-barrier trials prior to the combined procedure* met the extinction criterion in 30 trials or less, with a mean of 25.0 trials to meet the criterion of extinction. The eight dogs which had had *shock extinction,* and which *did* extinguish in the combined procedure, met the criterion for extinction in less than 70 trials of the combined procedure, with a mean of 43.8 trials.

The mean number of shocks required for extinction in the *combined* procedure was 8.2 shocks for the six dogs which had had the glass-barrier procedure and 14.3 shocks for the eight dogs which had had the shock-extinction procedure. This difference (14.3–8.2) is statistically significant (p=.01).

The results of this experiment indicate that the combined glass-barrier and shock-extinction procedure is highly effective in producing extinction of the jumping response. Again there is no difference between dogs trained for 10 and for 200 ordinary extinction trials. There are, however, several indications that this technique is more effective when preceded by the glass-barrier rather than by the shock-extinction technique. We have no data on how effective the present combined procedure would have been if introduced immediately after the ordinary extinction procedure, without the intervention of a special technique.

SUMMARY OF RESULTS

A summary of the findings of Experiments 2, 3, 4, and 5 is given in Table 2. From this table, several generalizations are possible: (*a*) The ordinary extinction procedure is ineffective. (*b*) Either the glass-barrier procedure or the shock-extinction procedure *can* produce extinction of jumping, but the proportions of animals extinguishing are small. There is little to choose between the two procedures. (*c*) The combined procedure is the most effective. It makes little difference whether it is preceded by 200 or by 10 ordinary extinction trials. There are, however, some indications that the combined procedure is more effective following the glass-barrier procedure than following the shock procedure. Not only is this trend detectable in the proportions of animals extinguishing with each sequence, but also the trend is clearly present in the data on (*a*) number of trials required to achieve the extinction criterion, and (*b*) number of shocks in the combined procedure required to produce extinction.

TABLE 2

PROPORTIONS OF ANIMALS EXTINGUISHING WITH THE VARIOUS EXTINCTION PROCEDURES

EXPERIMENT	SPECIAL PROCEDURE USED	AFTER 200 TRIALS ORDINARY EXTINCTION PROCEDURE	AFTER 10 TRIALS ORDINARY EXTINCTION PROCEDURE	TOTAL
2	None	0/13	0/10	0/23
3	Glass-barrier procedure only	1/5	1/4	2/9
4	Shock procedure only	1/7	2/6	3/13
5	Combined procedure, glass-barrier procedure first	4/4	2/2	6/6
5	Combined procedure, shock procedure first	5/6	3/4	8/10

THEORY

We had, in planning the extinction experiments, definite expectations about the relative effectiveness of the various extinction procedures. For the most part, these expectations were derived from the two-process theory outlined in another paper on the acquisition of traumatic avoidance learning (13).

We assumed that, in the course of acquisition, conditioned emotional reactions would be established through the process of classical conditioning, of CS-US contiguity in time. Further, we assumed that these emotional reactions would give rise to stimuli having drive properties. When an animal in our experiment was regularly avoiding the shock, he was in reality escaping from the CS, and each jump was, therefore, followed by a reduction in the intensity of the emotional response elicited by the CS. Thus, not only was the jumping response reinforced by shock termination during the escape phase of acquisition, but also by anxiety reduction during the early avoidance trials. Theoretically, we would expect that the jumping response would be reinforced on each occasion that the CS aroused anxiety and the jumping response terminated or reduced the anxiety. However, once the animal is regularly avoiding the shock, the CS is no longer followed by the US, and the temporal contiguity necessary to reinforce the conditioned emotional responses is no longer present. This dissociation of the CS and US should, according to classical conditioning principles, lead to extinction of the emotional responses in the presence of the CS. If this should occur, there should then be no basis for the reinforcement of the instrumental jumping responses; there would be no anxiety to be reduced. The instrumental response should then gradually extinguish. These formulations would predict a long course of extinction for avoidance responses because the conditioned emotional reactions would have to be extinguished before the extinction of the instrumental response would start to occur. However, extinction of the instrumental response should occur.

With such a theory in mind, we expected that the learned jumping response would continue to decrease in latency several trials after the last shock had been received. As long as the conditioned emotional response had not been extinguished, each jump in the presence of the CS would be followed immediately by anxiety reduction. According to the law of effect, the bond between the CS and the jumping response should be consistently strengthened, with a correlated progressive shortening of response latency. This, in fact, did occur. In Experiment 2, the latencies shortened over a period of about 100 trials following the trial on which the last shock was received. *But it is at this point that a real weakness in our interpretation appears.*

In a paper on the acquisition of traumatic avoidance responses we argued that the latency of the *conditioned emotional reaction* should be of the order of magnitude 1.5–2.5 seconds (13). But, after approximately 100 trials of ordinary extinction, the latency of the entire *jumping* response had in fact decreased to a mean value of about 1.7 seconds, and at the end of 200 trials of ordinary extinction the mean latency was about 1.6 seconds.

The instrumental jumping response *itself* usually has a latency of from 1.0 to 1.5 seconds at its asymptote. (Some individual animals produced latencies as short as 0.9 seconds, though very rarely.) Thus, if the animals' jumps were responses to their own emotional reactions, or if they were responding to drive arousal, the asymptote for the latency of the jumping response in the presence of the CS should have been approximately 2.4 to 3.4 seconds, i.e., the sum of the latencies for drive arousal and for jumping. But the animals were jumping faster than this within 10 to 20 trials after meeting the acquisition criterion (see Fig. 1), Obviously, the animals at this point *could not* have been responding to their own emotional reactions. The assumption that autonomic arousal serves as the stimulus for the jumping response is contradicted by the facts. The latency data indicate that, at this point, the jumping response has an extremely high habit strength and is activated by minimal drive. In any event, it cannot be argued that the jumping is energized by a *full-blown anxiety reaction to the CS.*

It is clear that either something was fundamentally wrong with our interpretation of acquisition (13), or that additional principles have been overlooked. For, once the animals are jumping in a period of time shorter than that required for the emotional reaction to take place in the presence of the CS, we can no longer argue that jumping continues to be reinforced by anxiety reduction. Then, since no further reinforcement occurs, the jumping response ought gradually to extinguish.

The data described in Experiment 2 indicate that the animals *did* become less and less emotional during the ordinary extinction trials. However, there is substantial evidence to indicate that the conditioned emotional responses which were established early in training had not, in reality, been *extinguished* at the end of 200 trials of ordinary extinction procedure. It will be remembered that most of the dogs in Experiment 3 were profoundly upset when introduction of the glass barrier first restrained them in the presence of the CS. Though, during ordinary extinction, with very short latencies of the instrumental act we see no emotional reactions, when the animals are later held in the presence of the CS by the glass barrier they demonstrate that the CS has *maintained its capacity to elicit anxiety.*

But if the emotional response has not been extinguished, what has happened to it? The effect of early reinforcements has shortened the latency of jumping to the point where, since the dog's jump removes the CS so quickly, the conditioned emotional reaction may not be elicited at all! This would save the emotional reaction from extinction, since it is no longer exercised. But, if the emotional reaction is not elicited, we are now confronted with the picture of a dog continuing to jump without further reinforcement!

The law of effect would account for the persistence of a *certain number* of jumping responses after the last reinforcement. The exact number would be a function of the amount of reinforcement received. The law of effect would then predict a lengthening of latency, leading toward extinction of the jumping response. But, *what will happen if the latency of jumping increases?* If, as we have hypothesized, the conditioned emotional reaction has not yet been extinguished, a longer latency will leave the dog in the presence of the CS for a time interval long enough for the emotional reaction to be elicited. (The glass-barrier procedure did, in fact, accomplish this.) The elicitation of the emotional response when a long latency occurs ought to lead to a jump which will in turn be followed immediately by anxiety reduction. This further reinforcement would again strengthen the bond between the CS and jumping. A cyclical mechanism of this sort, with the emotional reaction being elicited only rarely, but, when elicited, giving renewed strength to the CS-jumping bond, would account for the high resistance of the jumping response to the ordinary extinction procedure. The final extinction of jumping could begin only when the emotional reaction had been elicited often enough to be itself extinguished, or at a low strength.

We feel that the order of events which we have described is general to *all* learned avoidance responses. The picture is surprisingly akin to the clinical picture in compulsive neurosis. It contains the possibility of the organism "frightening itself" by remaining in the presence of the CS long enough for the CS to be effective, while no emotionality will be elicited with *short* latencies for *instrumental* acts. We feel that this interpretation can help us to circumvent the "dilemma of fear as a motivating force," which Eglash (1) feels is a stumbling block for anxiety-reduction theories.

This formulation suggests an interesting experiment. If the animal's jump can be delayed until *after* an emotional reaction has been elicited by the CS, and then the animal is allowed to jump, two opposing processes should go on. The habit strength of the jumping response should be augmented; but the elicitation of the emotional reaction without the presentation of the US (shock) should weaken the strength of the conditioned emotional response (anxiety). Of course, the best way to produce extinction of the emotional response would be to arrange the situation in such a way that an extremely intense emotional reaction takes place in the presence of the CS. This would be tantamount to a reinstatement of the original acquisition situation, and since the US is not presented a big decremental effect should occur. One of the main problems in extinguishing avoidance conditioning is, thus, keeping the animal in the presence of highly disturbing danger signals which are no longer followed by noxious stimulation. Usually, well-learned instrumental responses *prevent this from happening* and slow up the course of extinction. (It seems to us that the "reality-testing" procedures in psychotherapy are *partly* designed to face this problem.)

But there is a second principle, possibly just as important as the first. Extinction should be much easier to obtain if, in conjunction with the reality-testing procedure, the animal is not allowed to perform the instrumental avoidance response. Our glass-barrier procedure approximated this, since the CS was removed at the end of two minutes, and any anxiety reduction that took place was preceded in time by *responses other than jumping.* Thus, while only two out of nine of our animals extinguished in 10 days of the glass-barrier procedure, on the basis of our argument we would be forced to predict that this procedure would, in relatively few additional trials, have led to extinction for all animals. There is some reason to believe that the reality-testing procedure must approximate the acquisition situation more closely than we were able to approximate it, in order for this prediction to materialize. The behavior of the dogs was at first puzzling. It will be remembered that they were initially upset by the glass-barrier procedure, but they quickly quieted down with successive trials, and later most dogs remained in a relatively relaxed state when the glass barrier was encountered. It was surprising to see the dogs jump quickly on trials on which the glass barrier was not present when they seemed to have "learned to relax" in the presence of the CS plus the glass barrier. We feel that they learned to discriminate the two conditions, with and without glass barrier, and that each condition controlled its own response. This raises an interesting problem: after traumatic avoidance learning has taken place, how do we fashion reality-testing procedures that are indiscriminally different from the acquisition situation, so that the instrumental act can be removed? (Presumably the psychotherapist, when forcing reality testing on a patient, has learned some way of doing this, or else the patient would relapse completely on leaving the therapist's office.) We are forced to predict that the glass-barrier procedure would be far more efficient than the ordinary procedure. Furthermore, an additional experiment, in which the animal is forced to delay jumping long enough to elicit the emotional response, but is then allowed to jump, should definitely be *less* efficient than the glass-barrier procedure we have already used.

While a two-process theory forces us into fairly definite predictions about the glass-barrier, reality-testing procedure, its application to the shock-extinction procedure is by no means clear-cut. A simple Thorndikian interpretation of the shock-extinction procedure might maintain that punishment of the jumping response ought to stamp it out. The data do not support such an interpretation. (Neither do the data presented by Gwinn (2), on the use of punishment in the extinction of a running response in rats.) Only 3 of 13 dogs met the extinction criterion during 10 days of the shock-extinction procedure. The other 10 persisted in jumping, and on the average their latencies shortened. Gwinn has reported that his rats, when shocked for running, at first increased their running speed, but with additional shock trials some of them gradually extinguished, and the others were discontinued without reaching an extinction criterion.

From our point of view, the failure of the shock procedure to bring about extinction in 10 of our animals might be explained with the help of an additional assumption. In the shock-extinction procedure, we would clearly expect a few jumps on the basis of previous reinforcement of jumping. However, during the initial jumps into shock, the CS is once again followed by the US, though it is true that the jumping response intervenes between the two events. We assume that the renewed pairing of CS and US, by contiguity principles, drastically strengthens the fear reaction to the CS. At the same time, the habit strength of the jumping response should be weakened through Thorndikian action of punishment. We would then have to claim that the increase in drive more than counterbalances the decrease in the habit strength of jumping. We observed that emotionality between trials increased greatly during shock-extinction trials. The raised anxiety level probably affects the "operant level" for jumping in a shuttlebox situation. This phenomenon might overlap with the fear reaction specifically aroused by presentation of the CS.

This is clearly a dialectical argument in its present form, capable of accounting for *any* observable results! But it would not be difficult to test the formulation. Any technique which first reduces the strength of the anxiety reaction ought to increase the efficacy of the shock-extinction procedure. This follows, since the hypothetical events crucial for extinction are the lowering of drive and the lowering of habit strength. If, at the beginning of the shock-extinction procedure, drive (anxiety) is low, then the Thorndikian action of punishment can presumably take place *before* the new series of shocks can build up the intensity of anxiety in the presence of the

CS. We can then predict that if the use of punishment, increasing the anxiety level, *follows* a reality-testing procedure that is continued until the conditioned emotional reaction has diminished, it will be much more effective than if the punishment procedure *precedes* reality-testing. If, however, because of previous procedures the anxiety level elicited by the CS is fairly high at the beginning of the shock-extinction procedure, the new shock series will drive the anxiety level to a point where the decremental, Thorndikian effects of punishment cannot be observed. Thus, we can predict that *if extinction does take place with shock-extinction procedure, it must do so within very few trials, or else be extremely difficult to obtain.*

Our data are not in disagreement with these predictions. Of 10 animals who received the shock-extinction treatment before the combined treatment, two failed to extinguish during the combined procedure. Of six animals who received the glass-barrier procedure before the combined procedure, all extinguished during the combined procedure. In addition, the animals in the latter group required fewer trials of the combined procedure, as well as fewer shocks, to meet the extinction criterion. Using the shock procedure alone, those three animals which did extinguish did so in less than a dozen trials, while those who did not extinguish were jumping with short latencies after 100 trials. A more crucial experiment would have employed the shock and glass-barrier procedure in sequence, counterbalancing for order of presentation, without introducing the combined procedure. Then the sequence of glass barrier followed by shock should lead to rapid extinction.

Our interpretation of the shock-extinction procedure might be further sustained by an experiment in which the punishing shock is of low intensity. Here, the increment to conditioned anxiety might be less than the decrement to habit strength. Extinction might then take place quickly if the low-punishment procedure were preceded by the glass-barrier procedure.

The results of Experiment 5, on the combined extinction procedure, are not at odds with our theoretical discussion. This procedure, as we would expect, was more effective when it was preceded by glass-barrier treatment than when preceded by the punishment procedure. The combined procedure was effective in 14 out of 16 cases. Its effectiveness probably derives from the fact that the reality-testing procedure tends to reduce the strength of conditioned emotional reactions. This reduction offsets the increase of anxiety produced by the punishment shocks, at a time when the habit strength of jumping is being decreased through punishment. The result is extinction of jumping.

While the data of our five extinction experiments are fairly well fitted by our theoretical formulation, we are left with many points of difficulty, as well as some circular reasoning. Despite such shortcomings, the fact that the argument suggests new, independent lines of attack on the problem of extinction of traumatic avoidance learning is encouraging.

On most points, our interpretation of traumatic avoidance learning is in essential agreement with both Miller (5, 6) and Mowrer (7, 8). We agree with them that the development of the acquired anxiety drive is essential to the establishment of avoidance conditioning. We differ with Mowrer in including skeletal responses among those emotional reactions which are susceptible to classical conditioning. We differ with Miller because we feel that a one-process, S-R reinforcement theory is inadequate. We do not believe that anxiety is reinforced through the law of effect, but rather through principles of stimulus contiguity. In this respect, we tend to be more sympathetic with the views of Schlosberg (10) and of Skinner (11, 12).

In carrying out our analysis of the rather minute details of behavior of dogs in a traumatic avoidance learning situation, we were faced with several inadequacies in current theory. Our own theoretical formulations have been, for the most part, improvised to account for our own findings. They and those of Miller, Mowrer, and Schlosberg are not mutually exclusive. At present, all of the theoretical alternatives supplement each other and help to order the data on avoidance conditioning.

SUMMARY

We have described a series of five experiments in which several different procedures were used to bring about the extinction of traumatically induced avoidance responses in dogs. These were the main findings:

1. With an ordinary extinction procedure, cessation of responding was extremely difficult to obtain.

2. A glass-barrier, reality-testing procedure was moderately effective. In this procedure the animal was detained in the presence of the danger signal without being shocked and the instrumental avoidance response was not allowed to occur.

3. A shock-extinction or punishment procedure was approximately as effective as the glass-barrier procedure. In this procedure the animal was punished for making the instrumental response.

4. A combination of the reality-testing and punishment-for-responding procedures was very effective in producing extinction. This combination procedure was more effective when preceded by reality testing than it was when preceded by the punishment technique.

These findings were interpreted from the point of view of a modified two-process learning theory similar to Mowrer's (8).

REFERENCES

1. Eglash, A. The dilemma of fear as a motivating force. *Psychol. Rev.*, 1952, 59, 376–379.
2. Gwinn, G. T. The effects of punishment on acts motivated by fear. *J. exp. Psychol.*, 1949, 39, 260–269.
3. Maier, N. R. F., & Klee, J. B. Studies of abnormal behavior in the rat. XII. The pattern of punishment and its relation to abnormal fixations. *J. exp. Psychol.*, 1943, 32, 377–398.
4. Maier, N. R. F. Studies of abnormal behavior in the rat. XVII. Guidance versus trial and error in the alternation of habits and fixations. *J. Psychol.*, 1945, 19, 133–163.
5. Miller, N. E. Studies of fear as an acquirable drive: I. Fear as motivation and fear-reduction as reinforcement in the learning of new responses. *J. exp. Psychol.*, 1948, 38, 89–101.
6. Miller, N. E. Learnable drives and rewards. In S. S. Stevens (Ed.), *Handbook of experimental psychology*. New York: Wiley, 1951.
7. Mowrer, O. H. Anxiety-reduction and learning. *J. exp. Psychol.*, 1940, 27, 497–516.
8. Mowrer, O. H. On the dual nature of learning—A reinterpretation of "conditioning" and "problem solving." *Harv. educ. Rev.*, 1947, Spring, 102–148.
9. Mowrer, O. H. Learning theory and the neurotic paradox. *Amer. J. Orthopsychiat.*, 1948, 18, 571–609.
10. Schlosberg, H. The relationship between success and the laws of conditioning. *Psychol. Rev.*, 1937, 44, 379–394.
11. Skinner, B. F. Two types of conditioned reflex and a pseudo-type. *J. gen. Psychol.*, 1935, 12, 66–77.
12. Skinner, B. F. *The behavior of organisms.* New York: Appleton-Century-Crofts, 1938.
13. Solomon, R. L., & Wynne, L. C. Traumatic avoidance learning: acquisition in normal dogs. *Psychol. Monogr.*, 1953, 67, No. 4 (Whole No. 354).

Received November 21, 1952. Prior publication.

Editor's Comments on Papers 17, 18, and 19

17 PAVLOV
Excerpt from *Compound Stimuli*

18 GARCIA and KOELLING
Relation of Cue to Consequence in Avoidance Learning

19 KAMIN
Predictability, Surprise, Attention, and Conditioning

Discrimination learning has long been a major theme in the study of animal learning. An enormous variety of theoretical issues have been encompassed by the experimental study of discrimination learning, and the papers reproduced in this section touch on only a few of these. Fortunately, an excellent review and interpretation of animal discrimination learning (Sutherland and Mackintosh, 1971) is available to the interested reader.

As with so much else in this study of animal learning, many of the first, pace-setting experiments were performed by Pavlov. The brief selection reprinted here (Paper 17) is his report of an experiment in which conditioning was initially conducted with a compound stimulus, the components of which were subsequently tested to determine how much salivation they would induce. The results are the first demonstration of the phenomenon of "overshadowing." This phenomenon has proved especially important for the insight it gives us about selective attention in animals.

A somewhat different kind of selectivity is reported in Paper 18 by Garcia and Koelling. This paper, though relatively recent, has stimulated a mountain of research (see Riley and Baril, 1976, for a bibliography) and it has induced a rethinking of many assumptions about animal discrimination learning. The kind of selective discrimination learning demonstrated with rats in the paper reprinted here surprised many people, who had previously believed all detectable stimuli to be equivalent in terms of their roles in discrimination learning tasks. Garcia and Koelling challenged this belief, and their double-dissociation design implicated the relationship between the qualitative type of cue to the qualitative type of reinforcer as an important determinant of discrimination learning.

This finding precipitated a major rethinking of many of the assumptions that had guided the study of learning in animals. In so doing, this article has had a significance much broader than its immediate subject matter. For example, even though this specificity between cue and reinforcer was demonstrated with domesticated rats, the "wonder bread" of the animal kingdom, it led to a revival of interest in the evolutionary framework into which animals that learn fit. Calls for students of animal learning to pay more attention to evolutionary biology had (with a few exceptions) largely been ignored prior to the appearance of this paper. The experiment of Garcia and Koelling provided no solution to the conceptual problems of a comparative approach to the study of learning, but it sharply challenged the belief that all learning processes are equivalent.

Another type of selectivity is demonstrated in Paper 9 by Kamin. His experiments demonstrated that what a rat learns to a particular cue depends in large part on the context in which that cue occurs, particularly on the significance of other cues present at the same time. In interpreting his results, Kamin emphasized the importance of "surprise" in determining when conditioning occurs. The "surprise" concept has proved to be extremely helpful in analyses of conditioning. It has been formalized in a mathematical theory of Rescorla and Wagner (1972).

REFERENCES

Rescorla, R. A., and A. R. Wagner, 1972, A theory of Pavlovian conditioning: variations in the effectiveness of reinforcement and non-reinforcement, in *A. H. Black & W. F. Prokasy, eds., Classical Conditioning II: Current Research and Theory,* Appleton-Century-Crofts, New York, pp. 64–99.

Riley, A. L., and L. L. Baril, 1976, Conditioned taste aversions: a bibliography, *Anim. Learn. Behav.* **4**:suppl.

Sutherland, N. S., and J. J. Mackintosh, 1971, *Mechanisms of Animal Discrimination Learning,* Academic Press, New York.

17

Reprinted from pp. 141–144 of *Conditioned Reflexes: An Investigation of the Physiological Activity of the Cerebral Cortex*, G. V. Anrep, trans. and ed., Oxford University Press, London, 1927, 430 p.

COMPOUND STIMULI

I. P. Pavlov

[*Editor's Note:* In the original, material precedes this excerpt.]

With regard to the synthesizing activity of the nervous system, as compared with its analysing activity, little is known up to the present. It would, indeed, be futile for us to attempt to discuss the nature of its intimate mechanism : it can only be suggested that, in the future, synthesizing activity will be referred to the physico-chemical properties of synaptic membranes or anastomosing neuro-fibrils. Our immediate task must consist in accumulating experimental material concerning the synthesizing activity.

In addition to the formation of the conditioned reflex itself—which is, of course, primarily an expression of the synthesizing activity, and which has constantly formed the starting-point of our investigations—we have also examined the properties of compound conditioned stimuli. Compound stimuli were used with either simultaneous or successive action of their component parts.

In the case of compound simultaneous stimuli the following important relations have been observed :

When the stimuli making up the compound act upon different analysers, the effect of one of them when tested singly was found very commonly to overshadow the effect of the others almost completely, and this independently of the number of reinforcements of the compound stimulus. For example, a tactile component of a stimulatory compound was usually found to obscure a thermal component, an auditory component to obscure a visual component, and so on. Thus, in the following experiment by Dr. Palladin a conditioned reflex to acid was established to a simultaneous application of a thermal stimulus of 0° C. and a tactile stimulation of the skin. Tests were made both of the compound stimulus and of its individual components applied singly.

Time	Conditioned Stimulus	Salivary Secretion in c.cs. during 1 minute
11.15 a.m.	Tactile	0·8 c.c.
12.45 p.m.	Thermal	0·0 c.c.
1.10 p.m.	Tactile + thermal	0·7 c.c.

Another example may be taken from an experiment by Dr. Zeliony. A conditioned alimentary reflex was established to the simultaneous application of the tone of a pneumatic tuning-fork, which was considerably damped by being placed within a wooden box coated with wool, and of a visual stimulus of three electric lamps placed in front of the dog in the slightly shaded room.

Time	Conditioned Stimulus	Secretion of Saliva in drops during 30 seconds
3.37 p.m.	Tone + lamps	8
3.49 ,,	Lamps	0

In the above experiments the action of the thermal and visual components by themselves was ineffective, being completely overshadowed by the other respective components. It is obvious, however, that the ineffective components in the stimulatory compounds could easily be made to acquire powerful conditioned properties by independent reinforcement outside the combination. The true interpretation of the phenomenon which has just been described is revealed by experiments in which both components of a stimulus belong to one and the same analyser. For example, in one experiment there were used as components in a stimulatory compound two different tones, which appeared to the human ear to be of equal intensity. When the conditioned reflex to the compound became fully established, the tones sounded separately were found to produce an equal effect. In another experiment, a conditioned reflex was formed to a compound in which the two individual tones were of very different intensities. The effect of the tone of weaker intensity when tested singly was now very small or absent altogether. Such a case is illustrated in the following experiment by Dr. Zeliony. An alimentary reflex was established in a dog to a compound stimulus made up of the sound of a whistle and the sound of the tone d′ sharp of a pneumatic tuning-fork. Both these sounds appeared to the human ear to be of equal intensity, and both when tested separately elicited a secretion of 19 drops of saliva during one minute. In addition to this, another compound stimulus was established, made up of the same sound of the whistle plus the tone a′ of a tuning-fork of weaker intensity. When tested separately the whistle in this case elicited a secretion of seven drops of saliva during thirty seconds, and the tone only one drop.

It is evident from the above experiments that the obscuring of one stimulus by another belonging to the same analyser is determined by differences in their strength, and it is natural to assume that this explanation can be applied also to compound stimuli, the components of which belong to different analysers. On this assumption tactile cutaneous stimuli in our experiments should be regarded as being relatively stronger than thermal cutaneous stimuli, and auditory stimuli should be regarded as being relatively stronger than visual stimuli. The natural deductions from such an assumption are far-reaching, and it will be necessary at some future date to test its validity by the use of stimuli compounded from different analysers and varying as much as possible in their intensities, a very weak auditory stimulus being combined with a very strong visual one, and so on.

The phenomenon in which one stimulus is obscured by another in a simultaneous stimulatory compound, when the two stimuli belong to different analysers, presents several interesting features. The effect of the compound stimulus is found nearly always to be equal to that of the stronger component used singly, the weaker stimulus appearing therefore to be completely overshadowed by the stronger one. If, however, the stronger stimulus is even at long intervals of time, repeated singly without reinforcement by the unconditioned reflex, while the compound stimulus is constantly reinforced, the stronger stimulus by itself becomes completely ineffective, whereas in the stimulatory compound there is no diminution in its effect. It is evident, therefore, that although the effect of the weaker stimulus when tested singly is invisible, it nevertheless plays an important part in the stimulatory compound [experiments by Dr. Palladin].

Another feature of interest has been already described in the fourth lecture [experiments of Dr. Perelzweig, page 56]. If the weak component, which may be even quite ineffective when applied alone, is repeated at short intervals without reinforcement—*i.e.* is extinguished below zero—then both the compound and the stronger component undergo secondary extinction. In this experiment, therefore, the component which is normally of itself apparently ineffective becomes temporarily transformed by the process of experimental extinction into a strong inhibitory stimulus.

The following was observed in a single, but so far as the experimental conditions were concerned a perfect, experiment. When two stimuli belonging to different analysers were first separately made into conditioned stimuli, and only afterwards applied simultaneously to form a compound stimulus, which was repeatedly reinforced, the overshadowing of the one component by the other did not occur. From this it may be concluded that in the usual case where two hitherto neutral stimuli are used to form a compound stimulus, the stronger stimulus at once prevents the weaker from forming a corresponding connection with the centre for the unconditioned reflex. If, however, this connection has been established already, it is not disturbed during the subsequent establishment of the reflex to the compound stimulus. The mechanism on which the predominance of one component of a stimulatory compound over another depends is most probably a form of inhibition. This matter will be examined in detail in a subsequent lecture.

The cases mentioned above show that a definite interaction takes place between different cells of the cortex, resulting in a fusion or synthesis of their physiological activities on simultaneous excitation. In the case of a compound simultaneous stimulus made up of components of unequal strength belonging to the same analyser this synthesis is not so obvious. However, it comes out very clearly that even in these cases there is no summation of the individual reflex effect of each single component, the effect of the stronger component applied singly being equal to that of the compound stimulus.

The phenomenon of synthesis of stimuli belonging to the same analyser is much more evident in a modification of the experimental conditions which was first used in Dr. Zeliony's experiments, then again by Drs. Manuilov and Krylov, and has since been widely practised. It was noticed that if a conditioned reflex to a compound stimulus was established as described above, it was easy to maintain it in full strength and at the same time to convert its individual components, which gave a positive effect when tested singly, into negative or inhibitory stimuli. This result is obtained by constant reinforcement of the compound stimulus, while its components, on the frequent occasions when they are applied singly, remain without reinforcement. The experiment can be made with equal success in the reverse direction, making the stimulatory compound into a negative or inhibitory stimulus, while its components applied singly maintain their positive effect.

18

Reprinted from *Psychon. Sci.* **4**:123–124 (1966)

Relation of cue to consequence in avoidance learning[1]

JOHN GARCIA AND ROBERT A. KOELLING
HARVARD MEDICAL SCHOOL AND MASSACHUSETTS GENERAL HOSPITAL

An audiovisual stimulus was made contingent upon the rat's licking at the water spout, thus making it analogous with a gustatory stimulus. When the audiovisual stimulus and the gustatory stimulus were paired with electric shock the avoidance reactions transferred to the audiovisual stimulus, but not the gustatory stimulus. Conversely, when both stimuli were paired with toxin or x-ray the avoidance reactions transferred to the gustatory stimulus, but not the audiovisual stimulus. Apparently stimuli are selected as cues dependent upon the nature of the subsequent reinforcer.

A great deal of evidence stemming from diverse sources suggests an inadequacy in the usual formulations concerning reinforcement. Barnett (1963) has described the "bait-shy" behavior of wild rats which have survived a poisoning attempt. These animals utilizing olfactory and gustatory cues, avoid the poison bait which previously made them ill. However, there is no evidence that they avoid the "place" of the poisoning.

In a recent volume (Haley & Snyder, 1964) several authors have discussed studies in which ionizing radiations were employed as a noxious stimulus to produce avoidance reactions in animals. Ionizing radiation like many poisons produces gastrointestinal disturbances and nausea. Strong aversions are readily established in animals when distinctively flavored fluids are conditionally paired with x-rays. Subsequently, the gustatory stimulus will depress fluid intake without radiation. In contrast, a distinctive environmental complex of auditory, visual, and tactual stimuli does not inhibit drinking even when the compound stimulus is associated with the identical radiation schedule. This differential effect has also been observed following ingestion of a toxin and the injection of a drug (Garcia & Koelling, 1965).

Apparently this differential effectiveness of cues is due either to the nature of the reinforcer, i.e., radiation or toxic effects, or to the peculiar relation which a gustatory stimulus has to the drinking response, i.e., gustatory stimulation occurs if and only if the animal licks the fluid. The environmental cues associated with a distinctive place are not as dependent upon a single response of the organism. Therefore, we made an auditory and visual stimulus dependent upon the animal's licking the water spout. Thus, in four experiments reported here "bright-noisy" water, as well as "tasty" water was conditionally paired with radiation, a toxin, immediate shock, and delayed shock, respectively, as reinforcers. Later the capacity of these response-controlled stimuli to inhibit drinking in the absence of reinforcement was tested.

Method

The apparatus was a light and sound shielded box (7 in. x 7 in. x 7 in.) with a drinking spout connected to an electronic drinkometer which counted each touch of the rat's tongue to the spout. "Bright-noisy" water was provided by connecting an incandescent lamp (5 watts) and a clicking relay into this circuit. "Tasty" water was provided by adding flavors to the drinking supply.

Each experimental group consisted of 10 rats (90 day old Sprague-Dawley males) maintained in individual cages without water, but with *Purina Laboratory chow ad libidum.*

The procedure was: A. One week of habituation to drinking in the apparatus without stimulation. B. Pretests to measure intake of bright-noisy water and tasty water prior to training. C. Acquisition training with: (1) reinforced trials where these stimuli were paired with reinforcement during drinking, (2) nonreinforced trials where rats drank water without stimuli or reinforcement. Training terminated when there was a reliable difference between water intake scores on reinforced and nonreinforced trials. D. Post-tests to measure intake of bright-noisy water and tasty water after training.

In the x-ray study an audiovisual group and a gustatory group were exposed to an identical radiation schedule. In the other studies reinforcement was contingent upon the rat's response. To insure that both the audiovisual and the gustatory stimuli received equivalent reinforcement, they were combined and simultaneously paired with the reinforcer during acquisition training. Therefore, one group serving as its own control and divided into equal subgroups, was tested in balanced order with an audiovisual and a gustatory test before and after training with these stimuli combined.

One 20-min. reinforced trial was administered every three days in the x-ray and lithium chloride studies. This prolonged intertrial interval was designed to allow sufficient time for the rats to recover from acute effects of treatment. On each interpolated day the animals received a 20-min. nonreinforced trial. They were post-tested two days after their last reinforced trial. The x-ray groups received a total of three reinforced trials, each with 54 r of filtered 250 kv x-rays delivered in 20 min. Sweet water (1 gm saccharin per liter) was the gustatory stimulus. The lithium chloride group had a total of five reinforced trials with toxic salty water (.12 M lithium chloride). Non-toxic salty water (.12 M sodium chloride) which rats cannot readily distinguish from the toxic solution was used in the gustatory tests (Nachman, 1963).

The immediate shock study was conducted on a more orthodox avoidance schedule. Tests and trials were 2 min. long. Each day for four consecutive acquisition days, animals were given two nonreinforced and two reinforced trials in an NRRN, RNNR pattern. A shock, the minimal current required to interrupt drinking (0.5 sec. at 0.08-0.20 ma), was delivered through a floor grid 2 sec. after the first lick at the spout.

The delayed shock study was conducted simultaneously with the lithium chloride on the same schedule. Non-toxic salty water was the gustatory stimulus. Shock reinforcement was delayed during first trials and gradually increased in intensity (.05 to .30 ma) in a schedule designed to produce a drinking pattern during the 20-min. period which resembled that of the corresponding animal drinking toxic salty water.

Results and Discussion

The results indicate that all reinforcers were effective in producing discrimination learning during the acquisition phase (see Fig. 1), but obvious differences occurred in the post-tests. The avoidance reactions produced by

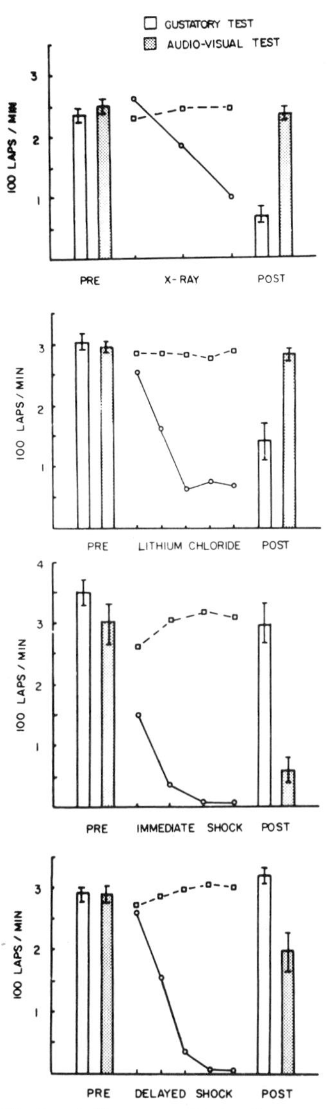

Fig. 1. The bars indicate water intake (± St. Error) during a gustatory test (a distinctive taste) and an audiovisual test (light and sound contingent upon licking) before and after conditional pairing with the reinforcers indicated. The curves illustrate mean intake during acquisition.

x-rays and lithium chloride are readily transferred to the gustatory stimulus but not to the audiovisual stimulus. The effect is more pronounced in the x-ray study, perhaps due to differences in dose. The x-ray animals received a constant dose while the lithium chloride rats drank a decreasing amount of the toxic solution during training. Nevertheless, the difference between post-test scores is statistically significant in both experiments ($p < 0.01$ by ranks test).

Apparently when gustatory stimuli are paired with agents which produce nausea and gastric upset, they acquire secondary reinforcing properties which might be described as "conditioned nausea." Auditory and visual stimulation do not readily acquire similar properties even when they are contingent upon the licking response.

In contrast, the effect of both immediate and delayed shock to the paws is in the opposite direction. The avoidance reactions produced by electric shock to the paws transferred to the audiovisual stimulus but not to the gustatory stimulus. As one might expect the effect of delayed shocks was not as effective as shocks where the reinforcer immediately and consistently followed licking. Again, the difference between post-test intake scores is statistically significant in both studies ($p < 0.01$ by ranks test). Thus, when shock which produces peripheral pain is the reinforcer, "conditioned fear" properties are more readily acquired by auditory and visual stimuli than by gustatory stimuli.

It seems that given reinforcers are not equally effective for all classes of discriminable stimuli. The cues, which the animal selects from the welter of stimuli in the learning situation, appear to be related to the consequences of the subsequent reinforcer. Two speculations are offered: (1) Common elements in the time-intensity patterns of stimulation may facilitate a cross modal generalization from reinforcer to cue in one case and not in another. (2) More likely, natural selection may have favored mechanisms which associate gustatory and olfactory cues with internal discomfort since the chemical receptors sample the materials soon to be incorporated into the internal environment. Krechevsky (1933) postulated such a genetically coded hypothesis to account for the predispositions of rats to respond systematically to specific cues in an insoluble maze. The hypothesis of the sick rat, as for many of us under similar circumstances, would be, "It must have been something I ate."

References

Barnett, S. A. *The rat: a study in behavior.* Chicago: Aldine Press, 1963.

Garcia, J., & Koelling, R. A. A comparison of aversions induced by x-rays, toxins, and drugs in the rat. *Radiat. Res.*, in press, 1965.

Haley, T. J., & Snyder, R. S. (Eds.) *The response of the nervous system to ionizing radiation.* Boston: Little, Brown & Co., 1964.

Krechevsky, I. The hereditary nature of 'hypothesis'. *J. comp. Psychol.*, 1932, 16, 99-116.

Nachman, M. Learned aversion to the taste of lithium chloride and generalization to other salts. *J. comp. physiol. Psychol.*, 1963, 56, 343-349.

Note

1. This research stems from doctoral research carried out at Long Beach V. A. Hospital and supported by NIH No. RH00068. Thanks are extended to Professors B. F. Ritchie, D. Krech and E. R. Dempster, U. C. Berkeley, California.

19

Reprinted from pp. 279–296 of *Punishment and Aversive Behavior,* B. A. Campbell and R. M. Church, eds., Appleton-Century-Crofts, New York, 1969, 597 p.
Reprinted by permission of Prentice-Hill, Inc., Englewood Cliffs, New Jersey

Predictability, Surprise, Attention, and Conditioning [1]

Leon J. Kamin
PRINCETON UNIVERSITY

The experiments to be described here have no special relevance to the problem of punishment. The studies to be reported do employ the CER procedure (Estes & Skinner, 1941). This procedure, within which an aversive US follows a warning signal regardless of the animal's behavior, has been contrasted to the arrangements employed in response-contingent punishment (Hunt & Brady, 1955). This type of comparison, however, is not germane to the present research. The kinds of results considered in this chapter derive from rats in a CER procedure, with shock as the US; but very similar results have been obtained in the McMaster laboratory by H. M. Jenkins, using pigeons in a food-reinforced operant discrimination. What appears to be involved in these studies is a concern with phenomena often referred to as examples of "selective attention." To the degree that punishment contingencies may be brought under stimulus control, the present work might be related to other contributions in this volume.

The present work arose from an interest in the possible role of attention in Pavlovian conditioning. The usual statement of the conditions sufficient for a Pavlovian CR asserts simply that a neutral, to-be-conditioned CS must be presented in contiguity with a US. What happens, however, when a compound CS consisting of elements known to be independently conditionable is presented in contiguity with a US? Are all elements of the CS effectively conditioned? Does the animal attend, and thus condition, more to some elements than to others? What kinds of experimental manipulations might direct the animal's attention to one or another element?

The first experimental approach to these questions was, in overview, as follows. First, condition an animal to respond to a simple CS, consisting of Element A. Then condition the animal to respond to a compound,

[1] The research reported here was supported by a research grant from the Associate Committee on Experimental Psychology, National Research Council of Canada.

consisting of Element A plus a superimposed Element B. Finally, test the animal with Element B alone. Will it respond to Element B? Put very naively, our primitive notion was that, because of the prior conditioning to Element A, that element might so "engage the animal's attention" during presentation of the compound that it would not "notice" the added Element B. The failure to notice the superimposed element might preclude any conditioning to it. To conclude that the prior conditioning to Element A was responsible for a failure to respond to Element B we must, of course, show that animals conditioned to the compound without prior conditioning to A do respond when tested with B. To control for amount of experience with the US, and variables correlated with it, we ought also to show that, if compound conditioning is followed by conditioning to A alone, the animal will respond when tested with B.

This relatively simple design has since expanded in a number of unexpected directions, and our original primitive notions about attention have been forcibly revised, if not refined. To date, we have utilized over 1200 rats as subjects in more than 110 experimental groups. There has been an earlier report of the first stages of this work (Kamin, 1968); in the present chapter, we shall review the basic preliminary findings, then focus on some of the more recent developments.

The basic CER procedure utilized in all these studies employs naive hooded rats as subjects, reduced to 75% of *ad libitum* body weight and maintained on a 24-hour feeding rhythm. The rats are first trained to press a bar for a food reward in a standard, automatically programmed operant conditioning chamber. The daily sessions are 2 hours in length, with food pellets being delivered according to a 2.5-minute variable-interval reinforcement schedule. The first five sessions (10 hrs.) produce stable bar-pressing rates in individual rats, and CER conditioning is then begun. During CER conditioning, the food-reinforcement schedule remains in effect throughout the daily 2-hour session, but four CS–US sequences are now programmed independently of the animal's behavior. The CS, typically, has a duration of 3 minutes and is followed immediately by a .5-second US, typically a 1-ma. shock. For each CER trial (four trials daily), a suppression ratio is calculated. The ratio is $B/A + B$, where B represents the number of bar presses during the 3-minute CS, and A the number of bar presses during the 3-minute period immediately preceding the CS. Thus, if the CS has no effect on the animal's bar pressing, the ratio is .50; but as the CS, with repeated trials, begins to suppress bar pressing, the ratio drops toward an asymptote very close to .00. We regard the learned suppression produced by the CS as an index of an association between CS and US, much as conditioned salivation to a metronome may be regarded as such an index.

The CS in the experiments to be described was either a white noise (typically 80 db), the turning on of an overhead house light (7.5-w. bulb diffused through milky plastic ceiling), or a compound of noise-plus-light presented simultaneously. The normal condition of the chamber is complete darkness. The various experimental groups received CER conditioning to various CS's, in different sequences. The precise sequences of CS's are detailed in the body of this report. Typically, following the CER conditioning, the animal was given a single test day, during which a nonreinforced CS was presented four times within the bar-pressing session. The data to be presented are suppression ratios for the first test trial. While no conclusions would be altered by including the data for all four test trials, the fact that the test CS is not reinforced means that test trials following the first contribute relatively little to differences between experimental groups.

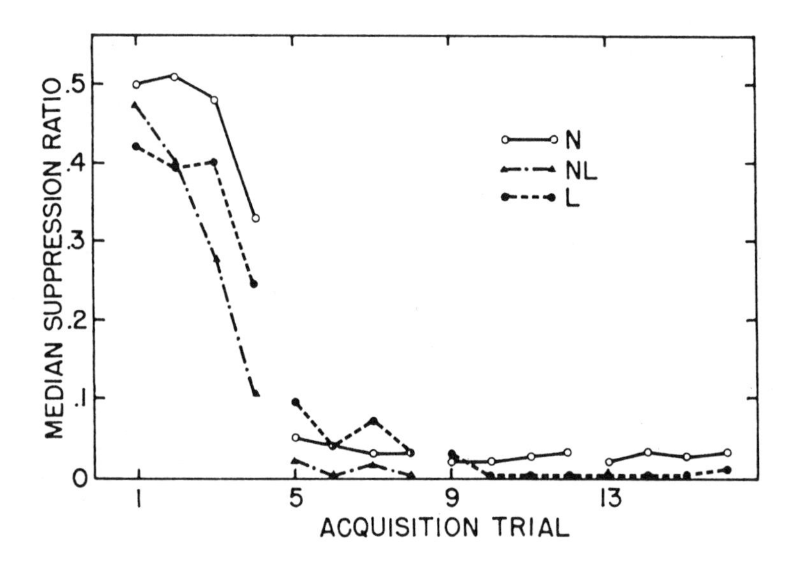

FIG 9–1. Acquisition of CER by trial, for three groups of rats, trained with either light, noise, or compound CS.

The characteristic outcome of our basic conditioning procedure is depicted in Fig. 9–1, which presents median suppression ratios, as a function of acquisition trial, for three representative groups of subjects. The groups have been conditioned with either noise, light, or the compound as a CS. The major point to note at present is that after a very few trials all groups approach asymptotic suppression. It can also be observed that light has a slightly suppressing effect on the very first trial so that the light group tends to acquire slightly more rapidly than the noise group. Finally, the compound group acquires significantly more rapidly than either of the others.

The first experimental approach to attention is illustrated in the design outlined below. The code letter for an experimental group is indicated at the left of the paradigm. Then the CS employed with that group during consecutive phases of CER conditioning is noted; L, N, and LN refer, respectively, to a light, a noise, or a compound CS. The number of reinforced trials with each type of CS is indicated in parentheses immediately following the CS notation; four reinforced trials are given daily. Finally, the CS employed during the test trial is indicated, together with the median suppression ratio for the group on the test trial. The number of animals per experimental group varies, in the studies to be reported, between 8 and 20.

Group A:	LN (8)	N (16)	Test L	.25
Group B:	N (16)	LN (8)	Test L	.45
Group G:	—	LN (8)	Test L	.05
Group 2-B:	—	N (24)	Test L	.44

There are a number of relevant comparisons which can be made within the above set of four experimental treatments. The basic comparison is that between Groups G and B. The test result for Group G indicates, as a kind of base line, the amount of control normally acquired by the light as a result of eight reinforced compound conditioning trials. This is very significantly different from the result for Group B, within which the same compound conditioning trials have been preceded by prior conditioning to the noise element. Thus, our speculation that prior conditioning to an element might block conditioning to a new, superimposed element receives support. When we next compare Groups A and B, we again observe a significant difference. These two groups have each received the same number of each type of CER conditioning trial, but in a different sequence. Group B, for whom the noise conditioning preceded compound conditioning, is less suppressed on the test trial than is Group A, for whom the noise conditioning followed compound conditioning. This again supports the notion that prior conditioning to A blocks conditioning to the B member of the compound. The further fact that Group A is not as suppressed as Group G is not to be regarded as produced by interpolation of noise conditioning after compound conditioning. It must be remembered that four days elapse for Group A between the last compound trial and the test; appropriate control groups have established that Group A's poor performance on the test, relative to Group G's, can be attributed to the passage of time. This *recency effect,* of course, works counter to the direction of the significant difference we have observed between Groups A and B. The failure of Group B to suppress to light as much as does Group A, even with a strong recency effect working to Group B's advantage, suggests a fundamental failure of conditioning to the light in Group B. This is confirmed when we compare the test results of Groups B and 2-B. These groups each experience 24 times noise followed by shock, but for Group B light is superimposed during the final eight trials. The fact that the test trial to light yields equivalent results for B and 2-B indicates that the superimpositions have produced literally no conditioning to the light. The test ratios for both these groups are slightly below .50, indicating again that, independent of previous conditioning, an initial presentation of light has a mildly disruptive effect on ongoing bar-pressing behavior.

The blocking effect demonstrated by the experimental treatments described above is not specific to the particular sequence of stimuli employed. When four new groups of rats were trained, reversing the roles of the light and noise stimuli, a total block of conditioning to the noise member of a compound was produced by prior conditioning to the light element (Kamin, 1968). Further, it should be pointed out that we have tested many rats, after *de novo* conditioning to the light–noise compound, to each element separately. We have never observed a rat which did not display some suppression to each element. Thus, granted the present intensity levels of light and noise, the blocking effect depends upon prior conditioning to one of the elements; when conditioned from the outset to the compound, no animal ignores completely one of the elements.

We should also note that animals conditioned to noise alone after previous conditioning to light alone acquire at the same rate as do naive animals conditioned to noise alone. Prior conditioning to noise alone also does not affect subsequent conditioning to light alone. It seems very probable that this lack of transfer between the two stimuli, as well as some degree of equivalence between the independent efficacies of the stimuli, are necessary preconditions for the kind of symmetrical blocking effect which we have demonstrated.

The results so far presented indicate that, granted prior conditioning to an element, no conditioning occurs to a new element which is now superimposed on the old. This might mean, as we first loosely suggested, that the animal does not notice (or perceive) the superimposed element; the kind of peripheral gating mechanism popularized by Hernandez-Peon (Hernandez-Peon et al., 1956) is an obvious candidate for theoretical service here. To speak loosely again, however, we might suppose that the animal does notice the superimposed stimulus but does not condition to it because the stimulus is redundant. The motivationally significant event, shock, is already perfectly predicted by the old element. The possible importance of redundancy and informativeness of stimuli in conditioning experiments has been provocatively indicated by Egger and Miller (1962). We thus decided to examine whether, in the case when the superimposed stimulus predicted something new (specifically, nonreinforcement), it could be demonstrated that the animal noticed the new stimulus. The following two groups were examined.

Group Y:	N (16)	LN, nonreinforced (8)	N, nonreinforced (4)
Group Z:	N (16)	N, nonreinforced (12)	

The results for both groups during nonreinforced trials are presented in Fig. 9–2.

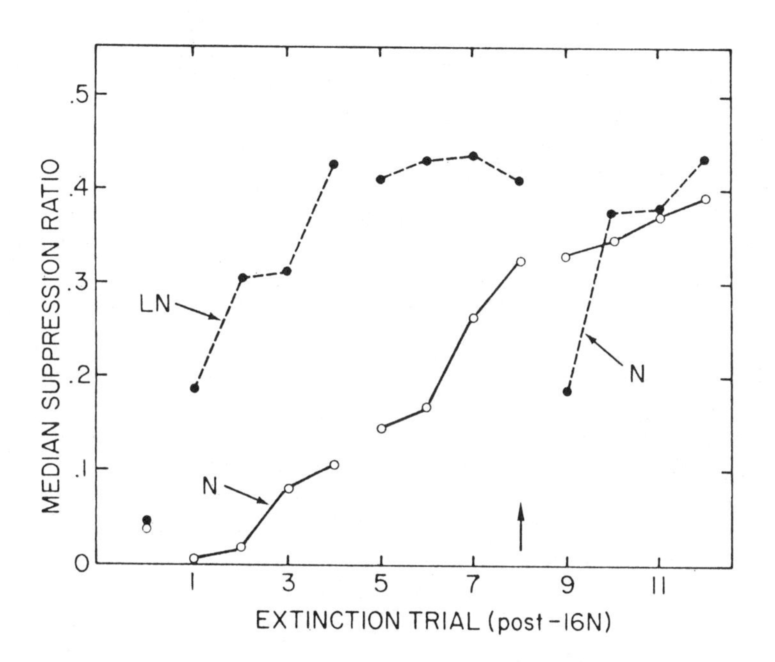

FIG. 9–2. Extinction of CER, by trial, following conditioning to noise. The groups were extinguished either to noise alone or to the compound. The arrow in the abscissa indicates point at which group extinguished to compound is switched to noise alone.

Through the first 16 CER conditioning trials these groups are treated identically, and on the sixteenth trial the median ratio to noise was .02 for each group. When Group Y was presented with the compound on its next trial, its ratio increased to .18; on the equivalent trial, Group Z, presented with the familiar noise, had a ratio of .01. The difference between groups on this trial fell short of significance, but it is certainly suggestive. The animals in Group Y seem to notice the superimposed light, even before the compound is followed by nonreinforcement. It must be remembered that, until the moment of nonreinforcement on Trial 17, Group Y is treated identically to the blocked Group B in the original experiment. Thus, if this result can be replicated, we have evidence that animals do notice the superimposed element, at least on the first trial of its introduction. The evidence is in the form of an attenuation of the suppression which would have occurred had not the new element been superimposed.

To return to the comparison between Groups Y and Z, on the second nonreinforced trial Group Y's ratio was .31, Group Z's was .02. This difference was significant. Thus a single nonreinforced presentation of the compound was sufficient for Group Y to discriminate between noise (always reinforced) and the compound (nonreinforced). Clearly, the light element had been perceived by Group Y. The very rapid extinction in Group Y cannot be attributed to the mere failure to reinforce the noise element, as Group Z's performance makes perfectly clear. The nature of the discrimination formed by Group Y is further illustrated by comparing performance of the two groups throughout the extinction phase of the experiment. By the eighth nonreinforced trial, the ratios were .41 for Group Y and .33 for Group Z. Then, on the next trial, the stimulus for Group Y was changed to noise alone. The Group Y ratio on this trial was .17, the Group Z ratio was again .33. This was a significantly lower ratio for Group Y than had been observed on the preceding trial. Thus, to some degree, animals in Group Y had learned that it was the compound which was nonreinforced; the noise element per se had been protected from extinction.

We now see that, if the superimposed element provides new information, the animal not only notices the element but can utilize the information which it provides with truly impressive efficiency. Further, the attenuated suppression noted on the transitional trial, when the new element is first superimposed on the old, suggested that, even in the earlier experiments in which the new element was redundant, the animals may have noticed it. This suggestion was confirmed by examining all of our data. We had at last count conditioned 153 animals with 16 trials of noise alone, followed by at least one trial of the compound. The median ratio of these animals on the sixteenth noise trial was .02; on the transitional trial (before reinforcement or nonreinforcement of the compound can exert any differential effect) the median ratio was .15. (When the transitional trial was reinforced, the median ratio on the second compound trial was again .02). There were 106 subjects which displayed higher ratios on the transitional trial than on the sixteenth noise trial; 17 which displayed lower ratios on the transitional trial; and 30 which had equal ratios on the two trials. This is a highly significant effect. There is thus no doubt that, at least on the first transitional trial, an animal previously conditioned to a single element notices the superimposition of a new element.

This observation is clearly fatal to our original theoretical notions. There remains the possibility, however, that in the case when the transitional trial proves the superimposed stimulus to be redundant, some gating mechanism is activated at that point such that the new element is not perceived on subsequent compound trials. Thus, it is at least conceivable that perceptual gating (deficient attention) provides the mechanism through which redundant stimuli are made nonconditionable. This view can be contrasted to the notion that redundant stimuli, though perceived in an intact manner, are simply not conditioned. We shall return to this problem a little later, after reviewing briefly some of the parameters of the blocking effect.

The data gathered to date, much of which has been more fully described elsewhere (Kamin, 1968), indicates such facts as the following. The blocking effect, granted prior conditioning to Element A, remains total even if the number of compound conditioning trials is very substantially increased; on the other hand, if conditioning to Element A is terminated before suppression has become asymptotic, a partial block of conditioning to the B member of the compound occurs. The amount of blocking is very smoothly related to the amount of prior conditioning to Element A. The block can be eliminated by extinguishing suppression to A prior to beginning compound conditioning; if suppression to A is extinguished following compound conditioning (A having been conditioned prior to the compound), the block remains. When blocking experiments were conducted with new groups of animals, holding constant the intensity value of Element B, while varying for different groups the intensity of Element A, the amount of blocking was a clear function of the relative intensities of the two elements. That is, more blocking of conditioning to B occurs if A is physically intense than if A is physically weak. This, however, is confounded with the fact that the level of suppression achieved by conclusion of the conditioning trials to A varies with the intensity of A; and we have already indicated that blocking varies with the level of suppression conditioned to A.

We have, as well, examined the blocking effect under a large number of procedural variations which have had no effect whatever on the basic phenomenon. Thus, for example, if the standard experiment is repeated employing a 1-minute, rather than a 3-minute, CS, a complete block is obtained. The same outcome is observed if the experiment is performed employing a 3-ma., rather than a 1-ma., US throughout. And again, complete blocking is obtained if the first CS, on which light onset is superimposed as a new element, is the turning off of a background 80-db noise, rather than the turning on of an 80-db noise. To put matters simply, the blocking phenomenon is robust, and easily reproducible.

We turn now to consideration of a classical phenomenon to which the blocking effect seems clearly related; we shall later return to a more detailed analysis of blocking itself. The blocking effect demonstrated in these studies seems in many ways reminiscent of the overshadowing of a weak element by a strong element in a compound CS. The basic observation reported by Pavlov (1927, pp. 141 ff.) was that if a compound CS was formed of two stimulus elements differing greatly in intensity or strength, the weaker element, when presented on test trials, failed to elicit any CR, despite repeated prior reinforcement of the compound. This was true although the weaker element was known to be independently conditionable. The major distinctions between the Pavlovian finding and the present blocking effect are: first, that overshadowing was said to occur without prior conditioning of the stronger element; and second, that overshadowing was reported to depend fundamentally on a substantial difference between the relative intensities of the two elements. The available summaries of Russian protocols from Pavlov's laboratory, however, indicate that at least in some of the overshadowing studies the dog had in fact, at an earlier time in its lengthy experimental history, been conditioned to the stronger stimulus. Thus it seemed possible to us that overshadowing might not be obtained if naive animals were, from the outset of an experiment, conditioned to a compound consisting of strong and weak elements.

The data already reported make it clear that complete overshadowing is not obtained when naive rats are conditioned to a compound of 80-db noise plus light. Following sixteen such reinforced compound trials, animals tested either to noise or to light each display clear conditioning; the ratios are .05 to light and .25 to noise. We wished now to see whether overshadowing might be observed if the relative intensities of the light and noise elements were radically changed. To test this, new groups were conditioned (this time for eight trials) to a compound consisting of our standard light plus 50-db noise. The group then tested to light displayed a ratio of .03, while the group tested to noise had a ratio of .42. The weak noise was thus almost completely overshadowed by light. Further, animals conditioned to 50-db noise alone, following conditioning to the compound, did not acquire significantly more rapidly than did naive rats conditioned from the outset to 50-db noise. These results are entirely corroborative of the Pavlovian reports. There remains the problem of relating overshadowing, which is not dependent on prior conditioning to one of the elements, to blocking, which is so dependent.

There is at least one obvious way of incorporating both phenomena within the same framework. We could assume that, during the early trials of conditioning to a compound, independent and parallel associations are being formed between each element and the US. With the further assumption that the association to the stronger element is formed more rapidly than that to the weaker, the overshadowing experiment becomes a case in which, implicitly, precisely the same sequence of events takes place which is explicitly produced in the blocking experiment. That is, in the overshadowing case an association to one element (the stronger) is substantially formed before conditioning to a second element takes place. Thus, conditioning of the second element is blocked.

These assumptions might be made more plausible if we examined the rates at which independent groups of animals acquire the CER when conditioned to either light, noise, or the compound. The relevant acquisi-

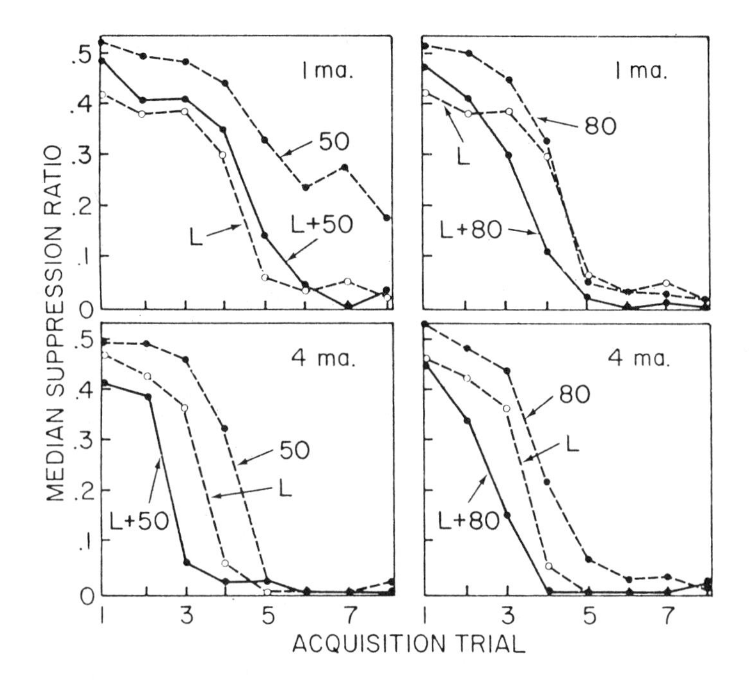

FIG. 9–3. Acquisition of CER, by trial, for independent groups of rats trained with either 50-db noise, 80-db noise, light, or compound CS. Two upper panels are for groups trained with 1 ma. US, two lower panels for groups trained with 4 ma. US.

tion curves for the first eight trials of conditioning are presented in Fig. 9–3. The upper left-hand panel of the figure presents curves for groups trained with light, 50-db noise, and the compound light plus 50 db, respectively. The group conditioned to light is asymptotically suppressed by Trial 5, before really substantial suppression is observed in the group conditioned to 50 db. The upper right-hand panel indicates that there is relatively little difference in the rates of conditioning to light and to 80-db noise. Thus, assuming the same rates of conditioning to each element within a compound as those observed when the elements are separately conditioned in independent groups, the overshadowing effect would be expected for the 50-db compound, but not for the 80-db compound.

There are further between-group comparisons possible within Fig.

9–3 which seem to support the argument. Within the upper right-hand panel, it can be observed that the compound group acquires significantly more rapidly than does either the light group or the 80-db group. That is, a clear summation of the two stimuli can be detected when conditioning to the compound. However, in the upper left-hand panel, there is clearly no summation; the compound group conditions at the same rate as the group trained to the stronger element, light. The 50-db element cannot be seen to affect in any way conditioning in the relevant compound group. Thus the presence or absence of overshadowing, measurable only after conditioning to a compound, is correlated with the presence or absence of a summation effect, detectable by comparing a compound group to other groups conditioned to single elements. This correlation of summation with overshadowing, it might be noted, seems relevant to Hull's (1943, Ch. 13) early interpretation of Pavlovian overshadowing. Basically, Hull regarded overshadowing as an extreme example of generalization decrement; the weaker member of the compound was assumed to be so dissimilar to the compound that it elicited no response. This view, which regards overshadowing as entirely dependent upon a postconditioning within-subject testing procedure, does not account for the association of overshadowing with the failure to observe summation in between-group comparisons made during conditioning. The very weak element in a compound CS really seems in some sense to be blotted out.

The weaker element in a compound, as has been noted, is one which, at least in independent groups, conditions less rapidly than the stronger element. The question thus arises whether overshadowing is a direct consequence of the relative intensities of the two elements, or whether the effect is mediated by the different rates of conditioning controlled by the separate elements. The finding that the effect depended directly upon relative intensities would be suggestive of perceptual and "attention-like" notions: for example, the weaker stimulus might not be noticed when compounded with a very strong stimulus. To fit overshadowing into the same framework as blocking, however, it would be convenient if the effect depended upon differential rates of conditioning. We have already reported that at least partial blocking of conditioning to a strong stimulus is obtained when the weak stimulus is conditioned prior to its compounding with the strong stimulus.

To decide between the two alternatives, we employed exactly the same pairs of CS elements utilized in the preceding studies, but manipulated the differential rates of conditioning controlled by the elements. This is quite easily done. When an intense US is employed in a CER procedure, differences in the rates of conditioning produced by CS's of different intensities are substantially reduced; all CS's are conditioned very rapidly (Kamin, 1965). We thus assumed that, by repeating the overshadowing studies already reported but now employing a 4-ma., rather than the standard 1-ma., US, the differences in rates of acquisition produced by light, by 50 db, and by 80 db would be reduced, with all groups tending to condition substantially in a very few trials. This in turn should mean that overshadowing, if it is dependent on the formation of a strong association to one element before substantial conditioning has occurred to the other, should be greatly reduced, if not eliminated.

The results were clear cut. The groups conditioned with a 4-ma. US to the compound light plus 80 db, when tested with, respectively, light or 80 db, displayed virtually total suppression. The same result was obtained when groups conditioned with a 4-ma. US to the compound light plus 50 db were tested with either light or 50 db. These CS elements, of course, are identical to those employed in the preceding overshadowing studies. The fact that light does not overshadow 50 db when an intense US is employed makes it clear that overshadowing is not a simple, direct consequence of the relative intensities of conditioned stimulus elements and seems to eliminate a simple attentional interpretation of overshadowing. The alternative interpretation seems quite well supported by examination of the lower two panels of Fig. 9–3. These panels present CER acquistion curves for new independent groups, analogous to the curves in the upper panels, but with US intensity now set at 4 ma. The new groups acquire more rapidly than do corresponding groups conditioned to 1 ma. More important, all new groups acquire rapidly, and none of the single element groups appears to have conditioned substantially before conditioning in another such group was well under way. We do not have enough data to make any precise guess about how much conditioning must occur to one element, in how many trials, before how much conditioning to another element, in order for overshadowing to occur in animals for whom the two elements are compounded. The results do indicate clearly, however, that overshadowing is not the result of a simple interaction of sensory events. They suggest as well that the occurrence of overshadowing can be predicted from examination of the rates of acquisition of independent groups conditioned to the separate elements. We might note, finally, that in each of the lower two panels of Fig. 9–3 clear summation effects are detectable, once again associated with the failure to observe overshadowing.

We return now to some further experimental analyses of the basic blocking effect. Within the work previously reported, substantial prior conditioning to an element has invariably given rise to no evidence of conditioning to the superimposed element. Thus the block has appeared to be a dramatically all-or-none affair. We now ask whether the total block which we observed in our basic Group B was in part an artifact of the relatively blunt measure of conditioning which we employed. The test trial to light, following compound conditioning, measures transfer from the compound to the element. The savings method is known to be extremely sensitive in demonstrating transfer, much more so than is the recall method represented by our test. We now repeated the basic experiment, but the test was no longer a single test trial to light; instead, all animals were given four reinforced conditioning trials to light at the end of the experiment. The focus of interest is on rate of acquisition during this conditioning to light. The two basic groups are outlined below.

Group 2-A:	N (16)	LN (8)	L (4)
Group 2-B:	—	N (24)	L (4)

While Groups 2-A and 2-B have each experienced noise followed by shock 24 times before the conditioning to light alone, the difference is of course that Group 2-A has on the last eight trials experienced the light superimposed on the noise. Will Group 2-A therefore show any savings, relative to Group 2-B, when conditioned to the light alone? Or have the eight superimpositions of light literally left no effect on the animal?

There was, as our earlier results would have suggested, no significant suppression to the light by either group on the first conditioning trial to light. However, Group 2-A displayed significantly more suppression on each of trials 2, 3, and 4 than did Group 2-B. Thus, it is clear that the eight light superimpositions did indeed leave some trace, which was manifested in a significant savings effect. However, we are reminded that our earlier data already demonstrated that, in groups conditioned similarly to Group 2-A, the animals did notice the superimposed light at least on the first, transitional trial. Can it be the case that the significant savings exhibited by Group 2-A is entirely attributable to the first trial on which light is superimposed? Or, do the compound trials following the first also contribute to the savings effect?

To answer this quetsion, Group 2-N was examined. The procedure is sketched below, and should be compared to those diagrammed in the immediately preceding paradigm.

Group 2-N:	N (16)	LN (1)	N (7)	L (4)

Group 2-N differs from Group 2-B only on the transitional trial; though the total number of reinforced experiences of noise is equated across Groups 2-A, 2-B, and 2-N, Group 2-N receives seven fewer light superimpositions than does Group 2-A. Nevertheless, the acquisition curves to light alone in the final phase of the experiment are virtually identical for Groups 2-N and 2-A; like Group 2-A, Group 2-N is significantly more suppressed than Group 2-B on each of Trials 2, 3, and 4. If we compute median suppression ratios over the four trials of light conditioning for each group, they are .28 for each of Groups 2-A and 2-N, but .38 for Group 2-B. Thus it is clear that the savings which we have demonstrated can be entirely attributed to the first, transitional trial. We had in any event independent evidence that the animal noticed the light on that trial, and it is now clear that the reinforcement at the termination of that trial does produce an increment in the associative connection between light and shock. There still, however, is nothing in the data which can allow us to conclude that the animal notices a redundant, superimposed element on any trial after the transitional trial; or at least, we have no indication that reinforced presentations of the superimposed element after the transitional trial in any way affect either the contemporaneous or the subsequent behavior of the animal. These results are obviously consistent with a perceptual gating concept, so long as the gating mechanism is not activated until after the transitional trial.

Where then do we stand now? The fact that the superimposed element proves to be redundant (that the US is already perfectly predicted by Element A) seems to be central to any interpretation of the blocking effect. Presumably, then, blocking would not occur if the superimposed element were made informative. We have earlier demonstrated that, if the compound is nonreinforced, the animal utilizes the information provided by Element B very efficiently. The strategy at this point was to perform a study within the blocking paradigm, reinforcing the compound trials, but at the same time making Element B informative. This was accomplished by radically increasing US intensity during the compound trials above the level employed during the prior conditioning to Element A, as with Group 2-M in the set of experimental treatments outlined below.

Group B:	N-1 ma. (16)	LN-1 ma. (8)	Test L	.45
Group 2-M:	N-1 ma. (16)	LN-4 ma. (8)	Test L	.14
Group 3-U:	N-4 ma. (8)	LN-4 ma. (8)	Test L	.36

The comparison between Groups B and 2-M is instructive. Here at last is a simple procedure which can virtually eliminate the blocking effect. Within Group 2-M, shock intensity is radically increased during the compound trials. The effect of this operation is to allow the formation of a clear association between the superimposed element and the US; Group 2-M, on the test trial, is significantly more suppressed than the standard Group B. This effect is not a simple consequence of employing an intense US during the compound trials. With Group 3-U, the same intense US is employed throughout the experiment, and a clear blocking effect is manifested: the test ratio of 3-U does not differ significantly from that of B, but does from that of 2-M. Thus, it is the change of shock intensity during the compound trials from that employed during prior conditioning which seems responsible for eliminating the block.

These results provide clear support for the assumption that blocking occurs because of the redundancy of the superimposed element. The question remains, how does redundancy prevent the formation of an association between a CS element and a US with which it is contiguously presented?

The most recent conception at which we have arrived seems capable of integrating all the data already presented. The notion is this: perhaps, for an increment in an associative connection to occur, it is necessary that the US instigate some mental work on the part of the animal. This mental work will occur only if the US is unpredicted, if it in some sense surprises the animal. Thus, in the early trials of a normal conditioning experiment, the US is an unpredicted, surprising event of motivational significance and the CS–US association is formed. Within the blocking experiment, the occurrence of the US on the first compound trial is to some degree surprising. This can be deduced, circularly, from the empirical observation that, on the transitional trial only, suppression is moderately attenuated; and some little learning about Element B can be demonstrated to have occurred on the transitional trial, but on no other compound trial. Finally, if in the blocking experiment US intensity is radically increased when compound training is begun, the new US is obviously surprising and no block is observed.

Precisely what mental work is instigated by a surprising US? The language in which these notions have been couched can be made more respectable, as well as more specific. Thus, as a first try, suppose that, for an increment in an associative connection to occur, it is necessary that the US provoke the animal into a backward scanning of its memory store of recent stimulus input; only as a result of such a scan can an association between CS and US be formed, and the scan is prompted only by an unpredicted US, the occurrence of which is suprising. This sort of speculation, it can be noted, leaves perception of the superimposed CS element intact. The CS element fails to become conditioned not because its input has been impeded, but because the US fails to function as a reinforcing stimulus. We have clearly moved some distance from the notion of attention to the CS, perhaps to enter the realm of retrospective contemplation of the CS.

These notions, whatever their vices, do suggest experimental manipulations. With the backward scan concept in mind, an experiment was performed which employed the blocking paradigm, but with an effort to surprise the animal very shortly after each presentation of the compound. Thus, animals were first conditioned, in the normal way, to suppress to the noise CS, with the usual 1-ma., .5-second US. Then, during the compound trials, the animal received reinforced presentations of the light-noise compound, again with a 1-ma., .5-second US. However, on each compound trial, 5 seconds following delivery of the US, an extra (surprising) shock (again 1 ma., .5 sec.) was delivered. When, after compound training, these subjects were tested with the light CS, they displayed a median ratio of .08. That is, the blocking effect was entirely eliminated by the delivery of an unpredicted shock shortly following reinforced presentation of the compound.

We have emphasized the close temporal relation between the unpredicted extra shock and the preceding compound CS. This emphasis is, of course, consistent with the backward scanning notion. There are, however, several alternative interpretations of the efficacy of the unpredicted shock in eliminating the blocking effect. There is the obvious possibility that the extra shock combines with the shortly preceding normal US to form, in effect, a US more intense than that employed during the prior conditioning to the noise element. We have already indicated that a radical increase of US intensity during the compound trials will eliminate the blocking effect. There is in the data, however, a strong indication that the extra shock functions in a manner quite different from that of an intense US. It is true that, if US intensity is increased from 1 ma. to 4 ma. during the compound trials, the blocking effect is eliminated; but it is also true that, if independent groups of naive rats are conditioned, with either a light, noise, or compound CS, paired with a 4-ma. US, they acquire the CER significantly more rapidly than do equivalent groups conditioned with a 1-ma. US. That is, acquisition of the CER is a clear positive function of US intensity. We have conditioned naive groups of animals, with either light or noise CS's, delivering the extra shock, 5 seconds after the normal US, from the outset of conditioning. In each case, the acquisition curve of rats conditioned with the extra shock was virtually superimposed on that of rats conditioned with the normal US. Thus, the extra shock does not appear to increase effective US intensity.

We have stressed the notion that the second, extra shock might cause the animal to scan the preceding sensory input, and that conditioning to the superimposed CS element occurs as a consequence of this scanning. There remains, however, the plausible alternative that the effect of the unpredicted, extra shock is to alert the animal in such a way that it is more attentive or sensitive to subsequent events; i.e., to the following compound trials. Thus, in this latest view, the extra shock does not increase the amount of conditioning taking place to the superimposed CS element on the first compound trial, but it does increase the amount of such conditioning taking place on all subsequent compound trials. Within the experiment already performed, there is unfortunately no way of deciding whether the extra shock facilitates conditioning to the CS which precedes it or to the CS which follows it. We do know, from appropriate control groups, that the extra shock does not cause the animal to suppress to extraneous exteroceptive stimuli which are subsequently presented.

There should be no great experimental difficulty in localizing the effect of the extra shock. We can, for example, deliver the extra shock to different groups at varying temporal intervals following the compound trials. Presumably, backward scanning should be less effective in forming an association when the extra shock is remote in time from the preceding trial. This approach, however, has the disadvantage that moving the extra shock away from the preceding trial moves it toward the subsequent trial. This problem in turn might be overcome by presenting only one compound trial a day. The sensitivity of the procedure seems to be such that, employing a savings technique, we might demonstrate the facilitating effect of a single extra shock, delivered on a single compound trial, with no subsequent compound conditioning. This effect in turn might be related to the temporal interval between the compound trial and the extra shock. There is no dearth of potential experiments to be performed, and not much sense in attempting to anticipate their outcomes.

To sum up, the blocking experiment demonstrates very clearly that the mere contiguous presentation of a CS element and a US is not a sufficient condition for the establishment of a CR. The question, very simply is: What has gone wrong in the blocking experiment? What is deficient? The experiment was conceived with a primitive hunch that attention to the to-be-conditioned stimulus element was a necessary precondition, and many of the results to date are consistent with the notion that the deficiency is perceptual, having to do with impeded input of the CS element. This blocked input was at first conceived as a consequence of a kind of competition for attention between the previously conditioned element and the new element. The results to date, however, make it clear that, if such an attentional deficit is involved, the redundancy of the new element is critical for producing it. The extra shock experiment, most recently, has suggested an alternative conception. The input of the new CS element can be regarded as intact, but the predictability of the US might strip the US of a function it normally subserves in conditioning experiments, that of instigating some processing of the memory store of recent stimulus input, which results in the formation of an association. There is also the possibility, of course, that the predictability of the US, by the time compound training is begun in the blocking experiment, strips the US of the function of alerting the animal to subsequent stimulus input.

There seems little doubt that, as experimentation continues, still other conceptions will be suggested. The experimental procedures are at least capable of discarding some conceptions and of reinforcing others. The progress to date might encourage the belief that ultimately these studies could make a real contribution toward answering the fundamental question toward which they are addressed: What are the necessary and sufficient conditions for the establishment of an association between CS and US within a Pavlovian paradigm?

REFERENCES

EGGER, M. C., & MILLER, N. E. Secondary reinforcement in rats as a function of information value and reliability of the stimulus. *Journal of Experimental Psychology*, 1962, **64**, 97–104.

ESTES, W. K., & SKINNER, B. F. Some quantitative properties of anxiety. *Journal of Experimental Psychology*, 1941, **29**, 390–400.

HERNANDEZ-PEON, R., SCHERRER, H., & JOUVET, M. Modification of electrical activity in cochlear nucleus during "attention" in unanesthetized cats. *Science*, 1956, **123**, 331–332.

HULL, C. L. *Principles of behavior*. New York: Appleton-Century, 1943.

HUNT, H. F., & BRADY, J. V. Some effects of punishment and intercurrent "anxiety" on a single operant. *Journal of Comparative and Physiological Psychology*, 1955, **48**, 305–310.

KAMIN, L. J. Temporal and intensity characteristics of the conditioned stimulus. In W. F. Prokasy (Ed.). *Classical conditioning*. New York: Appleton-Century-Crofts, 1965. Pp. 118–147.

KAMIN, L. J. "Attention-like" processes in classical conditioning. In M. R. Jones (Ed.), *Miami symposium on the prediction of behavior, 1967: Aversive stimulation*. Coral Gables, Fla.: University of Miami Press, 1968. Pp. 9–31.

PAVLOV, I. P. *Conditioned reflexes*. (Tr., G. V. Anrep.) London: Oxford University Press, 1927. (Reprinted, New York: Dover, 1960.)

Part III

COMPARATIVE STUDY OF LEARNING AND INTELLIGENCE

Editor's Comments on Papers 20, 21, and 22

For most of its history the study of learning in animals has been directed more toward understanding learning than toward understanding animals. Surprisingly little is known about the role of learned behavior in the lives of feral animals, and even less is known about how the various kinds of learning processes isolated in laboratory settings combine with one another to generate adaptive behavior in the wild. The hope that comparative study of learning in a wide range of species would provide an understanding of the adaptive significance, evolutionary history, and biological structure of learning has yet to be realized.

Although there have been thousands of studies of learning processes in animals, there has never developed a compelling, widely accepted methodology for comparing learning among different individual animals and across species. Unfortunately, it is far easier to decry the lack of such a methodology than to devise a satisfactory one. A variety of approaches to the comparative study of learning has been proposed and practiced, and each has proved vulnerable to an abundance of criticism. Some critics have even called for abandoning the enterprise, on grounds that it is doomed to failure.

Why has the quest for a sound comparative study of learning proved to be so difficult? Exemplars of species whose places in the phylogenetic tree bear potentially interesting relationships to one another have been tested in learning tasks and their performances have been compared. At first this may seem to be a forthright, even easy task. Problems become apparent when one recognizes that one wishes not merely to compare different animals, but to compare the learning processes exhibited by those animals. How can one be sure that learning processes observed in two different animals (or, for that matter, in the same animal at two different points in its development) are in fact comparable? Early investigators settled for making the procedures they employed in testing different animals as similar as possible, in the hope that similar testing procedures would tap comparable learning processes. The Hampton Court Maze, for example, can be produced in models that are appropriate for people, rats, turtles, monkeys, fish, and even cockroaches. What is to be made of the data generated with such devices? If a turtle learns a maze in fewer trials than does a human, does that mean the turtle is the "smarter" of the two? What makes two learning phenomena in different animals comparable?

This question has never received an altogether satisfactory answer. One influential attempt to deal with the question is found in the Thorndike article reprinted here (Paper 20). Thorndike believed that increasing efficiency and complexity were the hallmarks of evolving minds. By this view, what appear to be qualitative leaps in intellectual capacities in fact represent quantitative increases in the speed, complexity, and so forth of associative processes. Continuity is emphasized, as it had been earlier by Darwin (1871).

Some early attempts to validate Thorndike's hypothesis were based on a misguided and inaccurate view of the phylogenetic relationships among living species (see Hodos and Campbell, 1969). Can animals that share no common ancestors nevertheless be ranked in terms of the complexity of their learning processes? Razran (1971) has made an ambitious effort to order learning processes and to compare different animals in terms of where they lie in this ordering.

A somewhat different approach has been taken by M. E. Bitterman (1960, 1965). One of the first investigators to realize that superficial similarity of task requirements is not enough to ensure the comparability of observed learning phenomena, Bitterman has concentrated on qualitative differences in learning shown by different animals. Recognizing that different animals cannot be precisely "matched" with regard to such performance-affecting variables as motivation and alertness, he has advocated using a strategy of systematic variation of such variables, ensuring that animals are tested across a broad range of such factors. Some of the early generalizations that Bitterman and his colleagues made have been challenged by later experiments (see

Warren, 1973, for a concise review), but the realization that generalizations about "fishlike" versus "ratlike" behavior may not apply to all species of fish or strains of rat in no way invalidates the general approach (Paper 21). Perhaps the strategy would be used more widely if it were easier; systematic variation of all potential performance-affecting variables is difficult and arduous at best.

The entertaining article by the Brelands (Paper 22) offers another perspective on how similiar training procedures may produce very different results in different animals. Professional animal trainers, the Brelands used training techniques adapted from those that had proved sucessful in laboratory studies of animal learning. They discovered that their animals did not always behave as they had expected them to. By taking advantage of the "misbehavior" of the animals, the Brelands were able to produce commericially successful animal acts. More importantly, for our purposes, their keen observations of animal behavior provided a picture of how species-typical behavior patterns can override and modify the performance generated by standard training techniques.

REFERENCES

Bitterman, M. E., 1960, Toward a comparative psychology of learning, *Am. Psychologist,* **15**:704–712.

Bitterman, M. E., 1965, Phyletic differences in learning, *Am. Psychologist* **20**:396–410.

Darwin, C., 1871, *The Descent of Man,* Appleton, New York.

Hodos, W., and C. B. G. Campbell, 1969, *Scala naturae:* Why there is no theory in comparative psychology, *Psychol. Rev.* **76**:337–350.

Razran, G., 1971, *Mind in Evolution,* Houghton Mifflin, Boston.

Warren, J. M., 1973, Learning in vertebrates, in, *Comparative Psychology: A Modern Survery,* D. A. Desbury and D. A. Rethlingshafer, eds., McGraw-Hill, New York.

20

Reprinted from pp. 293–294 of *Animal Intelligence,* Macmillan, New York, 1911, 297 p.

THE EVOLUTION OF THE HUMAN INTELLECT

E. L. Thorndike

[*Editor's Note:* In the original, material precedes this excerpt.]

Nowhere in the animal kingdom do we find the psychological elements of reasoning save where there is a mental life made up of the definite feelings which I have called 'ideas,' but they spring up like magic as soon as we get in a child a body of such ideas. If we have traced satisfactorily the evolution of a life of ideas from the animal life of vague sense-impressions and impulses, we may be reasonably sure that no difficulty awaits us in following the life of ideas in its course from the chaotic dream of early childhood to the logical world-view of the adult scientist.

In a very short time we have come a long way, from the simple learning of the minnow or chick to the science and logic of man. The general frame of mind which one acquires from the study of animal behavior and of the mental development of young children makes our hypothesis seem vital and probable. If the facts did eventually corroborate it, we should have an eminently simple genesis of human faculty, for we could put together the gist of our contention in a few words. We should say: —

"The function of intellect is to provide a means of modifying our reactions to the circumstances of life, so that we may secure pleasure, the symptom of welfare. Its general law is that when in a certain situation an animal acts so that pleasure results, that act is selected from all those performed and associated with that situation, so that, when the situation recurs, the act will be more likely to follow than it was before; that on the contrary the acts which, when performed in a certain situation, have brought discomfort, tend to be dissociated from that situation. The intellectual evolution of the race consists in an increase in the number, delicacy, complexity, permanence and speed of formation of such associations. In man this increase reaches such a point that an apparently new type of mind results, which conceals the real continuity of the process. This mental evolution parallels the evolution of the cell structures of the brain from few and simple and gross to many and complex and delicate."

Nowhere more truly than in his mental capacities is man a part of nature. His instincts, that is, his inborn tendencies to feel and act in certain ways, show throughout marks of kinship with the lower animals, especially with our nearest relatives physically, the monkeys. His sense-powers show no new creation. His intellect we have seen to be a simple though extended variation from the general animal sort. This again is presaged by the similar variation in the case of the monkeys. Amongst the minds of animals that of man leads, not as a demigod from another planet, but as a king from the same race.

21

Reprinted from *Am. J. Psychol.* **71**:94–110 (1958)

SOME COMPARATIVE PSYCHOLOGY

By M. E. BITTERMAN, Bryn Mawr College, JEROME WODINSKY, American Museum of Natural History, and DOUGLAS K. CANDLAND, Princeton University

Although the capacity to learn has been demonstrated in a great variety of animals, there has been as yet no systematic comparative psychology of the learning process. Our highly generalized contemporary theory is based primarily on work with the rat, and the choice of subject for research on learning often is treated as a question of convenience. Is it true, as Skinner claims, that species "doesn't matter,"[1] or do the facts of comparative neurology attest the superficiality of any approach to learning that finds no place for phylogeny? Whatever the truth may be, an energetic search for the functional correlates of the wide structural differences that confront us in the animal series is long overdue.

How should such a search proceed? Taking the familiar rat as a point of reference, we might select for examination another animal, different enough from the rat to afford a marked phylogenetic contrast, yet similar enough that it may meaningfully be studied under analogous conditions. This is not, of course, to suggest that the two animals be compared in terms of their absolute scores in some standard apparatus. Work with the new animal, like work with the rat, would be directed to the discovery of functional relations. Its goal would be a theory of the new animal with which to compare the theory of the rat.[2]

On the assumption that a suitable subject for an enterprise of this kind might be found at the level of the fish, we have been doing some exploratory experiments with the African mouthbreeder (*Tilapia macrocephala*), an active, voracious animal, readily available, and easily maintained in the laboratory. Whether or not our choice of subject was a fortunate one remains to be determined. In the meantime, an account of several of our preliminary studies will give some notion of the problems to be met in comparative research and also perhaps of the returns to be anticipated.

* We are indebted to Dr. L. A. Aronson, of the American Museum of Natural History, and to Professor E. G. Wever, of Princeton University, who provided the facilities for these experiments. The work was done in the Princeton Laboratory while the senior author was a member of the Institute for Advanced Study.

[1] B. F. Skinner, A case history in scientific method, *Amer. Psychologist*, 11, 1956, 221-233.

[2] C. L. Hull, The place of innate individual and species differences in a natural-science theory of behavior, *Psychol. Rev.*, 52, 1945, 55-60.

I. Reversal-Learning

There seem to be many points at which the learning of rat and fish might be compared with profit. We chose to begin with a discriminative situation largely because of a long-standing interest in processes of discrimination in the rat, and we decided to study reversal-learning on the hunch that something like flexibility would prove to be an important functional correlate of phylogenetic level.

Our first experiment, with a simultaneous visual discrimination, already has been reported.[3] Two groups of mouthbreeders were studied, one by a method which provided opportunity for spontaneous correction of errors, and one by a method which did not. The animals were trained to strike at one of two simultaneously presented targets (horizontal *vs.* vertical stripes) irrespective of position, and the preference so established then was reversed repeatedly. The criterion of learning, both in the original problem and in each reversal, was 17 out of 20 errorless trials. The resulting curve of mean initial errors to criterion plotted against reversals looked like that found in higher forms, rising at first, and then falling gradually (with no appreciable decline after the sixth reversal). Further analysis of the improvement revealed, however, that its basis was unique. In Gatling's work with the rat, recovery from negative transfer was found to take place at all stages of learning in each reversal, yielding for successive reversals a series of essentially parallel learning curves which differed in over-all level.[4] In the fish, by contrast, performance on the first day of each reversal remained constant, and recovery from negative transfer appeared as an increase in the *rate* of error-elimination, yielding a series of learning curves for the successive reversals which differed primarily in slope.

A second experiment, now to be reported, was designed to examine the course of habit-reversal in the context of what was expected to be a simpler problem. The situation was the same as that of the first experiment, except that the fish were rewarded for response to spatial rather than to visual cues.

Subjects. Four naïve mouthbreeders, mature females, each about 2 in. long, were studied. They were drawn from the stock-tanks of the Department of Animal Behavior, American Museum of Natural History.

Method. The apparatus was that of the earlier experiment. A black metal housing which displayed two circular targets (each about 2 in. in diameter) was lowered

---

[3] Jerome Wodinsky and M. E. Bitterman, Discrimination-reversal in the fish, this Journal, 70, 1957, 569-576.

[4] F. P. Gatling, The effect of repeated stimulus reversals on learning in the rat, *J. compar. physiol. Psychol.*, 45, 1952, 347-351.

into S's (individual) living tank. The response of S, established by pretraining, was to strike at one of the targets. Displacement of a target by the animal activated a relay and signalled the response to E, who rewarded each correct choice with a pellet of food dropped into the water. After each trial, the housing was raised, and the lateral arrangement of the two targets changed (or not) in accordance with selected Gellermann-orders.[5] As in the first experiment, horizontally and vertically stripped targets (0.25-in. black and white stripes) were employed, although the animals were now reinforced for response to position irrespective of stripe. (It is customary in spatial training to use visually 'undifferentiated' alternatives, or to use visually differentiated alternatives in fixed positions. The method used here equates the perceptual conditions of visual and spatial problems.)

The animals were adapted to a 24-hr. feeding schedule for several weeks before work with them was begun; 20 pellets (of a specially prepared, highly attractive food) were given at the same time each day. During the experiment, there were 20 massed trials per day, and if less than 20 pellets were earned in the training period, the remainder were given soon after the animal was returned to the colony-room. The experiment consisted of a series of 15 problems, an original problem (Reversal 0) and 14 reversals, each learned to a criterion of 17 errorless trials on a single day. Two of the Ss were rewarded for going left in the original problem, and two for going right. The day after any animal reached criterion on the original problem, its training on the first reversal was begun, and so on, with each animal proceeding at its own pace. Training was by the correction-method which was used for one of the two groups of the first experiment. If the initial response on any trial was incorrect, S was permitted to make as many as four additional (repetitive) errors before the targets were withdrawn and the trial terminated. If the initial response was correct, or if any response thereafter was correct, reward was given and the trial terminated. To forestall the possibility of extinction early in training, a guidance-procedure was introduced after every series of 5 consecutive unreinforced trials (that is, after 25 consecutive unreinforced responses): the opening in the housing on the negative side was covered by a black disk, and the response which S sooner or later would make to the positive target (the only one available) was rewarded.

Results. Learning was rather rapid in this experiment. As Fig. 1 shows, the bulk of it took place on the first day of each problem; the criterion usually was reached on the second day, and only rarely were more than three days required. Mean errors, both to criterion and for the first and second days of each reversal separately, are plotted in Fig. 2. These curves show no evidence of progressive improvement over the series of problems. The increase in repetitive errors from Reversal 0 to Reversal 1—primarily a Day-1 effect—was consistent for all animals, but beyond that there was no recognizable trend. In the course of the 14 reversals, the average animal

[5] L. W. Gellermann, Chance orders of alternating stimuli in visual discrimination experiments, *J. genet. Psychol.*, 42, 1933, 356-360.

continued to make about 10 initial and 27 repetitive errors on Day 1, about 4 initial and 7 repetitive errors on Day 2, and so forth.

In the rat, progressive improvement in the reversal of spatial preferences is the rule. North found it with massed as well as with spaced practice, with

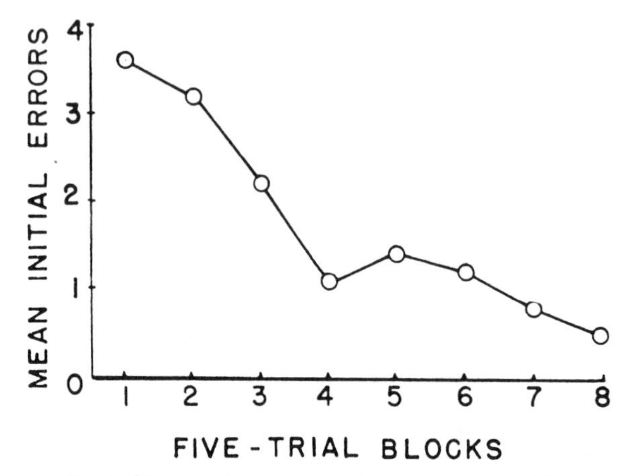

FIG. 1. SPATIAL REVERSAL TO CRITERION IN THE FISH
The curve shows within-problems improvement for the fourteen reversals combined.

correction as well as without correction, and with different degrees of training.[6] The rats of Dufort, Guttman, and Kimble, like those of Buytendijk 25 years before, were able to reverse a preference for either position with no more than a single error after a short series of problems.[7] Progressive improvement over a series of spatial reversals also has been found in other animals—not only in the chimpanzee, but in an eight-week-old pig, and in forms phylogenetically intermediate between fish and rat (newt and terrapin).[8] It is interesting to note, however, that Thompson was unable

[6] A. J. North, Improvement in successive discrimination reversals, *J. comp. physiol. Psychol.*, 43, 1950, 442-460.

[7] R. H. Dufort, Norman Guttman, and G. A. Kimble, One-trial discrimination reversal in the white rat, *J. compar. physiol. Psychol.*, 47, 1954, 248-249; F. J. J. Buytendijk, Über das Umlernen, *Arch. nëerl. Physiol.*, 15, 1930, 283-310.

[8] J. T. Cowles, Food tokens as incentives for learning by chimpanzees, *Comp. Psychol. Monogr.*, 14, 1935, (No. 71), 1-96; G. C. Myers, The importance of primacy in learning, *J. anim. Behav.*, 6, 1916, 64-69; Emanual Seidman, Relative ability of the newt and the terrapin to reverse a direction habit, *J. compar. physiol. Psychol.*, 42, 1949, 320-327. The over-all performance of Seidman's reptiles was superior to that of his amphibia. The reptilian performance was, in fact, unbelievably good, with many instances of *errorless* reversal in the later stages of the experiment.

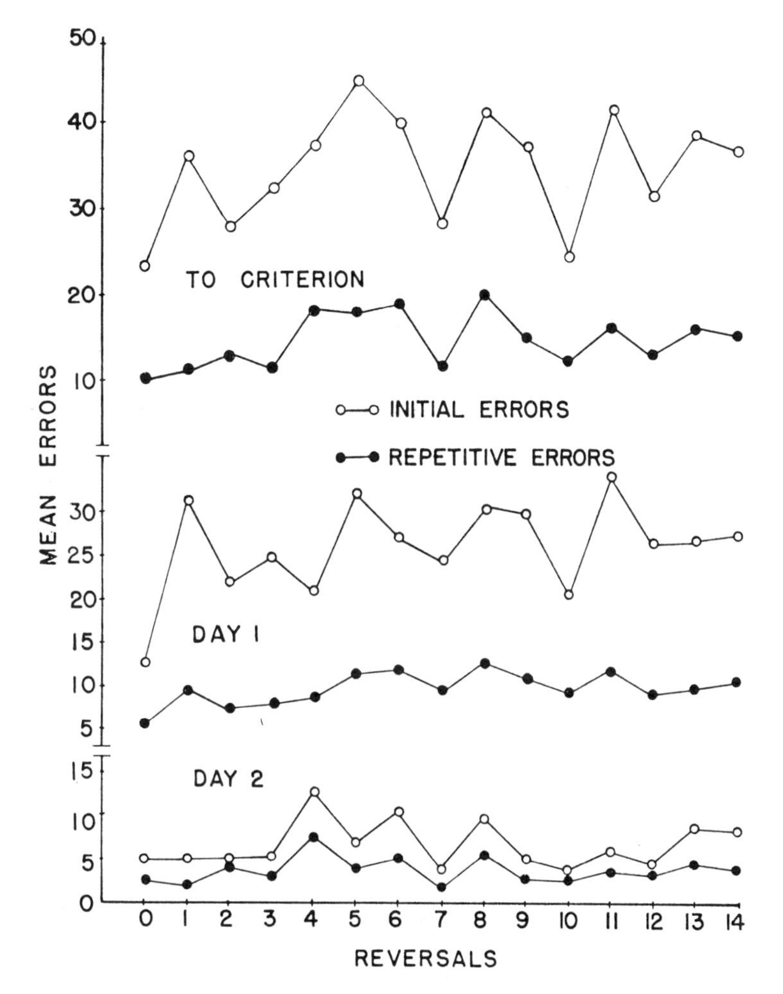

FIG. 2. SPATIAL REVERSAL TO CRITERION IN THE FISH

The curves show mean initial and repetitive errors per reversal, to criterion as well as for the first and second days on each reversal separately.

to find clear evidence of such improvement in an invertebrate (isopod), although he might have been more persistent in his search.[9]

Further information on the course of habit-reversal in the fish was

[9] Robert Thompson, Successive reversal of a position habit in an invertebrate, *Science*, 126, 1957, 163-164. *Cf.* an early study by R. M. Yerkes and G. E. Huggins, Habit formation in the crawfish, *Cambarus affinis*, *Harvard psychol. Studies*, 1, 1903, 565-577.

sought in a third experiment. One group of fish was trained, as in the first experiment, on the visual problem, and another, as in the second experiment, on the spatial problem, but in both cases the positive and negative stimuli were reversed each day. It was expected, of course, that the strength of the preference to be altered would be considerably less with daily reversal than with reversal to the criterion of 17/20. For purposes of comparison, two groups of rats were trained concurrently with the fish on analogous problems. Although some data already were available on spatial reversal in the rat after a small, fixed number of training trials, there were no corresponding visual data, and we thought it well to collect both kinds under similar conditions.

Subjects. The *S*s were 10 naïve mouthbreeders, male and female, comparable in every respect to those used earlier, and 10 naïve hooded rats, male and female, about 3 mo. old, bred in the laboratory. On the basis of adjustment in preliminary training, each group was divided into two matched subgroups of 5 *S*s each.

Method. The apparatus used for the fish was the same as that used in the earlier experiments. There were 20 trials per day with the horizontally and vertically striped targets, the lateral arrangement of which was varied from trial to trial in accordance with selected Gellermann-orders. For one (visual) group of fish, stripes were differentially reinforced; for the second (spatial) group, positions were differentially reinforced. Positive and negative stimuli were reversed daily. Some of the visual fish were reinforced for response to vertical stripes on odd days and for response to horizontal stripes on even days; others were reinforced for response to horizontal on odd days and vertical on even days. Similarly, the spatial fish were trained to one side (some left, some right) on odd days, and to the opposite side on even days. No spontaneous correction of errors was permitted in this experiment, but each incorrect response was followed by guidance: the housing was removed from the water, the incorrect stimulus was covered, the housing was reinserted, and the animal was rewarded for response to the positive stimulus. Every one of the 20 daily trials therefore terminated with a reinforced response. No other food was given the animals, which had been adapted before the beginning of the experiment to a 24-hr. feeding schedule of 20 pellets per day. There were 60 daily reversals, following which the fish on the visual problem were shifted to the spatial problem for 40 days, and conversely.

For work with the rat, a conventional model of Lashley's jumping apparatus was modified to eliminate punishment for errors. Narrow runways branched out from the starting platform to the two windows, which displayed horizontally and vertically striped cards (0.5-in. black and white stripes). The response of *S* (established in pretraining) was to run from the starting platform to one of the cards and to push against it; the incorrect card was locked in place, but the correct card gave access to a feeding platform in the rear. The two windows of the apparatus were equipped with hinged covers (of the same color as the surround) which were used for the purpose of guidance. After an incorrect response—that is, after the rat had approached the locked card and pushed against it—*E* returned the rat to the starting platform and simultaneously covered the window which displayed the negative card; the animal then went to the positive card (the only one visible), pushed it over, and found the food. The rats were maintained on a 24-hr. feeding schedule designed to keep them at 80% of their satiated bodily weights, and a large portion of each day's ration was consumed in the experimental situation. One group was put

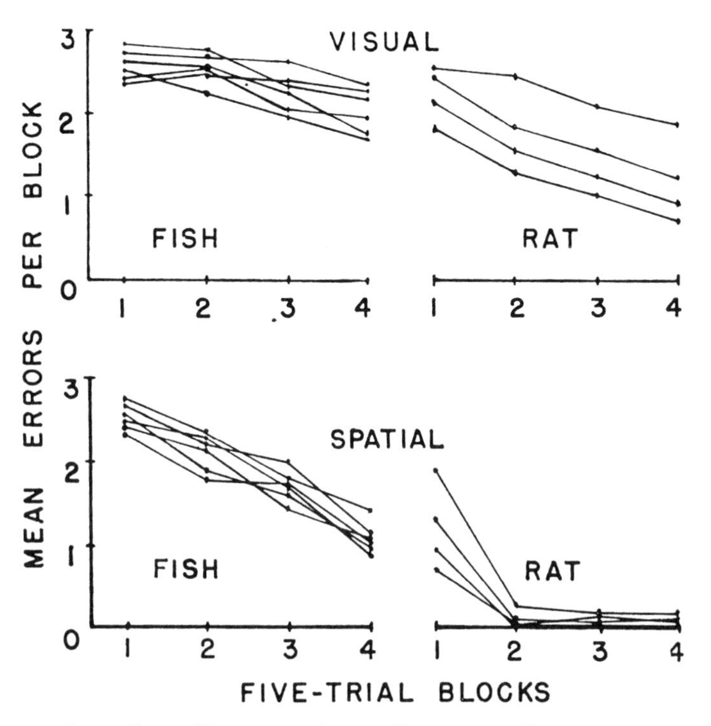

FIG. 3. DAILY VISUAL AND SPATIAL REVERSAL IN FISH AND RAT
The data plotted are for the first phase of the experiment. Each curve shows the average within-days performance of fish or rat during 10 days on a given problem. The relation among the curves in each cluster gives the change in performance from one 10-day period to the next. In neither case do the curves for the fish show any meaningful order. The curves for the rat are perfectly ordered in both cases—highest for the first 10-day period and lowest for the last.

on the visual problem and the other on the spatial problem. There were 20 trials per day, with daily reversal of the positive and negative stimuli, for 40 days, following which the visual group was put on the spatial problem, and the spatial group on the visual problem, for 40 days.

Results. The main results of the experiment are plotted in Fig. 3. Both in the visual problem and in the spatial problem, the fish showed significant within-days improvement, more in the spatial than in the visual, but in neither problem was there a consistent change from day to day. The curves for the six consecutive 10-day periods of training cross each other repeatedly and show no systematic variation in over-all level. The rat, too, improved significantly within days, again moreso in the spatial problem than in the visual. The difference between the two problems was much

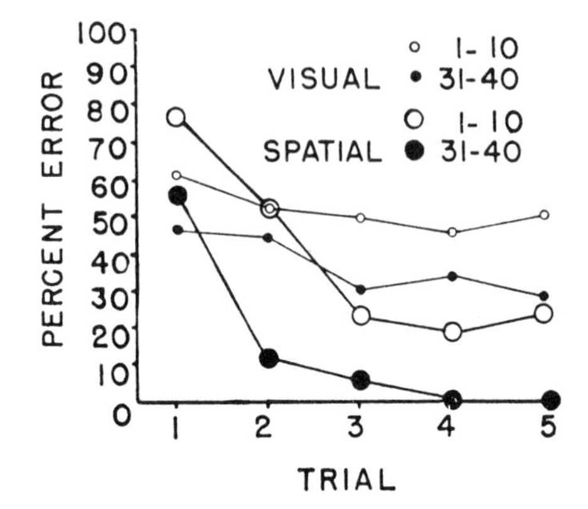

FIG. 4. DAILY VISUAL AND SPATIAL REVERSAL IN THE RAT
The curves show the performance of the rat on the first five trials of each problem in the first phase of the experiment. Performance on days 1-10 is compared with performance on days 31-40.

more marked for the rat than for the fish; the visual problem was equally difficult for the two forms in the beginning, but the spatial problem was far simpler for the rat than for the fish at the very outset. The most striking difference between the two forms, however, is to be found in the between-days change in the performance of the rat. The curves for the four consecutive 10-day periods on the visual problem are perfectly ordered, from the highest for Days 1-10 to the lowest for Days 31-40. Only the Block-1 points of the spatial curves are perfectly ordered, because errors were very rare after the first few trials of each day. As Fig. 4 shows, the between-days improvement in the performance of the rat appeared from the first trial of each day; that is, the parallel kind of improvement reported by Gatling could be observed in the very earliest stage of learning.[10] It is to be noted that the second-trial performance of the spatial rats became extremely good, but only one or two of the visual rats gave even sporadic evidence of one-trial reversal. Plots for the fish corresponding to those of Fig. 4 present four indistinguishable sets of points hovering about the 50% level.

[10] Gatling, *op. cit.*, 347-351.

The performance of fish on the spatial problem after training on the visual problem was indistinguishable from that of untrained fish on the spatial problem; that is, 60 days of visual reversal-training did not seem to affect spatial performance. Nor did 60 days of spatial reversal-training seem to affect performance on the visual problem. In the rat, by contrast, there was some indication of transfer from the first phase of the experiment to the second. Spatial reversal was largely unaffected by previous visual training, but there was a significant tendency for previous spatial training to interfere with visual performance, not only to begin with, but throughout the entire second phase of the experiment. The visual curve for Days 31-40 of the second phase was just about the same as the visual curve for Days 11-20 of the first phase. Previous spatial training was evidenced in the visual problem as a tendency to go on any given trial to the position which was correct on the preceding trial. In the second phase of the experiment, as in the first, the rat showed systematic between-days improvement in both problems, while the fish showed none.

One further bit of evidence on reversal-learning in the fish is available and should be considered.

Subjects. The Ss were five fish of the first experiment which had completed 15 visual reversals to a criterion of 17/20 errorless trials.[11]

Method. All Ss were trained on the visual problem, with daily reversal of positive and negative stimuli, for 20 days. There was no guidance, but there was opportunity for spontaneous correction of errors—each of the 20 daily trials terminated either with a correct response or with the fifth consecutive error. As in previous work with the correction-method, all pellets (up to a maximum of 20) unearned during training were given soon after the animal was returned to the colony-room after each experimental session.

Results. Under these conditions, the fish showed significant within-days improvement, but there was no between-days improvement, either in terms of initial or of repetitive errors. The level of initial error remained at about 10 and the level of repetitive error at about 20. This performance was, of course, clearly superior to the Day-1 performance of the same animals in the experiment on visual reversal to criterion (where the level of initial error was about 14 and the level of repetitive error about 30). The difference merely indicates that the preference to be reversed was stronger after training to criterion than after only a single day of training. Performance in daily visual reversal with correction but without guidance was somewhat inferior to performance in daily visual reversal with guid-

[11] Wodinsky and Bitterman, *op. cit.*, 569-576.

ance but without correction, although the difference did not quite reach statistical significance.

We now have studied habit-reversal in the fish under a variety of conditions: in visual and in spatial problems; with and without opportunity for correction; with and without systematic guidance; with training to a criterion and with reversal after a constant number of trials, in naïve *S*s and in *S*s with previous reversal-training under different circumstances. In these studies, progressive improvement in performance has been the exception rather than the rule; it was found only in a single experiment, and even then its pattern proved upon further analysis to be rather different from that found in the rat.[12] Attempts to account for progressive improvement in reversal-performance at the level of the rat, where it is the rule rather than the exception, have rested on very broad concepts—compounding of after-effects, receptor-orientation, mediated distinctiveness, and the like—which reflect the assumption that the phenomenon in question is a highly general one. Needless to say, these attempts have not been conspicuously successful. Our own results suggest that an account of progressive improvement in the course of reversal-training should be sought, not in terms of the broadest principles of learning, but in terms of processes unique to organisms of a certain class. A delineation of the class in which improvement occurs may provide a clue to the underlying processes.

II. Probability of Reinforcement

The experiments to be reported in this section were designed to study the effect of inconsistent reinforcement. The two-choice visual and spatial situations previously employed were used once more, but now one alternative was correct on only 75% of the trials and the other on the remaining 25% of trials. Our curiosity about the adjustment of the mouthbreeder to such problems was stimulated by a recent experiment of Bush and Wilson with the paradise fish, which, unfortunately, was terminated before a stable level of performance appeared.[13] Since corresponding data for the rat also were rather sketchy,[14] we trained some rats under conditions analogous to those used for the fish. At the human level, of course, atti-

[12] Wodinsky and Bitterman, *op. cit.*, 569-576.

[13] R. R. Bush and T. R. Wilson, Two-choice behavior of the paradise fish, *J. exp. Psychol.*, 51, 1956, 315-322.

[14] Egon Brunswik, Probability as a determiner of rat behavior, *J. exp. Psychol.*, 25, 1939, 175-197; another experiment, by Stanley, is summarized in R. R. Bush and Frederick Mosteller, *Stochastic Models for Learning*, 1955, 291-294.

tudes of a quite abstract kind have been found to influence behavior in such problems.[15]

Subjects. The *S*s were six experimentally naïve mouthbreeders, comparable in every respect to those used in our earlier studies, and 6 hooded rats, male, 90-120 days of age, bred in the laboratory.

Method. The apparatus used for each form was the same as that used in the comparative study of daily reversal in fish and rat which already has been described. Both groups were given 20 trials per day on the simultaneous horizontal-vertical discrimination, with response to one of the stimuli (horizontal stripes for half the animals in each group, vertical stripes for the remaining animals) reinforced on 70% of the trials, and response to the other stimulus reinforced on 30% of the trials. No spontaneous correction was permitted, but there was guidance after each unreinforced choice. The guidance, which was given by the method used in the comparative study of daily reversal, ensured that exactly 14 of each day's 20 reinforcements would be given for response to one of the stimuli and exactly 6 for response to the other stimulus. The reinforcements were balanced with respect to position, 7 for response to the 70% stimulus on the right, 7 for response to the 70% stimulus on the left, 3 for response to the 30% stimulus on the right, and 3 for response to the 30% stimulus on the left. The trials on which response to the 30% stimulus was to be reinforced were determined by the use of a table of random numbers, with the restriction that at least one such trial must occur in each block of five trials. The 70:30 training continued for 30 days, following which the ratio was changed to 100:0 (the previous 70% stimulus now being consistently reinforced). The fish, adapted to a 24-hr. feeding schedule of 20 pellets per day for some weeks prior to the beginning of the experiment, earned a pellet on each experimental trial (whether for an initial or for a guided response to the positive stimulus of that trial). The rats, on a 24-hr. feeding schedule which kept them at 80% of satiated bodily weight, were rewarded on each trial with wet mash, the small portion of each day's ration that was not consumed in the experimental situation being consumed after return to the home-cage.

Results. The curves of Fig. 5 show the percentage of each day's responses made to the more frequently reinforced stimulus during the 30 days of 70:30 training and the subsequent 10 days of 100:0 training. The fish went rapidly from a near-chance level of preference to about a 70% preference for the 70% stimulus which was maintained from Day 5 to Day 30. The percentages for the individual *S*s during the last 25 of the 30 days of 70:30 training were: 67.6, 70.6, 68.2, 70.0, and 69.8. (The results for only five fish are presented because one was lost on Day 16.) With the beginning of 100:0 training, the preference shifted rapidly upward to about the 95% level. The preference of the rats for the more frequently reinforced stimulus rose gradually from a near-chance level at the start

[15] J. J. Goodnow, Determinants of choice-distribution in two-choice situations, this JOURNAL, 68, 1955, 106-116.

of the 70:30 training to the 90% level at Day 30. There was no overlapping at all of the two groups (fish and rats) in terms of mean preference per animal for Days 26-30, during which time the percentages for the individual rats were 89, 92, 95, 91, 84, and 95. In the 10 days of 100:0

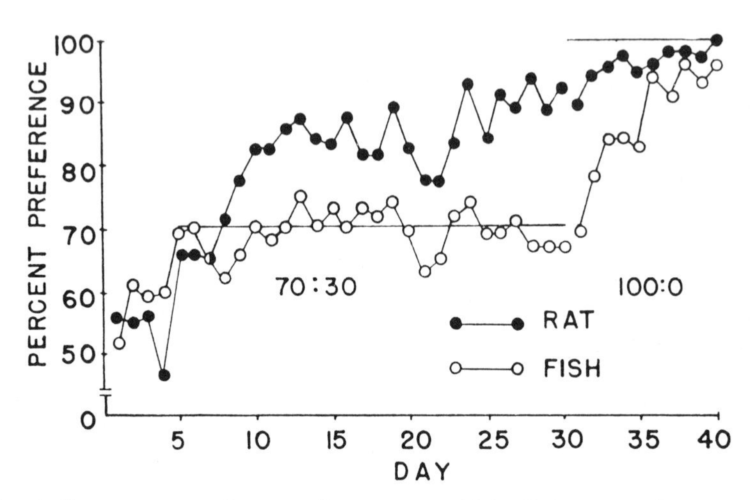

FIG. 5. PREFERENCE OF FISH AND RAT FOR THE MORE FREQUENTLY REINFORCED STIMULUS IN 70:30 AND 100:0 VISUAL PROBLEMS

training, the preference of the rats for the more frequently reinforced stimulus continued to rise gradually, as it might have done irrespective of the shift from inconsistent to consistent reinforcement.

Now a second comparison of the two groups was made in a 70:30 spatial problem.

Subjects. The Ss were the same as those used in the 70:30 visual problem.

Method. Two identical light-gray targets were used for the fish, and two identical light-gray cards for the rats. The animals were reinforced for going to one position (left for some Ss, right for the others) on 70% of the trials, and for going to the opposite position on 30% of the trials. In all other respects, the method employed in the first phase of spatial training was the same as that employed in the first phase of visual training. After 20 days, the method used for the fish was changed in only one respect: guidance following unreinforced choices was omitted (and all pellets not earned during each experimental session were given after return to the colony-room). After 10 days of 70:30 training, the rats were shifted to 60:40, still with guidance after each unreinforced choice.

Results. The curves of Fig. 6 show the performance of the two groups in terms of the percentage of response to the more frequently reinforced position. In the first phase of training, the curve for the fish rose rapidly to about the 75% level and remained there from Day 6 to Day 20. (The individual percentages for that 15-day period were: 73.3, 79.7, 78.0, 71.3, and 72.7—higher in each case, it may be noted, than for the visual problem.) Then, as soon as guidance was eliminated, the curve rose rapidly

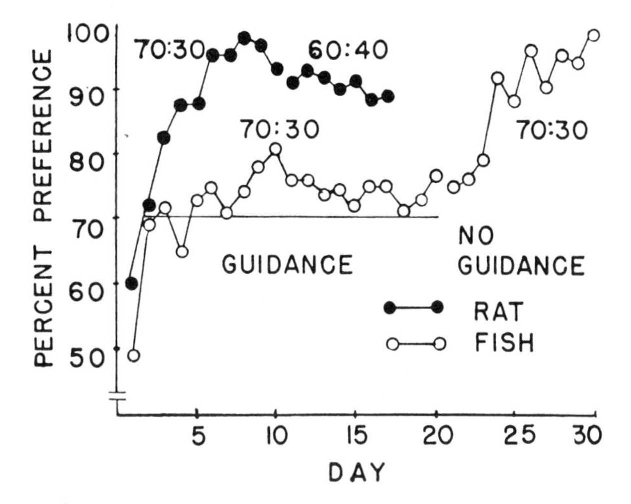

FIG. 6. PREFERENCE OF FISH AND RAT FOR THE MORE FREQUENTLY REINFORCED POSITION IN 70:30 AND 60:40 SPATIAL PROBLEMS

to the 100% level. The curve for the rats in the first phase rose rapidly to the 95% level, and it did not fall below 90% even when the ratio of reinforcement was changed from 70:30 to 60:40. During the last five days of 70:30 training, the percentages for individual rats were 94, 94, 95, 98, 92, and 99.

In both experiments, then—one with a 70:30 visual problem and a second with a 70:30 spatial problem—the fish rapidly established and maintained a level of preference for the more frequently reinforced stimulus which closely approximated the probability of reinforcement. The fact that 'matching' disappeared as soon as guidance was eliminated in the second phase of the spatial problem suggests that it was a product, not of unreinforced response to the 70% stimulus, but of reinforced response to the 30% stimulus. With guidance eliminated, rewarded response to the 30% stimulus became rare, preference for the 70% stimulus increased, rewarded response to the 30% stimulus became even more rare, and so forth, until the preference for the 70% stimulus reached the 100% level. (Unfortunately, the results of Bush and Wilson for the paradise fish—on a 75:25 spatial problem without guidance or opportunity for spontaneous correction—do not warrant comparison with our own, because the training was not carried far enough to establish an asymptote.[16]) By contrast, the rats of our two experiments gave no indication of matching, even under conditions which ensured that 30% of the reinforced responses would be made to the 30% stimulus. (In Brunswik's well-known experiment with the rat, frequency of reinforcement on each side of a T-maze was fixed—by the use of correction rather than guidance—but again the training was not carried far enough.[17] In Stanley's T-maze experiment, the rats of a 75:25 group developed a very strong preference for the 75% side, but under conditions such as those under which our fish also failed to match—that is, with no fixed level of reinforced response to each side.[18])

It is perhaps fruitless to speculate on the basis for the difference between fish and rat which is revealed in our two experiments on probability of reinforcement. The generality of the difference must first be tested, with other ratios of reinforcement and at other levels of drive. Perhaps the close correspondence between preference-ratio and reinforcement-ratio which here appears in the fish is mere coincidence; it may be, for example, that an 85:15 visual problem also would have given a 70:30 preference. One is tempted, however, to assume that the correspondence in the fish is indeed genuine, and to wonder why there is none in the rat. Does the difference in results for the two animals merely reflect the failure of the fish to discriminate between two-choice and guided trials? Ramond's data suggest that the rat makes such a discrimination readily. Its behavior on test-trials which afford a choice between two stimuli, both reinforced, soon tends to become independent of the relative frequency with which the stimuli have been reinforced previously in individual presentations.[19] If choice-trials and guided trials are functionally equivalent for the mouthbreeder, one may be certain that no mere sensory deficiency is responsible;

---

[16] Bush and Wilson, *op. cit.*, 315-322.

[17] Brunswik, *op. cit.*, 175-197.

[18] Bush and Mosteller, *op. cit.*, 291-294. W. K. Estes has reported some further data for the rat in a paper which appeared after the present one had gone to press (Of models and men, *Amer. Psychologist,* 12, 1957, 609-616). Over the last 28 days of a 56-day period of 75: 25 spatial training (correction, one trial per day), the mean preference of the 16 *Ss* for the 75% side closely approximated 75%, but individual performances gave no very good evidence of matching. In the final block of 16 trials, 6 *Ss* made 11-13 responses to the 75% side, 7 made 14-16 responses to the 75% side (of these at least 5 went always to that side), and 3 went predominantly to the 25% side. We are indebted to Dr. Estes for this supplementary information.

[19] C. K. Ramond, Performance in selective learning as a function of hunger, *J. exp. Psychol.,* 48, 1954, 265-270.

our exploration of the animal's visual capacities convinces us that they are quite superior to those of the rat. Another possibility worth considering is that different mechanisms of response-selection are at work in the two forms—an all-or-none mechanism in the rat and a continuous one in the fish.

III. Motivation

It very soon becomes necessary in an investigation of this kind to study the effects of drive-level. For one thing, our experimental control of the fish is based on the hunger-drive, and we should like, therefore, to know something about its properties. In all of our experiments we have utilized a highly arbitrary feeding schedule, selected intuitively, and retained simply because it seemed to yield stable levels of performance. The possibility must be faced that the schedule makes for a quite unrepresentative level of motivation and, consequently, for a quite unrepresentative kind of functioning. Furthermore, drive-level (at least in theory) is an important variable in the learning of the rat, our phylogenetic point of departure, and for that reason, too, we are led to study its effects in the fish. Here we describe a preliminary effort to achieve control of hunger in the fish. Taking as our measure the amount of food required for satiation, we attempted to map the relation between drive and deprivation-time.

Subjects. The *Ss* were mouthbreeders, male and female, six previously trained in the experiment on daily visual reversal to criterion, and six naïve animals of similar characteristics drawn from the same stock-tanks.

Method. The experienced animals were studied in an instrumental situation. The apparatus employed was that used for the experiments on discriminative learning, but one of the openings in the housing was permanently covered by a black disk. Effectively, then, we returned to the single-lever situation of our earliest work with the mouthbreeder.[20] There were several retraining sessions, with the target continuously in the water and each response rewarded with a pellet of food, until a stable rate of performance was attained. Then the experimental work was begun. The animals were satiated to a criterion of no response in 90 sec. on Day 0, and repeatedly thereafter at intervals of one, two, and four days, each of the six possible orders of deprivation-periods being used with a different *S*. For example, one *S* was satiated on Days 1, 3, and 7 (representing deprivation-periods of one, two, and four days, respectively); another *S* was satiated on Days 1, 5, and 7 (representing deprivation-periods of one, four, and two days, respectively); and so forth. In the second week, the order of deprivation-periods for each *S* was reversed. In the third week, all *Ss* were deprived for seven days and satiated on Day 21. In the fourth week, determinations were made once more at one and two days of deprivation, half the *Ss* being satiated on Days 22 and 24 (one and two days of deprivation), and the other half on Days 23 and 24 (two and one days of deprivation).

[20] J. V. Haralson and M. E. Bitterman, A lever-depression apparatus for the study of learning in fish, this Journal, 63, 1950, 250-256.

For the naïve fish, housed as were the experienced fish in individual tanks, a simple consummatory technique was used. After they had become adjusted to *E* and were readily taking pellets of food which he dropped into the water, the same deprivation-schedule as was used in the instrumental condition was begun. *E* would simply drop a pellet into the water in a given place, then a second pellet as soon as the first was taken, and so forth. If a pellet was untouched for 30 sec., a second was dropped; if no pellet was taken in the next 30 sec., a third was dropped; if no pellet was taken in the next 30 sec., the test was discontinued. In the simple consummatory condition as well as in the instrumental condition, therefore, the criterion of satiation was failure to respond in 90 sec. (Highly correlated with the criterion actually employed in the consummatory condition was a tendency on the part of the animals to eject pellets stored in the mouth.)

Results. The number of *S*s employed is so small that nothing much can be gained from analysis of individual satiation-curves. Most are negatively accelerated; others terminate rather abruptly after rising at a fairly constant rate. For some *S*s, initial rate of response increases with deprivation-time; for other *S*s, the initial rate tends to remain rather constant, with time of deprivation affecting only the point at which the curve begins to break.

Mean intake as a function of deprivation-time is plotted in Fig. 7. The one-, two-, and four-day points are based on the combined data of the first and second weeks, during which order of testing was perfectly balanced. The seven-day point is, of course, based only on the single test which followed. The results for one and two days of deprivation during the fourth week are, however, so close to the values plotted that the seven-day point probably is just about where it should be. We may conclude, in any case, that the major change in intake occurs during the first few days of deprivation. The pattern of change happens to be strikingly similar to that found by Stellar and Hill for water-intake as a function of water-deprivation in the rat.[21]

The opinion is common that the health of fish such as ours is impaired by 'overfeeding.' For that reason, we have sought in the past to control the motivation of our *S*s by a procedure which did not involve satiation. In this study, however, there was no obvious impairment of health in the course of repeated satiation, even daily satiation over a period of several weeks. Perhaps it will be possible, after all, to take advantage of the convenient point of reference which satiation constitutes in controlling the motivation of the fish. In any event, satiation-tests may provide a basis for evaluating the effects of other schedules of maintenance.

---

[21] Eliot Stellar and J. H. Hill, The rat's rate of drinking as a function of water deprivation, *J. compar physiol Psychol.*, 45, 1952, 96-102.

SUMMARY

This paper reports a series of experiments in comparative psychology. The animals studied were the fish and the rat. In the first set of experiments, on the course of habit-reversal (visual and spatial) under a variety of conditions, the progressive improvement in performance over a series of reversals which is the rule in the rat was found to be rare in the fish. Where progressive improvement did appear in the fish, furthermore, its pattern was quite different from that shown by the rat. A second set of experiments dealt with the effect of inconsistent reinforcement. When response to one of two stimuli (visual or spatial) was reinforced on 70%

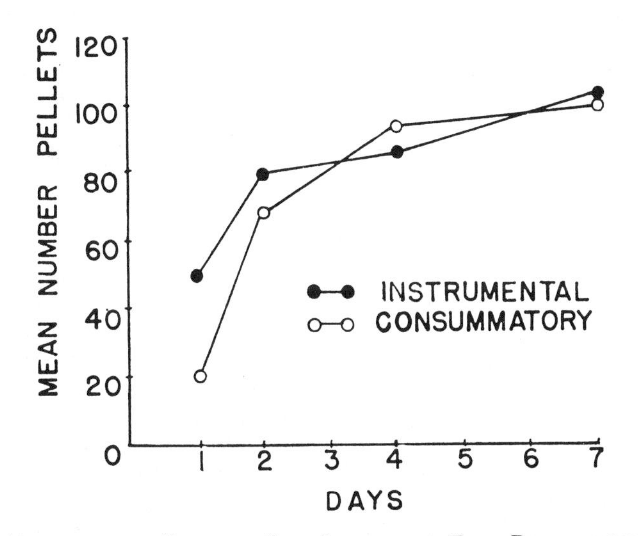

FIG. 7. THE RELATION BETWEEN FOOD-INTAKE AND FOOD DEPRIVATION IN THE FISH AS DETERMINED BY AN INSTRUMENTAL AND BY A SIMPLE CONSUMMATORY TECHNIQUE

of the trials and response to the other stimulus reinforced on 30% of the trials, the preference-ratio of the fish tended to match the reinforcement-ratio, but that of the rat did not. In both sets of experiments, then, marked functional differences between the two forms were found which seem to call for further investigation. A final experiment was designed to explore the relation between food-intake and food-deprivation in the fish as a prelude to an examination of the effect of drive-level. A negatively accelerated function was obtained which closely resembles that for water-intake and water-deprivation in the rat.

22

Reprinted from *Am. Psychologist* **16**:681–684 (1961)

THE MISBEHAVIOR OF ORGANISMS

KELLER BRELAND AND MARIAN BRELAND

Animal Behavior Enterprises, Hot Springs, Arkansas

THERE seems to be a continuing realization by psychologists that perhaps the white rat cannot reveal everything there is to know about behavior. Among the voices raised on this topic, Beach (1950) has emphasized the necessity of widening the range of species subjected to experimental techniques and conditions. However, psychologists as a whole do not seem to be heeding these admonitions, as Whalen (1961) has pointed out.

Perhaps this reluctance is due in part to some dark precognition of what they might find in such investigations, for the ethologists Lorenz (1950, p. 233) and Tinbergen (1951, p. 6) have warned that if psychologists are to understand and predict the behavior of organisms, it is essential that they become thoroughly familiar with the instinctive behavior patterns of each new species they essay to study. Of course, the Watsonian or neobehavioristically oriented experimenter is apt to consider "instinct" an ugly word. He tends to class it with Hebb's (1960) other "seditious notions" which were discarded in the behavioristic revolution, and he may have some premonition that he will encounter this bete noir in extending the range of species and situations studied.

We can assure him that his apprehensions are well grounded. In our attempt to extend a behavioristically oriented approach to the engineering control of animal behavior by operant conditioning techniques, we have fought a running battle with the seditious notion of instinct.[1] It might be of some interest to the psychologist to know how the battle is going and to learn something about the nature of the adversary he is likely to meet if and when he tackles new species in new learning situations.

Our first report (Breland & Breland, 1951) in the *American Psychologist*, concerning our experiences in controlling animal behavior, was wholly affirmative and optimistic, saying in essence that the principles derived from the laboratory could be applied to the extensive control of behavior under nonlaboratory conditions throughout a considerable segment of the phylogenetic scale.

When we began this work, it was our aim to see if the science would work beyond the laboratory, to determine if animal psychology could stand on its own feet as an engineering discipline. These aims have been realized. We have controlled a wide range of animal behavior and have made use of the great popular appeal of animals to make it an economically feasible project. Conditioned behavior has been exhibited at various municipal zoos and museums of natural history and has been used for department store displays, for fair and trade convention exhibits, for entertainment at tourist attractions, on television shows, and in the production of television commercials. Thirty-eight species, totaling over 6,000 individual animals, have been conditioned, and we have dared to tackle such unlikely subjects as reindeer, cockatoos, raccoons, porpoises, and whales.

Emboldened by this consistent reinforcement, we have ventured further and further from the security of the Skinner box. However, in this cavalier extrapolation, we have run afoul of a persistent pattern of discomforting failures. These failures, although disconcertingly frequent and seemingly diverse, fall into a very interesting pattern. They all represent breakdowns of conditioned operant behavior. From a great number of such experiences, we have selected, more or less at random, the following examples.

The first instance of our discomfiture might be entitled, What Makes Sammy Dance? In the exhibit in which this occurred, the casual observer sees a grown bantam chicken emerge from a retaining compartment when the door automatically opens. The chicken walks over about 3 feet, pulls a rubber loop on a small box which starts a repeated auditory stimulus pattern (a four-note tune). The chicken then steps up onto an 18-inch, slightly raised disc, thereby closing a timer switch, and scratches vigorously, round and round, over the disc for 15 seconds, at the rate of about two scratches per second until the automatic feeder fires in the retaining compartment. The chicken goes into the compartment to eat, thereby auto-

[1] In view of the fact that instinctive behaviors may be common to many zoological species, we consider *species specific* to be a sanitized misnomer, and prefer the possibly septic adjective *instinctive*.

matically shutting the door. The popular interpretation of this behavior pattern is that the chicken has turned on the "juke box" and "dances."

The development of this behavioral exhibit was wholly unplanned. In the attempt to create quite another type of demonstration which required a chicken simply to stand on a platform for 12–15 seconds, we found that over 50% developed a very strong and pronounced scratch pattern, which tended to increase in persistence as the time interval was lengthened. (Another 25% or so developed other behaviors—pecking at spots, etc.) However, we were able to change our plans so as to make use of the scratch pattern, and the result was the "dancing chicken" exhibit described above.

In this exhibit the only real contingency for reinforcement is that the chicken must depress the platform for 15 seconds. In the course of a performing day (about 3 hours for each chicken) a chicken may turn out over 10,000 unnecessary, virtually identical responses. Operant behaviorists would probably have little hesitancy in labeling this an example of Skinnerian "superstition" (Skinner, 1948) or "mediating" behavior, and we list it first to whet their explanatory appetite.

However, a second instance involving a raccoon does not fit so neatly into this paradigm. The response concerned the manipulation of money by the raccoon (who has "hands" rather similar to those of the primates). The contingency for reinforcement was picking up the coins and depositing them in a 5-inch metal box.

Raccoons condition readily, have good appetites, and this one was quite tame and an eager subject. We anticipated no trouble. Conditioning him to pick up the first coin was simple. We started out by reinforcing him for picking up a single coin. Then the metal container was introduced, with the requirement that he drop the coin into the container. Here we ran into the first bit of difficulty: he seemed to have a great deal of trouble letting go of the coin. He would rub it up against the inside of the container, pull it back out, and clutch it firmly for several seconds. However, he would finally turn it loose and receive his food reinforcement. Then the final contingency: we put him on a ratio of 2, requiring that he pick up both coins and put them in the container.

Now the raccoon really had problems (and so did we). Not only could he not let go of the coins, but he spent seconds, even minutes, rubbing them together (in a most miserly fashion), and dipping them into the container. He carried on this behavior to such an extent that the practical application we had in mind—a display featuring a raccoon putting money in a piggy bank—simply was not feasible. The rubbing behavior became worse and worse as time went on, in spite of nonreinforcement.

For the third instance, we return to the gallinaceous birds. The observer sees a hopper full of oval plastic capsules which contain small toys, charms, and the like. When the S_D (a light) is presented to the chicken, she pulls a rubber loop which releases one of these capsules onto a slide, about 16 inches long, inclined at about 30 degrees. The capsule rolls down the slide and comes to rest near the end. Here one or two sharp, straight pecks by the chicken will knock it forward off the slide and out to the observer, and the chicken is then reinforced by an automatic feeder. This is all very well—most chickens are able to master these contingencies in short order. The loop pulling presents no problems; she then has only to peck the capsule off the slide to get her reinforcement.

However, a good 20% of all chickens tried on this set of contingencies fail to make the grade. After they have pecked a few capsules off the slide, they begin to grab at the capsules and drag them backwards into the cage. Here they pound them up and down on the floor of the cage. Of course, this results in no reinforcement for the chicken, and yet some chickens will pull in over half of all the capsules presented to them.

Almost always this problem behavior does not appear until after the capsules begin to move down the slide. Conditioning is begun with stationary capsules placed by the experimenter. When the pecking behavior becomes strong enough, so that the chicken is knocking them off the slide and getting reinforced consistently, the loop pulling is conditioned to the light. The capsules then come rolling down the slide to the chicken. Here most chickens, who before did not have this tendency, will start grabbing and shaking.

The fourth incident also concerns a chicken. Here the observer sees a chicken in a cage about 4 feet long which is placed alongside a miniature baseball field. The reason for the cage is the interesting part. At one end of the cage is an automatic electric feed hopper. At the other is an opening through which the chicken can reach and pull a loop on a bat. If she pulls the loop hard enough the bat (solenoid operated) will swing, knocking a small baseball up the playing field. If it gets past the miniature toy players on the field and hits the back fence, the chicken is automatically reinforced with food at the other end of the cage. If it does not go far enough, or hits one of the players, she tries again. This results in behavior on an irregular ratio. When the feeder sounds, she then runs down the length of the cage and eats.

Our problems began when we tried to remove the cage for photography. Chickens that had been well conditioned in this behavior became wildly excited when the ball started to move. They would jump up on the playing field, chase the ball all over the field, even knock it off on the floor and chase it around, pecking it in every direction, although they had never had access to the ball before. This behavior was so persistent and so disruptive, in spite of the fact that it was never reinforced, that we had to reinstate the cage.

The last instance we shall relate in detail is one of the most annoying and baffling for a good behaviorist. Here a pig was conditioned to pick up large wooden coins and deposit them in a large "piggy bank." The coins were placed several feet from the bank and the pig required to carry them to the bank and deposit them, usually four or five coins for one reinforcement. (Of course, we started out with one coin, near the bank.)

Pigs condition very rapidly, they have no trouble taking ratios, they have ravenous appetites (naturally), and in many ways are among the most tractable animals we have worked with. However, this particular problem behavior developed in pig after pig, usually after a period of weeks or months, getting worse every day. At first the pig would eagerly pick up one dollar, carry it to the bank, run back, get another, carry it rapidly and neatly, and so on, until the ratio was complete. Thereafter, over a period of weeks the behavior would become slower and slower. He might run over eagerly for each dollar, but on the way back, instead of carrying the dollar and depositing it simply and cleanly, he would repeatedly drop it, root it, drop it again, root it along the way, pick it up, toss it up in the air, drop it, root it some more, and so on.

We thought this behavior might simply be the dilly-dallying of an animal on a low drive. However, the behavior persisted and gained in strength in spite of a severely increased drive—he finally went through the ratios so slowly that he did not get enough to eat in the course of a day. Finally it would take the pig about 10 minutes to transport four coins a distance of about 6 feet. This problem behavior developed repeatedly in successive pigs.

There have also been other instances: hamsters that stopped working in a glass case after four or five reinforcements, porpoises and whales that swallow their manipulanda (balls and inner tubes), cats that will not leave the area of the feeder, rabbits that will not go to the feeder, the great difficulty in many species of conditioning vocalization with food reinforcement, problems in conditioning a kick in a cow, the failure to get appreciably increased effort out of the ungulates with increased drive, and so on. These we shall not dwell on in detail, nor shall we discuss how they might be overcome.

These egregious failures came as a rather considerable shock to us, for there was nothing in our background in behaviorism to prepare us for such gross inabilities to predict and control the behavior of animals with which we had been working for years.

The examples listed we feel represent a clear and utter failure of conditioning theory. They are far from what one would normally expect on the basis of the theory alone. Furthermore, they are definite, observable; the diagnosis of theory failure does not depend on subtle statistical interpretations or on semantic legerdemain—the animal simply does not do what he has been conditioned to do.

It seems perfectly clear that, with the possible exception of the dancing chicken, which could conceivably, as we have said, be explained in terms of Skinner's superstition paradigm, the other instances do not fit the behavioristic way of thinking. Here we have animals, after having been conditioned to a specific learned response, gradually drifting into behaviors that are entirely different from those which were conditioned. Moreover, it can easily be seen that these particular behaviors to which the animals drift are clear-cut examples of instinctive behaviors having to do with the natural food getting behaviors of the particular species.

The dancing chicken is exhibiting the gallinaceous birds' scratch pattern that in nature often precedes ingestion. The chicken that hammers capsules is obviously exhibiting instinctive behavior having to do with breaking open of seed pods or the killing of insects, grubs, etc. The raccoon is demonstrating so-called "washing behavior." The rubbing and washing response may result, for example, in the removal of the exoskeleton of a crayfish. The pig is rooting or shaking—behaviors which are strongly built into this species and are connected with the food getting repertoire.

These patterns to which the animals drift require greater physical output and therefore are a violation of the so-called "law of least effort." And most damaging of all, they stretch out the time required for reinforcement when nothing in the experimental setup requires them to do so. They have only to do the little tidbit of behavior to which they were conditioned—for example, pick up the coin and put it in the container—to get reinforced immediately. Instead, they drag the process out for a matter of minutes when there is nothing in the contingency which forces them to do this. Moreover, increasing the drive merely intensifies this effect.

It seems obvious that these animals are trapped by strong instinctive behaviors, and clearly we have here a demonstration of the prepotency of such behavior patterns over those which have been conditioned.

We have termed this phenomenon "instinctive drift." The general principle seems to be that wherever an animal has strong instinctive behaviors in the area of the conditioned response, after continued running the organism will drift toward the instinctive behavior to the detriment of the conditioned behavior and even to the delay or preclusion of the reinforcement. In a very boiled-down, simplified form, it might be stated as "learned behavior drifts toward instinctive behavior."

All this, of course, is not to disparage the use of conditioning techniques, but is intended as a demonstration that there are definite weaknesses in the philosophy underlying these techniques. The pointing out of such weaknesses should make possible a worthwhile revision in behavior theory.

The notion of instinct has now become one of our basic concepts in an effort to make sense of the welter of observations which confront us. When behaviorism tossed out instinct, it is our feeling that some of its power of prediction and control were lost with it. From the foregoing examples, it appears that although it was easy to banish the Instinctivists from the science during the Behavioristic Revolution, it was not possible to banish instinct so easily.

And if, as Hebb suggests, it is advisable to reconsider those things that behaviorism explicitly threw out, perhaps it might likewise be advisable to examine what they tacitly brought in—the hidden assumptions which led most disastrously to these breakdowns in the theory.

Three of the most important of these tacit assumptions seem to us to be: that the animal comes to the laboratory as a virtual *tabula rasa*, that species differences are insignificant, and that all responses are about equally conditionable to all stimuli.

It is obvious, we feel, from the foregoing account, that these assumptions are no longer tenable. After 14 years of continuous conditioning and observation of thousands of animals, it is our reluctant conclusion that the behavior of any species cannot be adequately understood, predicted, or controlled without knowledge of its instinctive patterns, evolutionary history, and ecological niche.

In spite of our early successes with the application of behavioristically oriented conditioning theory, we readily admit now that ethological facts and attitudes in recent years have done more to advance our practical control of animal behavior than recent reports from American "learning labs."

Moreover, as we have recently discovered, if one begins with evolution and instinct as the basic format for the science, a very illuminating viewpoint can be developed which leads naturally to a drastically revised and simplified conceptual framework of startling explanatory power (to be reported elsewhere).

It is hoped that this playback on the theory will be behavioral technology's partial repayment to the academic science whose impeccable empiricism we have used so extensively.

REFERENCES

BEACH, F. A. The snark was a boojum. *Amer. Psychologist*, 1950, **5**, 115–124.

BRELAND, K., & BRELAND, M. A field of applied animal psychology. *Amer. Psychologist*, 1951, **6**, 202–204.

HEBB, D. O. The American revolution. *Amer. Psychologist*, 1960, **15**, 735–745.

LORENZ, K. Innate behaviour patterns. In *Symposia of the Society for Experimental Biology*. No. 4. *Physiological mechanisms in animal behaviour*. New York: Academic Press, 1950.

SKINNER, B. F. Superstition in the pigeon. *J. exp. Psychol.*, 1948, **38**, 168–172.

TINBERGEN, N. *The study of instinct*. Oxford: Clarendon, 1951.

WHALEN, R. E. Comparative psychology. *Amer. Psychologist*, 1961, **16**, 84.

Editor's Comments on Papers 23, 24, and 25

Although the study of learning in animals began as a search for animal intelligence, major effort has been devoted to showing that seemingly complex learning processes can be broken down into simpler components. What appears to be supremely intelligent activity may in fact represent species-typical responses to particular stimuli, the concatenation of relatively simple learning processes, and/or the misinterpretation by a human observer of the relationship between an animal and its environment. Have investigators of animal behavior figuratively slit their throats with Occam's razor by allowing their skepticism to blind them to important instances of animal intelligence? The debate over this issue is, as it has been for many years, lively, loud, and laden with emotion.

Wolfgang Köhler's book *The Mentality of Apes*, from which Paper 23 is reprinted, provided a variety of observations of the apparently intelligent activity of chimpanzees. For Köhler, a pacesetter among Gestalt psychologists, many of the tasks accomplished sucessfully by his chimps were demonstrations of "insight," a sudden reorganization of the perceptual field. As with most other attributions of a high level of intelligence to nonhumans, Köhler's interpretation has been challenged (for example Schiller, 1952). However, more recent observations of chimpanzees in the wild suggest that there are aspects of chimp behavior that cannot be reduced to implementation of species-typical motor responses or trial-and-error learning (see Warren, 1976, for an excellent review and critique).

The influential learning theorist and behaviorist Edward C. Tolman believed that animals do more than react to stimuli in their immediate environment, and that they learn about orderly relations in their worlds on the basis of confirmation or disconfirmation of expectancies. He argued that animals do not merely react to stimuli in their sensory field, but, rather, that they are guided by representations they form of the world with which they commerce. Tolman believed these representations are largely spatial in nature, comprising "cognitive maps." Paper 24 by Tolman and Honzik reports a demonstration of "insight" in rats, one to which the cognitive map theory is easily applied. Tolman's contributions to our understanding of animal learning were substantial and his 1932 book, *Purposive Behavior in Animals and Men,* remains an entertaining, informative, and stimulating source.

In his early studies of cats learning to escape from puzzle-boxes (Paper 6), Thorndike noted that cats that had previously learned to escape from several different boxes learned to escape from a new box more rapidly than did cats without such previous experience. He argued that the animal's "tendency to pay attention to what it is doing gets strengthened, and this is something which may properly be called a change in degree of intelligence." A similar phenomenon was discovered by Harry Harlow and is described in Paper 25 reprinted here. Harlow studied rhesus monkeys trained in a series of discrimination tasks (he also reports data from some human children). The improvement his animals showed over successive tasks suggested that learning sets, like simpler learned habits, are acquired gradually. Since Harlow's initial work, learning-set formation has been extensively studied. There are large species differences in learning-set formation (Warren, 1965), but the pattern of differences is a complicated one. For example, while some birds do very poorly in learning-set tasks, others do quite well (Kamil and Hunter, 1970; Hunter and Kamil, 1971).

Among Harlow's many other contributions to our understanding of how animals change with experience are his studies of attachment behavior. For a summary of much of this important work, see Harlow (1971).

REFERENCES

Harlow, H. G., 1971, *Learning to Love,* Albion, San Francisco.

Hunter, M. W., III, and A. C. Kamil, 1971, Object discrimination learning set and hypothesis behavior in the northern bluejay (*Cyanocitta* aristata), *Psychon. Sci.* **22**:271–273.

Kamil, A. C., and M. W., Hunter III, 1970, Performance on object-discrimination learning set by the greater hill myna (*Gracula religiosa*), *J. Comp. Physiol. Psychol.* **73**:68–73.
Schiller, P. H., 1952, Innate constituents of complex responses in primates. *Psychol. Rev.* **59**:177–191.
Tolman, E. C., 1932, *Purposive Behavior in Animals and Men,* Appleton-Century-Crofts, New York.
Warren, J. M., 1965, The comparative psychology of learning, *Ann. Rev. Psychol.* **16**:95–118.
Warren, J. M., 1976, Tool use in mammals, in *Evolution of Brain and Behavior in Vertebrates,* R. B. Masterton, et al., Lawrence Erlbaum Associates, Hillsdale, N. J., pp. 407–424.

23

Reprinted from pp. 103–104, 130–138 of *The Mentality of Apes,* E. Winter, trans., Harcourt, Brace & Co., New York, 1925, 336 p.

THE MAKING OF IMPLEMENTS

W. Köhler

IN all intelligence tests of the kind applied here, one circumstance is always repeated: if one *single part* of the "solutions" in the proceedings we have discussed (e.g. the beginning) be considered by itself and without any relation to the remaining parts, it represents behaviour which, in the face of the task, i.e. the attaining of the objective, seems to be either quite irrelevant or else to lead in the opposite direction. It is only when we consider the *whole course* of the experiment, or, as later, at least considerable parts instead of those sections, that this whole seems to have some significance, and each of the sections previously isolated takes on a meaning as a part of *this whole.*

[Only for *one* section is this not true, namely, the last, in which each time, as a result of all the preceding acts, the objective is just grabbed. This section, of course, has significance even when considered alone.]

The above is not philosophy, it is not even a theory of the actual proceedings, but a simple statement with which anyone must agree who distinguishes between "significant in relation to a task" and "non-significant", or "meaningless", and who will consider examples objectively.

When a man or an animal takes a roundabout way (in the ordinary sense of the word) to his objective the beginning—considered by itself only and regardless of the further course of the experiment—contains at least one component which must seem *irrelevant; in very indirect routes there are usually some parts of the way which, when considered alone, seem in contradiction to the purpose of the task*, because they lead away from the goal. If the subdivision in thought be dropped, the *whole* detour, and each part of it considered separately, becomes full of meaning in *that* experiment.

If, with the aid of a stick, I fetch an objective, otherwise unattainable, the same applies. The act of picking up the stick lying near me, considered separately, and without reference to the further conduct of the experiment, the use of the stick as an implement, is quite irrelevant in relation to the objective. It does not, keeping always of course this supposed isolation in mind, bring me any nearer to my objective, and is, therefore, meaningless in this situation. But, on the other hand, considered as part of the total proceeding, it has the significance of a necessary part of a meaningful whole.

The same reflections applied to other "indirect ways" (used figuratively) illustrate the same circumstances, and that is why we call them all "indirect ways".

Thus matters stand from the point of view of a purely objective consideration. How the chimpanzee arrives at his actual solutions in such cases is another question, which we cannot investigate until later. But the purpose of all further experiments is to set up situations in which the possible solution becomes more complicated. Thus the subjective consideration of the course of the experiment in parts will show up, ever more clearly and in greater number, sections which, *taken separately, are meaningless in relation to the task*, but which become *significant again*, when they are considered *as a part of the whole.* How does the chimpanzee behave in such situations?

[*Editor's Note:* Material has been omitted at this point.]

Are the two sticks ever combined so as to become technically useful? This time Sultan is the subject of experiment (20.4). His sticks are two hollow, but firm, bamboo rods, such as the animals often use for pulling along fruit. The one is so much smaller than the other, that it can be pushed in at either end quite easily. Beyond the bars lies the objective, just so far away that the animal cannot reach it with either rod. They are about the same length. Nevertheless, he takes great pains to try to reach it with one stick or the other, even pushing his right shoulder through the bars[1]. When everything proves futile, Sultan commits a "bad error", or, more clearly, a great stupidity, such as he made sometimes on other occasions. He pulls a box from the back of the room towards the bars; true, he pushes it away again at once as it is useless, or rather, actually in the way. Immediately afterwards, he does something which, although practically useless, must be counted among the "good errors": he pushes one of the sticks out as far as it will go, then takes the second, and with it pokes the first one cautiously towards the objective, pushing it carefully from the nearer end and thus slowly urging it towards the fruit. This does not always succeed, but if he has got pretty close in this way, he takes even greater precaution; he pushes very gently, watches the movements of the stick that is lying on the ground, and actually touches the objective with its tip. Thus, all of a sudden, for the first time, the contact "animal-objective" has been established, and Sultan visibly feels (we humans can sympathize) a certain satisfaction in having even so much power over the fruit that he can touch and slightly move it by pushing the stick. The proceeding is repeated; when the animal has pushed the stick on the ground so far out that he cannot possibly get it back by himself[2], it is given back to him. But although, in trying to steer it cautiously, he puts the stick in his hand exactly to the cut (i.e. the opening) of the stick on the ground, and although one might think that doing so would suggest the possibility of pushing one stick into the other, there is no indication whatever of such a practically valuable solution. Finally, the observer gives the animal some help by putting one

[1] This is not in contradiction to the statement made above; in order not to discourage the animal from the very beginning, I put the objective only just out of reach of the single stick.

[2] The way in which he does that is reported on p. 180.

finger into the opening of one stick under the animal's nose (without pointing to the other stick at all). This has no effect; Sultan, as before, pushes one stick with the other towards the objective, and as this pseudo-solution does not satisfy him any longer, he abandons his efforts altogether, and does not even pick up the sticks when they are both again thrown through the bars to him. The experiment has lasted over an hour, and is stopped for the present, as it seems hopeless, carried out like this. As we intend to take it up again after a while, Sultan is left in possession of his sticks; the keeper is left there to watch him.

Keeper's report: "Sultan first of all squats indifferently on the box, which has been left standing a little back from the railings; then he gets up, picks up the two sticks, sits down again on the box and plays carelessly with them. While doing this, it happens that he finds himself holding one rod in either hand in such a way that they lie in a straight line; he pushes the thinner one a little way into the opening of the thicker, jumps up and is already on the run towards the railings, to which he has up to now half turned his back, and begins to draw a banana towards him with the double stick. I call the master: meanwhile, one of the animal's rods has fallen out of the other, as he has pushed one of them only a little way into the other; whereupon he connects them again"[1].

The keeper's report covers a period of scarcely five

[1] The keeper's tale seems acceptable to me, especially as, upon inquiries, he emphasized the fact that Sultan had first of all connected the sticks in play and without considering the objective (his task). The animals are constantly poking about with straws and small sticks in holes and cracks in their play, so that it would be more astonishing if Sultan had never done this, while playing about with the two sticks. There need be no suspicion that the keeper quickly "trained the animal"; the man would never dare it. If anybody continues to doubt, even that does not matter, for Sultan continually not only performs this act but shows that he realizes its meaning.

PLATE III. SULTAN MAKING A DOUBLE-STICK

minutes, which had elapsed since stopping the experiment. Called by the man, I continued observation myself: Sultan is squatting at the bars, holding out one stick, and, at its end, a second bigger one, which is on the point of falling off. It does fall. Sultan pulls it to him and forthwith, with the greatest assurance, pushes the thinner one in again, so that it is firmly wedged, and fetches a fruit with the lengthened implement. But the bigger tube selected is a little too big, and so it slips from the end of the thinner one several times; each time Sultan rejoins the tubes immediately by holding the bigger one towards himself in the left and the thinner one in his right hand and a little backwards, and then sliding one into the other.[1] (Plate III.) The proceeding seems to please him immensely; he is very lively, pulls all the fruit, one after the other, towards the railings, without taking time to eat it, and when I disconnect the double-stick he puts it together again at once, and draws any distant objects whatever to the bars.

The next day the test is repeated; Sultan begins with the proceeding which is in practice useless, but after he has pushed one of the tubes forward with the other for a few seconds, he again takes up both, quickly puts one into the other, and attains his objective with the double stick.

(1.5) The objective lies in front of the railings, still farther away; Sultan has three tubes to resort to, the two bigger ones fitting over either end of the third. He tries to reach his objective with two tubes, as before; as the outer one keeps falling off, he takes distinct pains to push the thinner stick farther into the bigger one. Contrary to expectation, he actually attains his objective with the

[1] The illustration is culled from a cinematograph film that was taken a month later in other surroundings, more suitable for photographs.

double tube, and pulls it to him. The long tool sometimes gets into his way when doing this, by its farther end getting caught between the railings, when being moved obliquely, so the animal quickly separates it into its parts, and finishes the task with one tube only. From now on, he does this every time when the objective is so close that *one* stick is sufficient, and the double-stick awkward. The new objective is placed still farther away. In consequence, Sultan tries which of the bigger tubes is more useful when joined to the thin one; for they do not differ very much in length (64 and 70 cms.), and, of course, the animal does not lay them together in order to compare their lengths. *Sultan never tries to join the two bigger tubes;* once he puts them opposite to each other for a moment, not touching, and looks at the two openings, but puts one aside directly (without trying it) and picks up the third thinner one; the two wide tubes have openings of the same size[1]. The solution follows quite suddenly: Sultan fishes with a double-stick, consisting of the thinner one and one of the bigger ones, holding, as usual, the end of the smaller one in his hand. All of a sudden he pulls the double-stick in, turns it round, so that the thin end is before his eyes and the other towering up in the air behind him, seizes the third tube with his left hand, and introduces the tip of the double-stick into its opening. With the triple pole he reaches the objective easily; and when the long implement proves a hindrance in pulling the objective to him, it is disconnected as before.

According to observations in this experiment, Sultan never attempted to join tubes which would not have

[1] It can be shown that when the chimpanzee connects the double-stick he is guided by the relation between the two thicknesses of the tubes (compare *Nachweis Einfacher Strukturfunktionen*, *etc. Abh. d. Preuss. Akad. d. Wiss.* 1918, *Phys.-Math. Kl.*, No. 2, p. 56 seqq.).

fitted together[1]. Once, when an experiment was to be shown to visitors, I put down the objective outside, and at the same time threw two different-sized tubes, which happened to be at hand, through the bars to Sultan. He took hold of them at once, the bigger one, as usual, in his left hand, the thinner, in his right, and was already lifting his right hand to connect the tubes, when he suddenly stopped, without carrying out his intention, turned the thick tube round, looked at its other end, and immediately dropped both tubes to the ground. I let him pass them out to me and discovered that both ends of the wider one happened to have a nodule, thus closing the opening; in these circumstances Sultan did not even attempt to connect the tubes. When I cut away this closed part, he made the trial at once.

(6.8) The wide tube is closed at one end; a wooden block is put into the other end before the experiment; it sticks out just a little, being somewhat narrower than the tube, so that a space is left between it and the tube. Sultan seizes the tubes, looks for a moment at the block in the hole, tries to squeeze the thinner tube into the narrow opening between the block and the side of the tube, fails, and straightway pulls out the stopper, throws it aside, and connects the tubes.

Sometimes, however, he experiences a difficulty, where one would least expect it. Holding both tubes in his hand and wanting to proceed as usual to connect them, he hesitates for a few minutes and seems strangely uncertain; this is when the tubes lie in his hand in certain positions, namely, almost parallel, or else across each other

[1] In those cases in which mere observation does not lead to a definite conclusion, a trial, is, of course, made. Compare experiment of 6.8.

in the shape of a very narrow "X". This difficulty has now almost disappeared, but at first it occurred frequently. When Chica had, later on, adopted the same procedure, she showed exactly the same embarrassment when the two tubes were in this position, and her embarrassment was even more striking than Sultan's. As soon as the animals have again optically separated one tube from the other, the action proceeds quite smoothly. The optical factor in the situation, which at other times is a sure guide to the chimpanzee, making his acts and his behaviour appear the direct result of it, must in this case be so changed that it does not determine the motor factor quite so definitely. We humans can always see the relative positions of two tubes like this clearly enough not to be thus embarrassed, but if slightly complicated conditions (unfolding a folding-couch) be introduced, we too take some seconds to readjust, before our vision can dictate our movements.

In cases of pure *alexia* (Wertheimer) this uncertainty seems to be greatly increased. Gradually it becomes obvious that to understand the capacities and mistakes of chimpanzees in visually given situations is quite impossible without a theory of visual functions, especially of shapes in space.

In another experiment, further manufacture of implements is demanded of Sultan. (17.6) Besides a tube with a large opening, he has at his disposal a narrow wooden board, just too broad to fit into the opening. Sultan takes the board and tries to put it into the tube. This is not a mistake; the different *shapes* of the board and the tube would tempt even a human being to try it, because the difference in thickness of both these objects is not obvious at first sight. When he is not successful, he bites the end of the tube and breaks off a long splinter from its side, obviously because the side of the tube was in the way of the wood ("good error"). But as soon as he has his splinter, he tries to introduce it into the still intact end of the tube; a surprising turn, which should lead to the solution, were not the splinter a little too big. Sultan seizes the board once more, but now works at it with his teeth, and correctly too, from both edges at one end towards the middle, so that the board becomes narrower. When he has chewed off some of the (very hard) wood, he tests whether the board now fits into the sound opening of the tube, and continues working thus (here one must speak of real "work") until the wood goes about two centimetres deep into the tube. Now he wishes to fetch the objective with his implement, but two centimetres is not deep enough, and the tube falls off the top of the wood over and over again. By this time Sultan is plainly tired of biting at the wood; he prefers to sharpen the wooden splinter at one end and actually succeeds so far as to get it to stick firmly in the sound end of the tube, thus making the double stick ready for use. In connexion with this treatment of the wood it must be remarked that, contrary to my expectation, Sultan bit away wood almost exclusively from *one* end of the board, and, even if he took the other end between his teeth for a moment, he never gnawed blindly first at one, and then at the other. His way of dealing with the tube was also very satisfactory. The one opening of the tube that had been spoiled by breaking its side is thereafter left unheeded. I had some anxiety for the other opening during the further experiment, but although Sultan, when the wood and splinter did not fit in, put his teeth into it several times, he never really bit into the side of the tube, so that the opening could still be used. I could not guarantee that each repetition of the experiment would turn out so well. Sultan evidently had a specially bright day.

They have often sharpened wood, moreover, before this experiment. For instance, if Grande wants to poke somebody through the bars, she will swiftly bite a board in two and thus get the splinters she needs. Sultan too, if there is no key about, will occasionally sharpen a piece of wood in order to poke about in the keyhole; a fact noticed over and over again in the literature of this subject. But I was never quite clear about this sharpening business, and, therefore, we now investigated whether Sultan would proceed rationally with the very hard wood, which he would never have separated into serviceable splinters in mere play or by accident, but which he would have to work at somewhat methodically.

It will be obvious, after all the foregoing, that the double stick is made as promptly, when the objective is too high up to be knocked down with *one* stick only, and also that Chica, having adopted the new method, will apply it on occasion to the jumping process.

24

Reprinted from *Univ. California Publ. in Psychol.* **4**:215–232 (1930)

"INSIGHT" IN RATS[1]

BY

E. C. TOLMAN AND C. H. HONZIK

INTRODUCTION

Using a maze suggested by the senior writer, H. H. Hsiao[(4)] conducted an experiment designed to show whether or not rats were capable of grasping "a material, inner relation of two things to each other." More specifically, Hsiao's object was to discover whether a rat can get the "insight" that two paths have a common section—that, if the common section is closed, both of these paths are useless and that only a third, alternative, path not including this common section remains as the proper one whereby to reach the goal. Hsiao obtained what seemed to be positive results, but since these were based on only three rats, it was thought desirable to repeat the experiment with a larger number of animals.

EXPERIMENT I (MAZE I)

Apparatus and method.—The maze used in the first experiment is shown in figure 1. In principle and in general shape it is essentially the same as that used by Hsiao. It presents three paths to food, numbered 1, 2, and 3 in order of increasing length. Further, paths 1 and 2 have a common final section which is not common to path 3. The principal difference between the present maze and Hsiao's is that the final common section is now longer.

---

1 The cost of this investigation was met in large part by grants to the Department of Psychology from the Research Board of the University of California.

The maze was constructed of unpainted redwood. The alleys were 4 inches wide, the walls 6 inches high. The tops of the alleys were covered with ½-inch mesh hardware cloth. The five gates indicated in the figure were also made of the same hardware cloth. These gates were pivoted at the bottom and slanted upward away from the rat. The upper edges of the gates were held by rubber bands against the covers of the alleys. The tension of the rubber bands was such that the rat by stepping on a

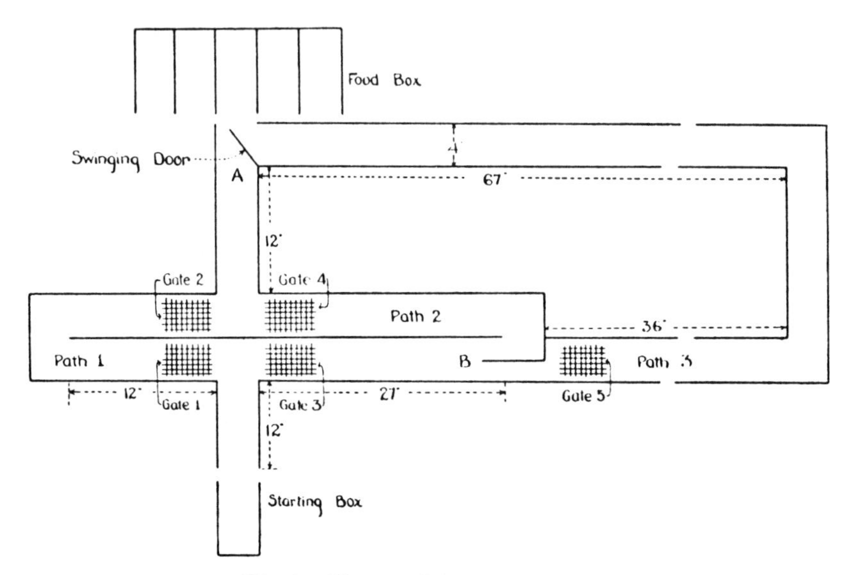

Fig. 1. Maze used in Experiment I

gate would, by the force of his weight, be able to push the gate down and pass over it. Once over, the gate would swing up behind him and retracing was prevented.[2]

The entrance to each path could be closed by a sliding door, made of beaver board.

General procedure.—The general course of the experiment can be divided into two periods: (1) a *preliminary training period* so as to acquaint the rat with the features of the maze and

[2] For a more detailed description of this type of gate, see Tolman, Tryon, and Jeffress.(7)

develop a strong preference for Path 1, a somewhat less strong preference for Path 2, and a weak, or no preference for Path 3;[3] and (2) a *final test period* in which to discover whether, when the swinging door at the end of the common section to paths 1 and 2 is locked, the rats, upon retracing out of Path 1, will then have the "insight" to avoid Path 2 and take Path 3 directly—and this in spite of the fact that, in the preliminary training, when the first gate to Path 1 was locked, the rats were in the habit of going directly by way of Path 2 and avoiding Path 3.

For each run, whether in the preliminary training period or in the final test period, the rat was given about 2 grams of a modified Steenbock mash. After the last run for the day the remainder of the daily ration was given in the rat's living cage.

The experiment was carried out with two groups of rats, Group A and Group B, the special conditions and results of which will be presented separately and in more detail below.

GROUP A

Animals.—Group A consisted of 10 male rats of a mixed breed obtained from the Department of Anatomy of the University of California. Nine of the rats were white, one hooded. They were from three to four months old at the beginning of the experiment. None of them had had previous training of any sort.

Preliminary training.—The preliminary training consisted of 6 runs each day for 15 days, making a total of 90 runs for each rat. The first three runs each day were "forced runs," i.e., by blocking successive pairs of the three paths the rat was forced down each of the three paths once. The remaining three of the six daily runs were "free," i.e., by leaving all the paths open the rat could choose a path according to his preference.[4]

---

[3] Many previous experiments have established that, when given an alternative between a long and a short path to food, a rat will come to choose consistently the shorter path: DeCamp(2); Kuo(5); Yoshioka(10); Blodgett(1); Hsaio.(4)

[4] For the invention of this method of "forced," followed by "free," runs we are indebted to Yoshioka.(9)

The results of the preliminary training were as follows:

Forced runs: These were distributed equally, by virtue of the method, into the three paths, 150 runs in each path.

Free runs: Path 1—309
Path 2— 81
Path 3— 60

When the rats had a free choice, it was evident that by far the greatest preference was for Path 1. The preference for Path 2 as against Path 3 was not great, and is roughly measured by the proportion 81 to 60.

Test period.—Six "insight" or test runs were given on the sixteenth day. The two gates in Path 1 were let down, permitting return. Each rat, as would be expected from the training series, immediately took Path 1. Finding himself blocked at *A* (see fig. 1), he was forced to return to the starting point (the second gate in Path 2 was not let down and this prevented his return by way of Path 2). He was then free to take either Path 2 or Path 3.

Test results.—On the *first* "insight" run, 4 of the 10 rats after returning out of Path 1 avoided Path 2. On the *second* "insight" run, 3 of the 10 rats after returning out of Path 1 avoided Path 2. On the *third* "insight" run, 6 of the 10 rats after returning out of Path 1 avoided Path 2. On the remaining three runs all the rats had learned not to take Path 1, but ran immediately into Path 3, or else in a few cases ran first into Path 2 and then into Path 3.

Since only 4 of the 10 rats avoided Path 2 on the first "insight" run, it was obvious that the results from this group were negative, particularly when we consider that not all these 4 rats avoided Path 2 on their second test run.

GROUP B

Animals.—Group B consisted of 11 male rats, also of varying color, three to four months of age, and without previous training of any sort.

Preliminary training.—The preliminary training consisted of 6 runs a day for 14 days. The first three runs each day, as in Group A, were "forced," that is, one in each path. The next three runs were "free." In the last three of the nine daily runs the entrance to Path 1 was closed by the sliding door at its entrance, so that the rats had to choose directly between Path 2 and 3.

The results of the preliminary training were as follows:

"Free" runs (second three): Path 1—204
Path 2—211
Path 3— 47

Last three runs (choice between paths 2 and 3):

Path 2—343
Path 3— 68

The preference for Path 2 as against Path 3 is quite evident in both types of runs, the "free" and those where choice was restricted to Path 2 or 3.

Test period.—The "insight" runs were given on the fifteenth day. As in Group A the two gates in Path 1 were let down to permit return. To prevent the rat from taking Path 3 or 2 immediately without going first into Path 1, the entrance to Path 2 was closed and not opened until the rat had entered Path 1 as far as the block at A.

The results of the test runs were as follows:

"Insight" run	No. of the 11 rats that avoided Path 2
1	1
2	1
3	2
4	6
5	7
6	9
7	9

The results are again negative.

The fact that all but 2 of the 11 rats avoided Path 2 on the sixth and seventh runs does not indicate "insight" in the sense in which we wish to use the term. For it appears obvious that, after having several times found themselves blocked by taking Path 2, the rats would *learn* to avoid this path. Their subsequent avoiding of Path 2 is, then, mere evidence of "trial and error" learning and does not indicate any grasp of the fact of the common section to paths 1 and 2 previous to their first experience of the block in Path 2.

EXPERIMENT II (MAZE II)

Apparatus and conditions.—In view of the negative results obtained on Maze I it was decided to try a maze in which the choice between paths 2 and 3 would require a larger movement, that is, a turning through 90° to the right or left (fig. 2).

A group of 11 male rats, also of mixed breed and without previous maze training, was run in this maze.

Preliminary training.—The preliminary training consisted of 10 runs a day for 13 days. For the first two days the first three of the ten daily runs were "forced," as explained above; the remaining seven runs were "free." By the end of the second day's running all the rats had begun to show a decided preference for the shortest path, viz., Path 1. Therefore, after the second day, only two runs daily (and these not in immediate succession) were permitted over Path 1; in the remaining eight of the ten runs Path 1 was closed by a wire netting set 9 inches back from the entrance, the wire gate at this entrance being let down by removing the rubber band. Thus the rat, entering Path 1 according to his first preference, found himself blocked, had to turn and retrace, and then to choose Path 2 or 3. This sort of retracing from Path 1 had the advantage of being similar to the "insight" run itself in which, however, the retracing is from the block at *A*. Thus both in the preliminary training runs and in the test runs the rat retraces out of Path 1, but, although in the training runs the rat has learned then to take Path 2, in the test run he must take Path 3 in order to show "insight."

It is not meant to imply that the rats invariably entered Path 1 when the wire-netting block was in place. On the contrary they learned to look into the alley without entering and to choose Path 2 or 3 immediately. This propensity, however, was somewhat counteracted by the two runs daily into Path 1, which had the effect of keeping the rat familiar with this path and inducing him to try it frequently.

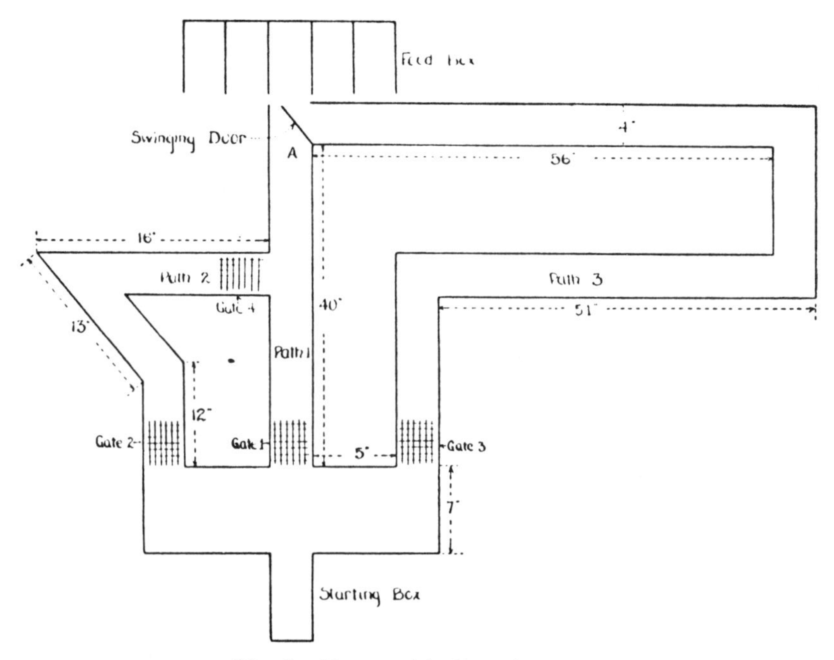

Fig. 2. Maze used in Experiment II.

Results of training period.—There was a total of 901 runs with Path 1 blocked as described above. Of these runs, 752 were made by way of Path 2 and 149 by way of Path 3. We find a considerable preference for Path 2 as against Path 3, and it is this preference or habit which must be overcome on the "insight" run, when the block is put at *A*, if "insight" is to be demonstrated.

Test period and its results.—On the fourteenth day 5 "insight" runs were given. The results of these runs are as follows:

"Insight" run	No. of the 11 rats that avoided Path 2
1	5
2	4
3	8
4	8
5	6

All the 5 rats that avoided Path 2 on the first run blundered into Path 2 on the second run. Of the 6 rats who did not avoid Path 2 on the first run, 4 avoided Path 2 on the second run. No rat avoided Path 2 consistently. It is clear that again "insight" cannot be inferred from the results.

EXPERIMENT III (MAZE III)

It now seemed possible to the writers that the situations presented by the two mazes I and II were not simple enough for the animals to show "insight." A third maze was therefore built.

Apparatus and methods.—This maze (fig. 3) was of the elevated type first used by Miles.[6] The rails, or runways, were 1½ inches wide, and 30 inches above the floor. The gate in Path 2 was inserted in a short tunnel and was similar to the gates used in the other mazes. Its purpose was to prevent return by way of Path 2. The blocks used to prevent entrance upon any path were of wire netting mounted upon a strip of wood fitted over the rail.

GROUP A

Animals.—The rats, 15 in number, used in this group on this maze were all males of mixed breed, from five to eight months of age, and had been previously run in an experiment on inheritance[8] for 21 days in a 17 unit T-maze. An interval of six weeks elapsed between that running and the present experiment. It may be well to emphasize that the maze in which the rats had their first training was an automatic self-recording maze[7] of the ordinary T-type, set on the floor, with 5-inch walls and 4-inch alleys, and was thus of an entirely different kind from the elevated one used in the present experiment.

Preliminary training period.—Days 1–8 (twelve runs a day). By the end of the first day preference for Path 1 was evident, so that thereafter 10 of the 12 daily runs were with Path 1 blocked at *A* (see fig. 3). As occurred in the case of Maze II, the rats in

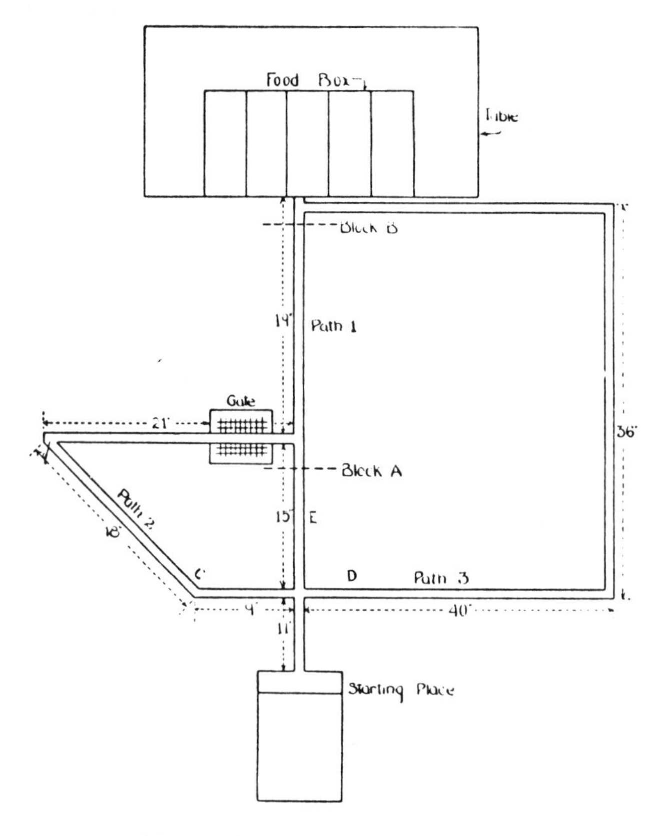

Fig. 3. Maze used in Experiment III.

this maze also learned merely to look toward block *A* and make the choice between paths 2 and 3 immediately without entering Path 1 at all.

Special effort was made to prevent such immediate choices, for the following reason. In the "insight" or test runs (with the block at *B*) the rat is forced to turn and head back toward the starting point; the behavior of the rat on Path 1 in the training runs should be similar to his behavior in the "insight" runs in this respect, that the rat, in the training runs, should make his choice of Path 2 or Path 3 while headed back toward the starting point. In this way the training runs are made to differ from the "insight" runs only in the position of the blocks. In the training runs the block is at *A*, in the test runs it is at *B*. The choice between paths 2 and 3 is made, therefore, in both training and test runs after the rat has been stopped and has been turned back, but the choice in the test run must be different from that of the training run if it is to show "insight." In order to induce the rats to attempt Path 1, at least two runs each day were given over Path 1, with no block at *A*. Knowing that he had made a successful run over Path 1 recently the rat was more likely to try it again. But this was not entirely successful. Consequently, still another method was tried. For the wire-netting block at *A*, one made of a piece of clear window glass was substituted on the fourth day. This was considerably less visible than the wire block. This method was effective for a time, but again the rats learned to see the glass and not to approach it. As a last resort, rats that did not approach the block at *A* voluntarily were forced to do so by blocks placed at *C* and *D*. Finding paths 2 and 3 blocked the rat ran to block *A*. The blocks at *C* and *D* were then quickly and quietly removed while he was still facing toward block *A*, and on his return he chose either Path 2 or Path 3.

Days 9–13 (twelve runs a day). On the ninth day of the training period a new type of run was introduced. This consisted of moving the block *A* on Path 1 forward from the point *A* to the point *E*, and setting up also block *B*. With this arrangement the rats ran to the block at *E*, had to retrace and then choose Path 2 or Path 3. If Path 2 was chosen the rat found himself blocked again at *B*. To permit him to return to the starting point again the block at *E* was quietly removed while he was facing block *B*. Having returned now to the starting point the rat could then take Path 3 or blunder into Path 2 by way of which he had just found himself blocked. It is interesting to note that the first time this type of run was given all the 15 rats chose Path 3 on the second return to the starting point; and thereafter very few mistakes of taking Path 2 a second time were made.

The defect inherent in this sort of run is obvious; it is very much like the "insight" run which was to follow later and may be considered direct training for that later run. It is therefore important to emphasize that each rat had only 12 runs of this sort, that these runs were distributed over 5 days (average of 2 a day), and that the runs were interspersed among 65 other runs which each rat had during these 5 days. The fact that the runs were few in number and were scattered among other runs led the experimenter to believe that their training effect could not be great. Further, it should be noted that block *B*, on this run, is encountered not by way of Path 1 (as in the "insight" run) but by way of Path 2.[5]

The advantage of this run and the only reason for its introduction is this: If the rat had never before encountered the block at *B*, his first encounter with it in the test or "insight" run might conceivably cause considerable confusion, a factor we wished to minimize. Therefore, some acquaintance with block *B* was thought advisable, not by way of Path 1 but by way of Path 2, as just said.

Results of training period.—There was a total in all 13 days of training of 1357 runs with block *A* in place. We ask now how many times was Path 2 chosen and how many times Path 3, since this will tell us how strong a preference there was, or how strong a habit was formed, for Path 2. Of the 1357 runs 1229 (90.57 per cent) were by way of Path 2, 128 runs (9.43 per cent) by way of Path 3. The preference for Path 2 could hardly be much stronger, especially when we consider that more than half of the

[5] A later control group (see below, Group B) without the introduction of this special run gave exactly the same sort of results as did the present group.

runs over Path 3 were made in the early part of the training while the animals were incompletely familiar with the maze. There were also 171 "forced" runs over Path 3, paths 1 and 2 being blocked.

The number of times each rat took paths 2 and 3, when the block was at *A*, is given in table 1.

TABLE 1

Path \ Rat	H58	H65	W56	W53	W59	H64	W74	W50	W60	B51	H53	W22	W29	W55	W16	Total
2	84	83	87	78	73	78	84	80	86	81	73	73	88	84	82	1214
3	5	6	10	8	13	8	3	3	3	14	27	15	1	6	6	128

Test period.—The "insight" runs were given on the fourteenth day. The results are as follows:

"Insight" run	No. of the 15 rats that avoided Path 2
1	14
2	13
3	10
4	12
5	11
6	12
7	13

The first "insight" run must be considered crucial since after this first run the element of training enters. And for the first "insight" trial, 14 of the 15 rats responded correctly, taking Path 3 immediately in spite of the strong preference and habit for Path 2.

Taking the number of errors made by individual rats during the seven test runs, the results are as follows:

In 7 runs: 1 rat made 5 errors
3 rats made 4 errors
3 rats made 1 error
8 rats made 0 errors

More than half the rats made 7 perfect runs. Only one rat made an error on the first test run. The others made errors on later runs, notably the third. The fact that 6 rats chose the right path, viz., Path 3, on the first test run and made errors on later runs may be explained by a temporary reassertion of, or lapse into, an old habit.

If the taking of Path 2 or Path 3, after return from block *B*, were a matter of pure chance similar to the tossing of a coin, the probability[6] of 14 out of 15 rats taking paths 3 and 1 rat taking Path 2 would be .00046. The actual result of 14 rats taking Path 3 and 1 rat taking Path 2 on the first test run indicates, therefore, a very decided "loading" in favor of Path 3, in spite of the fact that the "loading" during the training period was just the reverse, viz., in favor of Path 2.

Retraining and alternated runs.—On the day following the test runs, viz., the fifteenth, the ordinary training runs were resumed, 10 of the 12 runs on this day being with the block at *A*. On the sixteenth day 11 runs were given, and these were as follows: The first four runs were "insight" or test runs, the next two were training runs with block at *A*, then four more "insight" runs, and finally one more training run. Thus, for the entire group of 15 rats there was a total of 120 "insight" runs, and a total of 45 training runs on this day. For the training runs with block at *A* the taking of Path 3 was called an error, since Path 2 was open and was the shorter path. In the "insight" runs the taking of Path 2 was as usual called an error. The errors were as follows:

In the 120 "insight" runs—18 errors
In the 45 training runs—4 errors

In the "insight" runs we may say that each rat made an average of one error in 8 runs, while in the training runs each rat made an average of one-fourth of an error in 3 runs.

This remarkable ability to adjust correctly to changing conditions suggested the idea of alternating the blocks between *A*

[6] This has been calculated by the point binomial: see Holzinger.(3)

and *B* on successive runs, i.e., one training run with block at *A*, the next an "insight" run with block at *B*, the third a training with block at *A*, and so on. This was tried on the seventeenth day. Eleven runs were given, 6 being training runs and 5 test runs. Counting errors as above, the results were as follows:

	"Insight" errors	Training-run errors
10 rats	0	0
1 rat	2	1
1 rat	0	2
1 rat	1	0
1 rat	3	0
1 rat	1	0

In these alternated runs both the glass and the wire-netting blocks were used, and their positions were alternated between points *A* and *B*. For one run the glass block was at *B*, the wire block at *A*, and for the succeeding run the glass block was put at *A*, the wire block at *B*. This was to check the possibility that the rats might have become conditioned to the nature of the blocks, the glass block being a sign for Path 2, the wire block for Path 3. The alternation of blocks caused no change or confusion in the behavior of the rats. Their behavior gave unmistakable evidence that they had grasped so well the relation of the paths to each other that they were able to take the right path wherever either block was placed.

GROUP B

To ascertain the possible effect of the twelve special runs per rat which were introduced during Days 9–13 with Group A, viz., the runs with blocks at *E* and *B*, which might perhaps have been direct training for the "insight" trials, another group of 10 male rats of mixed breed from five to eight months of age and without previous training of any sort was given training for 14 days, but without these blocks-at-*E*-and-*B* runs. Otherwise the general conditions were the same as for Group A.

Results of training period.—There were for the entire group of 10 rats 1178 runs with Path 1 blocked at *A*. Of these 1178 runs, 1081 (or 91.77%) were over Path 2, and 97 (8.23%) over

Path 3. The preference for Path 2 as against Path 3 is obvious and is even stronger in this group than in the previous Group A.

The number of times each rat took paths 2 and 3, with the block at *A*, is given in table 2.

TABLE 2

Path \ Rat	H82	W11	H52	W51	H5	H55	W65	W31	W17	H1	Total
2	107	104	106	108	108	109	110	107	111	111	1081
3	8	17	13	5	15	6	1	20	4	8	97

Results of test period.—Day 15. The results of the "insight" runs were as follows:

"Insight" run	No. of the 10 rats that avoided Path 2
1	7
2	9
3	all
4	8
5	all
6	9
7	9

Taking the number of errors made by individual rats during the seven test runs, the results were as follows:

In 7 runs: 1 rat made 3 errors
1 rat made 2 errors
3 rats made 1 error
5 rats made 0 errors

The probability of 7 out of 10 rats taking Path 3, and 3 rats taking Path 2, assuming that the choice of these paths is a matter of pure chance, would be 0.117. That is, assuming pure chance, we should expect a distribution of this sort about twelve times in one hundred. But actually there was a heavy "loading" the other way, viz., in favor of Path 2, built up during the training runs, as is indicated by the proportion of 1081 runs in Path 2 to 97 runs in Path 3.

RESULTS

1. "Insight," in the sense in which we have used the term here, seems to be definitely proved for Maze III with both Group A and Group B. That is, under the conditions of this elevated maze and of the kinds and amounts of preliminary training given, the rats of both groups, upon *first* finding a block in the final common path, as a result of taking one "entering" path (Path 1) *immediately* (i.e., without any trial and error learning) also avoided the taking of the other "entering" path (Path 2). And this result was obtained in spite of the fact that these rats had shown a very strong propensity to take this second "entering" path, when the first "entering" path (and not the final common path) was blocked.

2. No evidence of such "insight" was obtained under our conditions for either Maze I or Maze II.

DISCUSSION

To explain the fact that we obtained no evidence of insight with our Maze I, although Hsaio[(4)] did obtain evidence of such insight with a very similar maze, two points are to be noted (1) Our maze was not exactly identical in shape with Hsaio's and (2) we did not give the amount and distribution of preliminary training that he did. Either of these two points may have been enough to explain the difference between his results and ours. For the factors which actually govern the appearance or non-appearance of "insight" are in the present state of our knowledge still quite uncertain.

To explain the fact that no insight was obtained in Maze II although it was obtained in Maze III which had an identical ground pattern, it would seem important that Maze III had no side walls as did Maze II and hence the rats were able in Maze III to "see" the situation as a whole. Or, even if the rats in Maze III were not able to "see" all the paths at any one moment, they might still have been better able to grasp the connections between the paths, owing perhaps to the open space on all sides of the runways, which may have served to accentuate the relations between the paths.

LITERATURE CITED

[1] Blodgett, H. C.

1929. The effect of introduction ot reward upon the maze performance of rats. *Univ. Calif. Publ. Psychol.*, 4:113–134.

[2] DeCamp, J. E.

1920. Relative distance as a factor in the white rat's selection of a path. *Psychobiol.*, 2:245–253.

[3] Holzinger, K. J.

Statistical methods for students in Education. (Boston, Ginn & Co., 1928), chap. 11.

[4] Hsiao, H. H.

1929. Experimental study of the rat's insight within a spatial complex. *Univ. Calif. Publ. Psychol.*, 4:57–70.

[5] Kuo, Z. Y.

1922. The nature of unsuccessful acts and their order of elimination. *Jour. Comp. Psychol.*, 2:1–27.

[6] Miles, W. R.

1927. The narrow-path elevated maze. *Proc. Soc. Exp. Biol. and Med.*, 24:454–456.

[7] Tolman, E. C., Tryon, R. C., and Jeffress, L. A.

1929. A self-recording maze with an automatic delivery table. *Univ. Calif. Publ. Psychol.*, 4:99–112.

[8] Tryon, R. C.

1929. The genetics of learning ability in rats. *Univ. Calif. Publ. Psychol.*, 4:71–89.

[9] Yoshioka, J. G.

1928. Preliminary study of discrimination of maze patterns by the rat. *Univ. Calif. Publ. Psychol.*, 4:1–18.

[10] Yoshioka, J. G.

1929. Weber's Law in the discrimination of maze distance by the white rat. *Univ. Calif. Publ. Psychol.*, 4:155–184.

25

Reprinted by permission from *Psychol. Rev.* **56**:51–65 (1949)

THE FORMATION OF LEARNING SETS[1, 2]

BY HARRY F. HARLOW

University of Wisconsin

In most psychological ivory towers there will be found an animal laboratory. The scientists who live there think of themselves as theoretical psychologists, since they obviously have no other rationalization to explain their extravagantly paid and idyllic sinecures. These theoretical psychologists have one great advantage over those psychological citizens who study men and women. The theoreticians can subject their subhuman animals, be they rats, dogs, or monkeys, to more rigorous controls than can ordinarily be exerted over human beings. The obligation of the theoretical psychologist is to discover general laws of behavior applicable to mice, monkeys, and men. In this obligation the theoretical psychologist has often failed. His deductions frequently have had no generality beyond the species which he has studied, and his laws have been so limited that attempts to apply them to man have resulted in confusion rather than clarification.

One limitation of many experiments on subhuman animals is the brief period of time the subjects have been studied. In the typical problem, 48 rats are arranged in groups to test the effect of three different intensities of stimulation operating in conjunction with two different motivational conditions upon the formation of *an isolated* conditioned response. A brilliant Blitzkrieg research is effected—the controls are perfect, the results are important, and the rats are dead.

If this *do and die* technique were applied widely in investigations with human subjects, the results would be appalling. But of equal concern to the psychologist should be the fact that the derived general laws would be extremely limited in their application. There are experiments in which the use of naive subjects is justified, but the psychological compulsion to follow this design indicates that frequently the naive animals are to be found on both sides of the one-way vision screen.

The variety of learning situations that play an important rôle in determining our basic personality characteristics and in changing some of us into thinking animals are repeated many times in similar form. The behavior of the human being is not to be understood in terms of the results of single learning situations but rather in terms of the changes which are affected through multiple, though comparable, learning problems. Our emotional, personal, and intellectual characteristics are not the mere algebraic summation of a near infinity of stimulus-response bonds. The learning of primary importance to the primates, at least, is the formation of learning sets; it is the *learning how to learn efficiently* in the situations the animal frequently encounters. This learning to learn transforms the organism from a creature that adapts to a changing environment by trial and error to one that adapts by seeming hypothesis and insight.

The rat psychologists have largely ignored this fundamental aspect of learning and, as a result, this theoretical

[1] This paper was presented as the presidential address of the Midwestern Psychological Association meetings in St. Paul, May 7, 1948.

[2] The researches described in this paper were supported in part by grants from the Special Research Fund of the University of Wisconsin for 1944–48.

domain remains a *terra incognita.* If learning sets are the mechanisms which, in part, transform the organism from a conditioned response robot to a reasonably rational creature, it may be thought that the mechanisms are too intangible for proper quantification. Any such presupposition is false. It is the purpose of this paper to demonstrate the extremely orderly and quantifiable nature of the development of certain learning sets and, more broadly, to indicate the importance of learning sets to the development of intellectual organization and personality structure.

The apparatus used throughout the studies subsequently referred to is illustrated in Fig. 1. The monkey responds by displacing one of two stimulus-objects covering the food-wells in the tray before him. An opaque screen is interposed between the monkey and the stimulus situation between trials and a one-way vision screen separates monkey and man during trials.

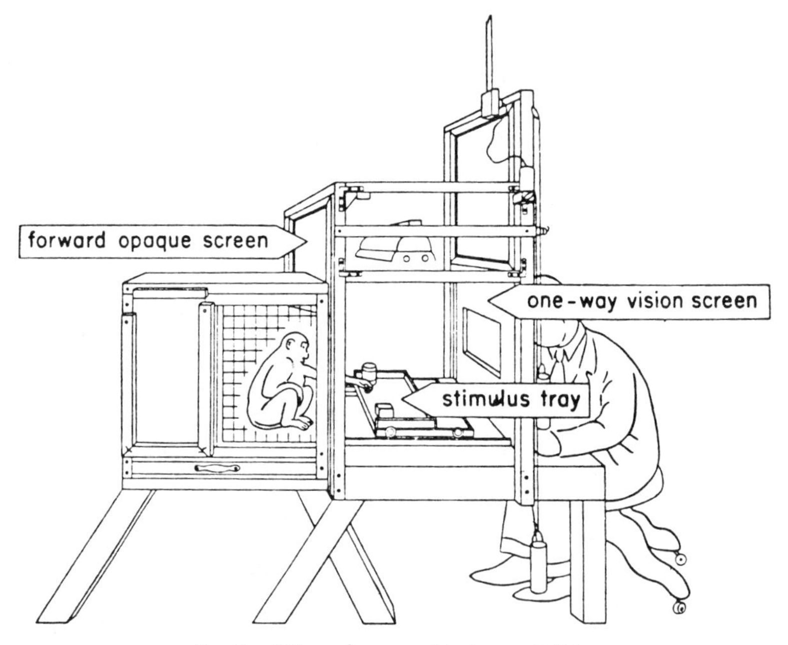

FIG. 1. Wisconsin general test apparatus.

The first problem chosen for the investigation of learning sets was the object-quality discrimination learning problem. The monkey was required to choose the rewarded one of two objects differing in multiple characteristics and shifting in the left-right positions in a predetermined balanced order. A series of 344 such problems using 344 different pairs of stimuli was run on a group of eight monkeys. Each of the first 32 problems was run for 50 trials; the next 200 problems for six trials; and the last 112 problems for an average of nine trials.

In Fig. 2 are presented learning curves which show the per cent of correct responses on the first six trials of these discriminations. The data for the first 32 discriminations are grouped for blocks of eight problems, and the remaining discriminations are arranged in blocks of 100, 100, 56, and 56 problems. The data indicate that the subjects progressively improve in their ability to learn object-quality discrimination problems. The monkeys *learn how to learn* individual problems with a minimum of errors. It is this *learning how to learn a kind of problem* that we designate by the term *learning set.*

The very form of the learning curve changes as learning sets become more efficient. The form of the learning curve for the first eight discrimination problems appears S-shaped: it could be described as a curve of 'trial-and-error' learning. The curve for the last 56 problems approaches linearity after Trial 2. Curves of similar form have been described as indicators of 'insightful' learning.

We wish to emphasize that this *learning to learn,* this *transfer from problem to problem* which we call the formation of a learning set, is a highly *predictable, orderly* process which can be demonstrated as long as controls are maintained over the subjects' experience and the difficulty of the problems. Our subjects, when they started these researches, had no previous laboratory learning experience. Their entire discrimination learning set history was obtained in this study. The stimulus pairs employed had been arranged and their serial order determined from tables of random numbers. Like nonsense syllables, the stimulus pairs were equated for difficulty. It is unlikely that any group of problems differed significantly in intrinsic difficulty from any other group.

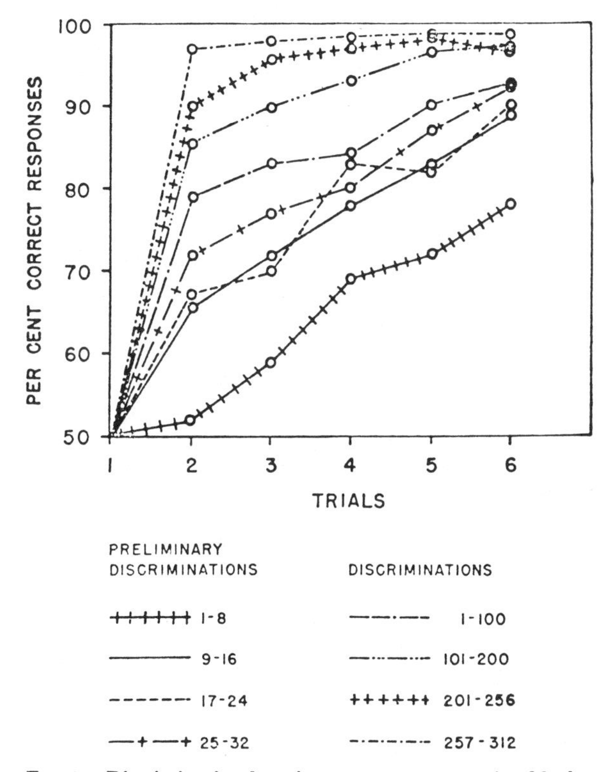

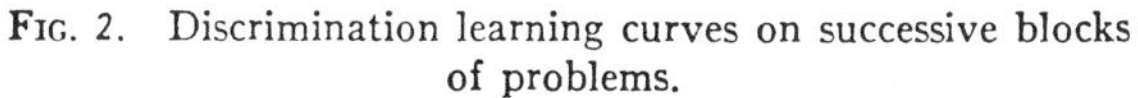

FIG. 2. Discrimination learning curves on successive blocks of problems.

In a conventional learning curve we plot change of performance over a series of *trials;* in a learning set curve we plot

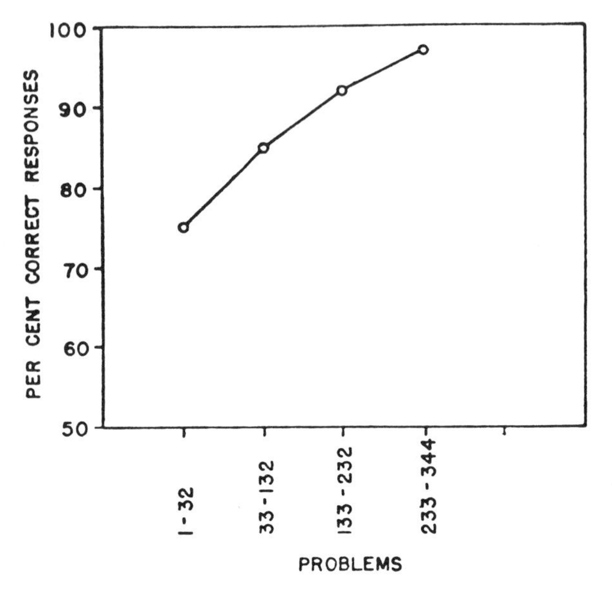

FIG. 3. Discrimination learning set curve based on Trial 2–6 responses.

change in performance over a series of *problems.* It is important to remember that *we measure learning set in terms of problems* just as *we measure habit in terms of trials.*

Figure 3 presents a discrimination learning set curve showing progressive increase in the per cent of correct responses on Trials 2–6 on successive blocks of problems. This curve appears to be negatively accelerated or possibly linear.

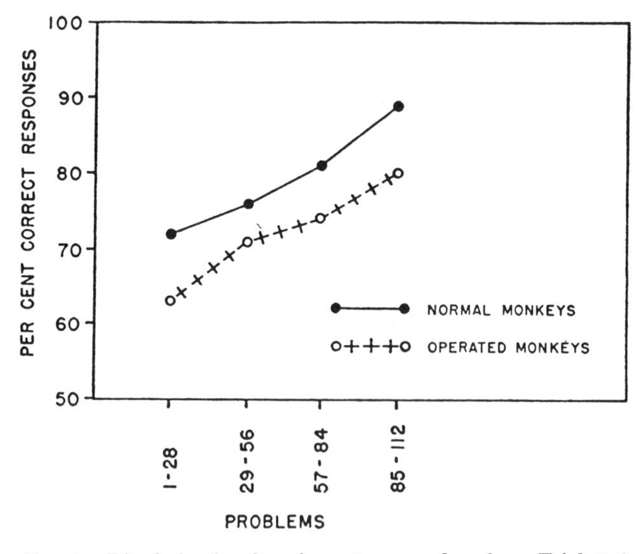

FIG. 4. Discrimination learning set curves based on Trial 2–6 responses: normal and operated monkeys.

Discrimination learning set curves obtained on four additional naive normal monkeys and eight naive monkeys with extensive unilateral cortical lesions, are shown in Fig. 4. Brain-injured as well as normal monkeys are seen to form effective discrimination learning sets, although the partial hemidecorticate monkeys are less efficient than the normal subjects. Improvement for both groups is progressive and the fluctuations that occur may be attributed to the small number of subjects and the relatively small number of problems, 14, included in each of the problem blocks presented on the abscissa.

Through the courtesy of Dr. Margaret Kuenne we have discrimination learning set data on another primate species. These animals were also run on a series of six-trial discrimination problems but under slightly different conditions. Macaroni beads and toys were substituted for food rewards, and the subjects were tested sans iron-barred cages. The data for these 17 children, whose ages range from two to five years and whose intelligence quotients range from 109 to 151, are presented in Fig. 5. Learning set curves are plotted for groups of children attaining a predetermined learning criterion within differing numbers of problem blocks. In spite of the small number of cases and the behavioral vagaries that are known to characterize this primate species, the learning set curves are orderly and lawful and show progressive increase in per cent of correct responses.

Learning set curves, like learning curves, can be plotted in terms of correct responses or errors, in terms of responses on any trial or total trials. A measure which we have frequently used is per cent of correct Trial 2 responses—the behavioral measure of the amount learned on Trial 1.

Figure 6 shows learning set curves measured in terms of the per cent correct Trial 2 responses for the 344-problem series. The data from the first 32 preliminary discriminations and the 312 subsequent discriminations have been plotted separately. As one might expect, these learning set curves are similar to those that have been previously presented. What the curves show with especial clarity is the almost unbelievable

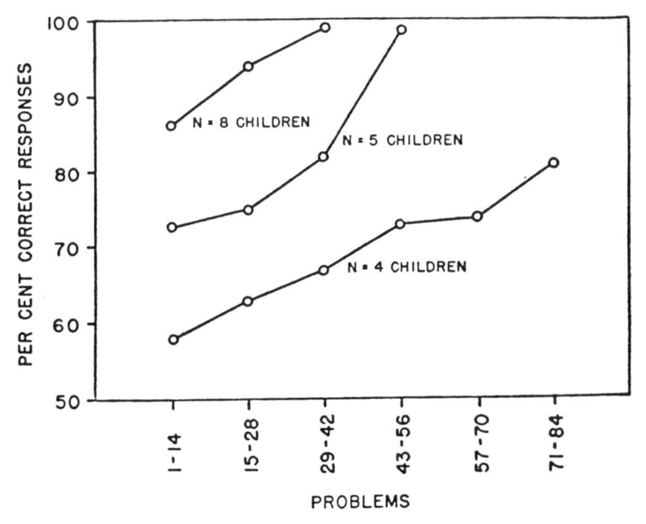

FIG. 5. Discrimination learning set curves based on Trial 2–6 responses: children.

change which has taken place in the *effectiveness of the first training trial.* In the initial eight discriminations, this single paired stimulus presentation brings the Trial 2 performance of the monkeys to a level less than three per cent above chance; in the last 56 discriminations, this first training trial brings the performance of the monkeys to a level *less than three per cent* short of perfection. Before the formation of a discrimination learning set, a single training trial produces negligible gain; after the formation of a discrimination learning set, *a single training trial constitutes problem solution.* These data clearly show that *animals can gradually learn insight.*

In the final phase of our discrimination series with monkeys there were subjects that solved from 20 to 30 consecutive problems with no errors whatsoever following the first blind trial,—and many of the children, after the first day or two of training, did as well or better.

These data indicate the function of learning set in converting a problem which is initially difficult for a subject into a problem which is so simple as to be immediately solvable. The learning set is the mechanism that changes the problem from an intellectual tribulation into an intellectual triviality and leaves the organism free to attack problems of another hierarchy of difficulty.

For the analysis of learning sets in monkeys on a problem that is ostensibly at a more complex level than the discrimination problem, we chose the discrimination reversal problem. The procedure was to run the monkeys on a discrimination problem for 7, 9, or 11 trials and then to reverse the reward value of the stimuli for eight trials; that is to say, the stimulus previously correct was made incorrect and the stimulus previously incorrect became correct.

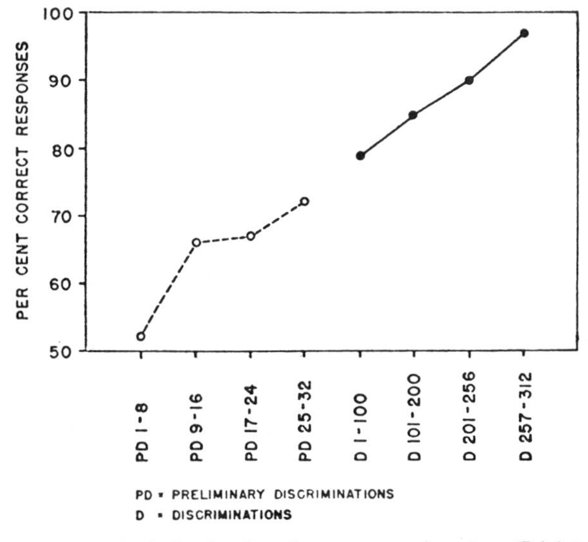

FIG. 6. Discrimination learning set curve based on Trial 2 responses.

The eight monkeys previously trained on discrimination learning were tested on a series of 112 discrimination reversal problems. Discrimination reversal learning curves for successive blocks of 28 problems are shown in Fig. 7. The

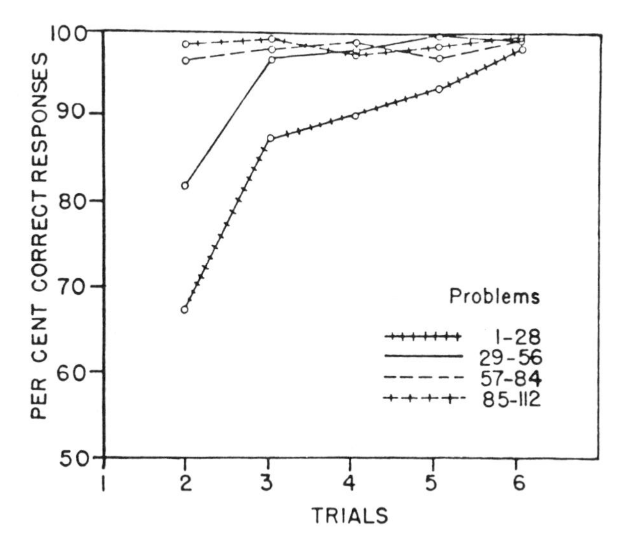

FIG. 7. Discrimination reversal learning curves on successive blocks of problems.

measure used is per cent of correct responses on Reversal Trials 2 to 6. Figure 8 presents data on the formation of the discrimination reversal learning set in terms of the per cent of correct responses on Reversal Trial 2 for successive blocks of 14 problems. Reversal Trial 2 is the first trial following the 'informing' trial, *i.e.*, the initial trial reversing the reward value of the stimuli. Reversal Trial 2 is the measure of the effectiveness with which the single informing trial leads the subject to abandon a reaction pattern which has proved correct for 7 to 11 trials, and to initiate a new reaction pattern to the stimulus pair. On the last 42 discrimination reversal problems the monkeys were responding as efficiently on Reversal Trial 2 as they were on comple-

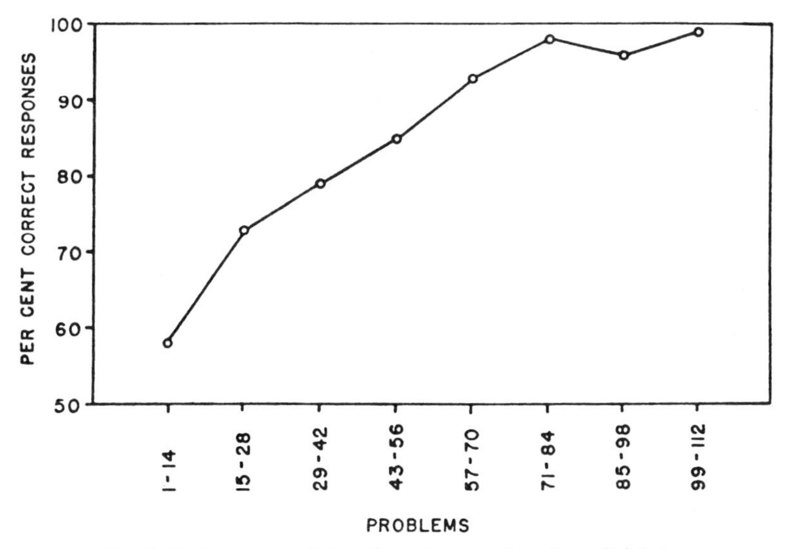

FIG. 8. Discrimination reversal learning set curve based on Trial 2 responses.

mentary Discrimination Trial 2, *i.e.*, they were making over 97 per cent correct responses on both aspects of the problems. The eight monkeys made from 12 to 57 successive correct second trial reversal responses. Thus it becomes perfectly obvious that at the end of this problem the monkeys possessed sets both to learn and to reverse a reaction tendency, and that this behavior could be consistently and immediately elicited with hypothesis-like efficiency.

This terminal performance level is likely to focus undue attention on the one-trial learning at the expense of the earlier, less efficient performance levels. It should be kept in mind that this one-trial learning appeared only as the end result of an orderly and progressive learning process; insofar as these subjects are concerned, the insights are only to be understood in an historical perspective.

Although the discrimination reversal problems might be expected to be more difficult for the monkeys than discrimination problems, the data of Fig. 9 indicate that the discrimination reversal learning set was formed more rapidly than the previously acquired discrimination learning set. The explanation probably lies in the nature of the transfer of training from the discrimination learning to the discrimination reversal problems. A detailed analysis of the discrimination learning data indicates the operation throughout the learning series of certain error-producing factors, but with each successive block of problems the frequencies of errors attributable to these factors are progressively decreased, although at different rates and to different degrees. The process might be conceived of as a learning of response tendencies that counteract the error-producing factors. A description of the reduction of the error-producing factors is beyond the scope of this paper, even though we are of the opinion that this type of analysis is basic to an adequate theory of discrimination learning.

Suffice it so say that there is reason to believe that there is a large degree of transfer from the discrimination series to the reversal series, of the learned response tendencies counteracting the operation of two of the three primary error-producing factors thus far identified.

The combined discrimination and dis-

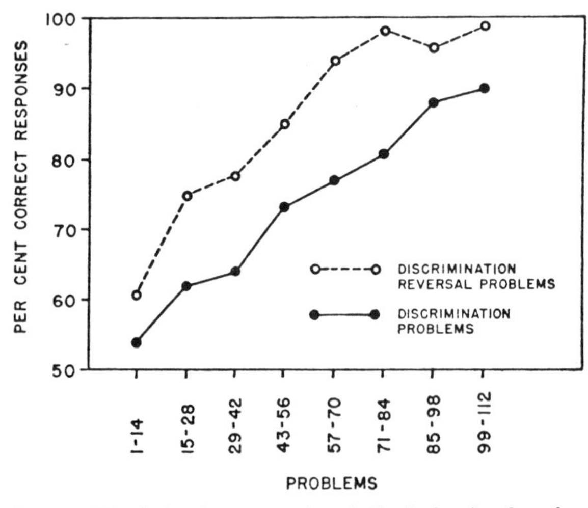

FIG. 9. Discrimination reversal and discrimination learning set curves based on Trial 2 responses.

crimination reversal data show clearly how the learning set delivers the animal from Thorndikian bondage. By the time the monkey has run 232 discriminations and followed these by 112 discriminations and reversals, he does not possess 344 or 456 specific habits, bonds, connections or associations. We doubt if our monkeys at this time could respond with much more than chance efficiency on the first trial of any series of the previously learned problems. But the monkey does have a generalized ability to learn *any* discrimination problem or *any* discrimination reversal problem with the greatest of ease. Training on several hundred specific problems has not turned the monkey into an automaton exhibiting forced, sterotyped, reflex responses to specific stimuli. These several hundred habits have, instead, made the monkey an adjustable creature with an *increased capacity* to adapt to the ever-changing demands of a psychology laboratory environment.

We believe that other learning sets acquired in and appropriate to the monkey's natural environment would enable him to adapt better to the changing conditions there. We are certain, moreover, that learning sets acquired by man in and appropriate to his environment have accounted for his ability to adapt and survive.

Before leaving the problem of discrimination reversal learning we submit one additional set of data that we feel merits attention. Nine of the children previously referred to were also subjected to a series of discrimination reversal problems. The outcome is partially indicated in Fig. 10 which shows the per cent of correct Reversal Trial 2 responses made on successive blocks of 14 problems. It can be seen that these three to five-year-old children clearly bested the monkeys in performance on this series of problems. Trial 2 responses approach perfection in the second block of 14 discrimination reversal

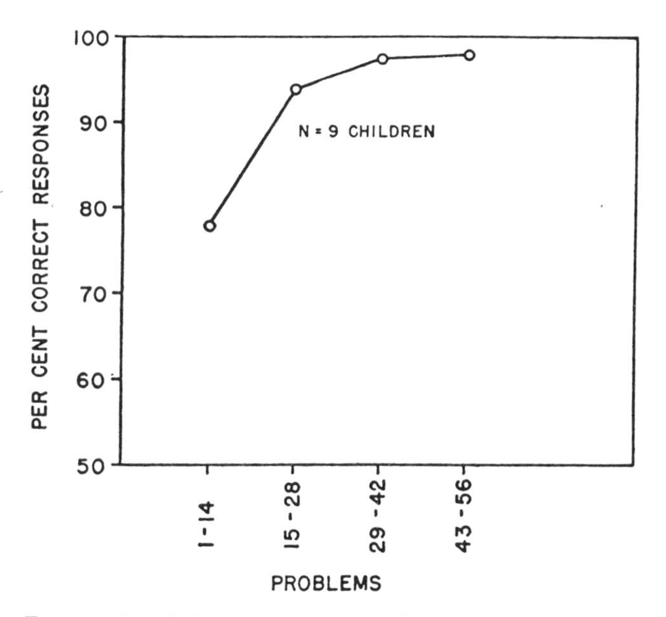

FIG. 10. Discrimination reversal learning set curve based on Trial 2 responses: children.

problems. Actually, over half of the total Trial 2 errors were made by one child.

These discrimination reversal data on the children are the perfect illustration of set formation and transfer producing adaptable abilities rather than specific bonds. Without benefit of the monkey's discrimination reversal set learning curves we might be tempted to assume that the children's data indicate a gulf between human and subhuman learning. But the *extremely rapid* learning on the part the children is not unlike the *rapid* learning on the part of the monkeys, and analysis of the error-producing factors shows that the same basic mechanisms are operating in both species.

Following the discrimination reversal problem the eight monkeys were presented a new series of 56 problems designed to elicit alternation of unequivocally antagonistic response patterns. The first 7, 9, or 11 trials of each problem were simple object-quality discrimination trials. These were followed immediately by ten right-position discrimination trials with the same stimuli continuing to shift in the right-left positions in predetermined orders. In the first 7 to 11 trials, a particular object was correct regardless of its position. In the subsequent 10 trials, a particular position—the experimenter's right position—was correct, regardless of the object placed there. Thus to solve the problem the animal had to respond to object-quality cues and disregard position cues in the first 7 to 11 trials and, following the failure of reward of the previously rewarded object, he had to disregard object-quality cues and respond to position cues.

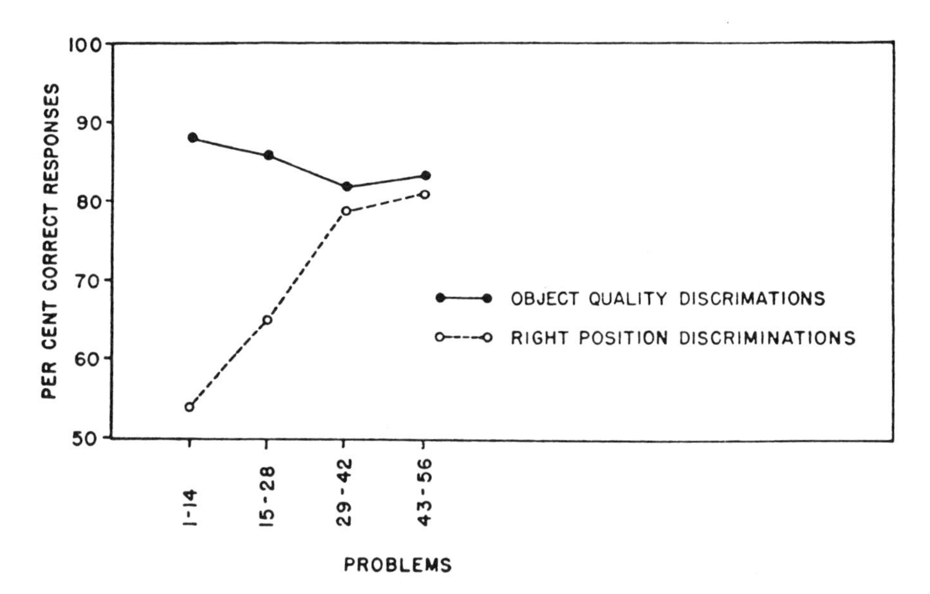

FIG. 11. Learning set curves for problem requiring shift from object-quality discrimination to right-position discrimination.

The learning data on these two antagonistic tasks are presented in Fig. 11. It is to be noted that the object-quality curve, which is based on Trials 1 to 7, begins at a very high level of accuracy, whereas the position curve, plotted for Trials 1 to 10, begins at a level little above chance. This no doubt reflects the operation of the previously well-established object-quality discrimination learning set. As the series continues, the object-quality curve shows a drop until the last block of problems, while the position curve rises progressively. In the evaluation of these data, it should be noted that chance performance is 50 per cent correct responses for the object-quality discriminations and 45 per cent for the position discriminations, since each sequence of 10 position trials includes an error "informing" trial. It would appear that the learning of the right-position discriminations interferes with the learning of the object-quality discriminations to some extent. In spite of this decrement in object-quality discrimination performance for a time, the subjects were functioning at levels far beyond chance on the antagonistic parts of the problems during the last half of the series. We believe that this behavior reflects the formation of a right-position learning set which operates at a high degree of independence of the previously established object-quality discrimination learning set.

The precision of the independent operation of these learning sets throughout the last 14 problems is indicated in Fig. 12. Since the right-position part of the problem was almost invariably initiated by an error trial, these data are limited to those problems on which the first trial object-quality discrimination response was incorrect. The per cent of correct Trial 7 responses to the 'A' object, the correct stimulus for the object-quality discriminations, is 98. The initiating error trial which occurs when the problem shifts without warning to a right-position problem, drops this per cent response to the 'A' object to 52—a level barely above chance. The per cent of Trial 7 responses to the right position during the object-quality discriminations is 52. The single error trial initiating the shift of the problem to a right-position discrimination is followed by 97 per cent right-position responses on the next trial. In other words, *it is as though* the outcome of a single *push of an object* is adequate to switch off the 'A'-object choice reaction tendency and to switch on the right-position choice reaction tendency.

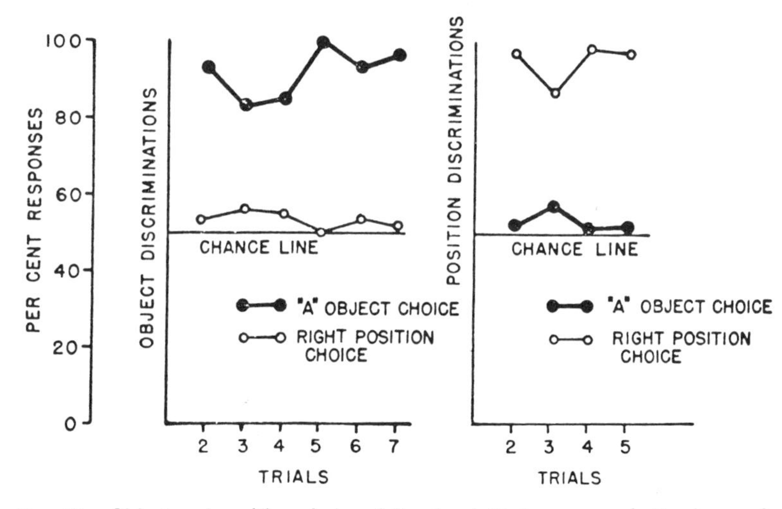

FIG. 12. Object and position choices following initial errors on both phases of object-position shift series, based on problems 42–56.

The cue afforded by a single trial produces at this point almost complete discontinuity of the learning process. The only question now left unsettled in the controversy over hypotheses in subhuman animals is whether or not to use this term to describe the behavior of a species incapable of verbalization.

Again, it should be remembered that both the object-quality discrimination learning set and the right-position discrimination learning set developed in a gradual and orderly manner. Only after the learning sets are formed do these phenomena of discontinuiy in learned behavior appear.

Further evidence for the integrity of learning sets is presented in an additional experiment. Six monkeys with object-quality discrimination learning experience, but without training on reversal problems or position discriminations, were given seven blocks of 14 problems each, starting with a block of 25-trial object-quality discriminations, followed by a block of 14 25-trial positional discriminations composed of right-position and left-position problems presented alternately. The remaining five blocks of problems continued the alternate presentation of 14 object-quality discrimination problems and 14 right-left positional discrimination problems. Figure 13 presents curves showing the per cent of correct responses on total trials on these alternate blocks of antagonistic discriminations. The complex positional discrimination learning set curve shows progressive improvement throughout the series, whereas the object-quality discrimination curve begins at a high level of accuracy, shows decrement on the second block, and subsequently recovers. By the end of the experiment the two basically antagonistic learning sets had 'learned' to live together with a minimum of conflict. These data are the more striking if it is recalled that between each two blocks of object-quality discriminations there were 350 trials in which no object was differentially rewarded, and between

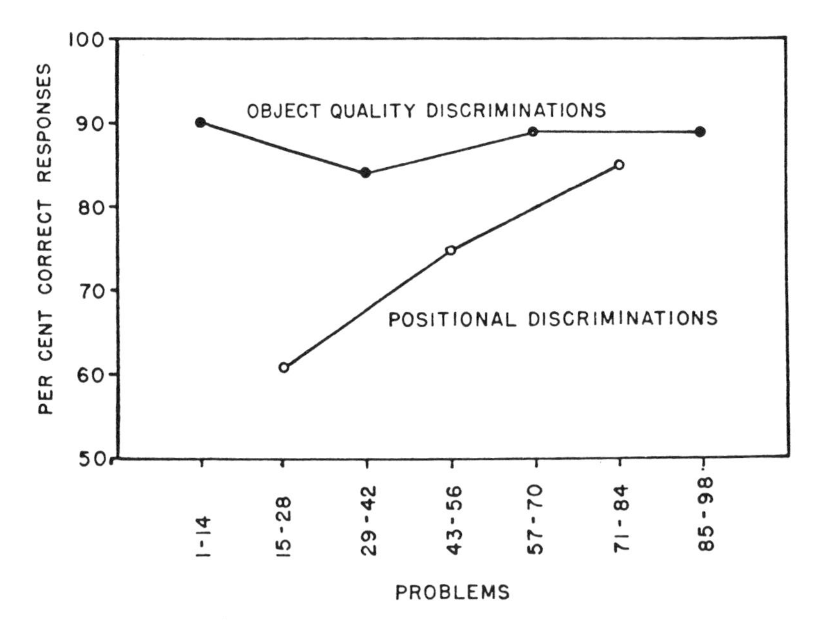

FIG. 13. Learning set curves for problem series with alternating object-quality and positional discriminations, based on total trial responses.

each two blocks of 14 positional discriminations there were 350 trials in which no position was differentially rewarded.

In Fig. 14 we present additional total-trial data on the formation of the positional learning set. These data show the change in performance on the first and last seven positional discriminations in each of the three separate blocks of positional discriminations. The interposed object-quality discrimination problems clearly produced interference, but

112 six-trial discriminations. The lower curves show total errors on an additional group of 56 discriminations presented one year later. In both situations the full-brained monkeys make significantly better scores, but one should note that the educated hemidecorticate animals are superior to the uneducated unoperated monkeys. Such data suggest that half a brain is better than one if you compare the individuals having appropriate learning sets with the individuals lacking them.

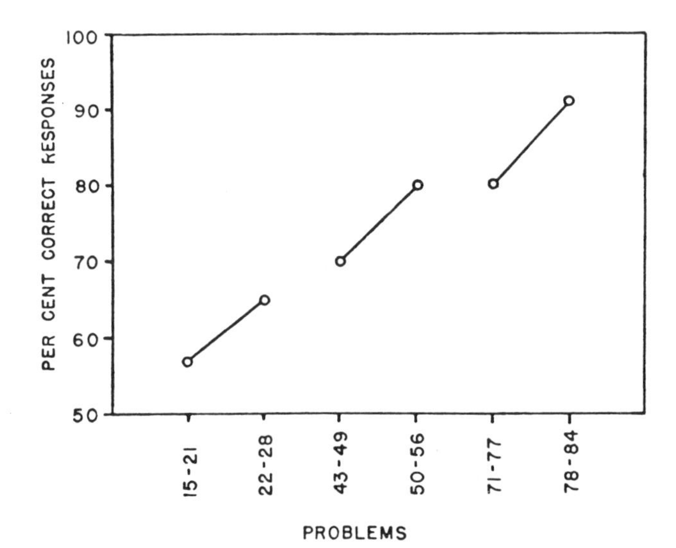

FIG. 14. Right-left positional discrimination learning set curve based on total trial responses. (Data on antagonistic object-quality discrimination problems omitted.)

they did not prevent the orderly development of the positional learning sets, nor the final attainment of a high level of performance on these problems.

We have data which suggest that the educated man can face arteriosclerosis with confidence, if the results on brain-injured animals are applicable to men. Figure 15 shows discrimination learning set curves for the previously described groups of four normal monkeys and eight monkeys with very extensive unilateral cortical injury. The upper curves show total errors on an initial series of

More seriously, these data may indicate why educated people show less apparent deterioration with advancing age than uneducated individuals, and the data lend support to the clinical observation that our fields of greatest proficiency are the last to suffer gross deterioration.

Although our objective data are limited to the formation of learning sets which operate to give efficient performance on intellectual problems, we have observational data of a qualitative nature on social-emotional changes in our animals. When the monkeys come to us they are wild and intractable but within a few years they have acquired, from the experimenter's point of view, good personalities. Actually we believe that one of the very important factors in the development of the good personalities of our monkeys is the formation of social-emotional learning sets organized in a manner comparable with the intellectual learning sets we have previously described. Each contact the monkey has with a human being represents a single specific learning trial. Each person represents a separate problem. Learning to react favorably to one person is followed by learning favorable reactions more rapidly to the next person to whom the monkey is socially introduced. Experience with additional individuals enables the monkey to learn further how to behave with human beings, and eventually the monkey's favorable reactions to new people are acquired so rapidly as to appear almost instantaneous.

The formation of social-emotional learning sets is not to be confused with mere stimulus generalization, a construct applied in this field with undue freedom. Actually a learning set once formed determines in large part the nature and direction of stimulus generalization. In the classic study in which Watson conditioned fear in Albert, the child developed a fear of the rat and generalized this fear, but failed to develop or generalize fear to Watson, even though Watson must have been the more conspicuous stimulus. Apparently Albert had already formed an affectional social-emotional learning set to people, which inhibited both learning and simple Pavlovian generalization.

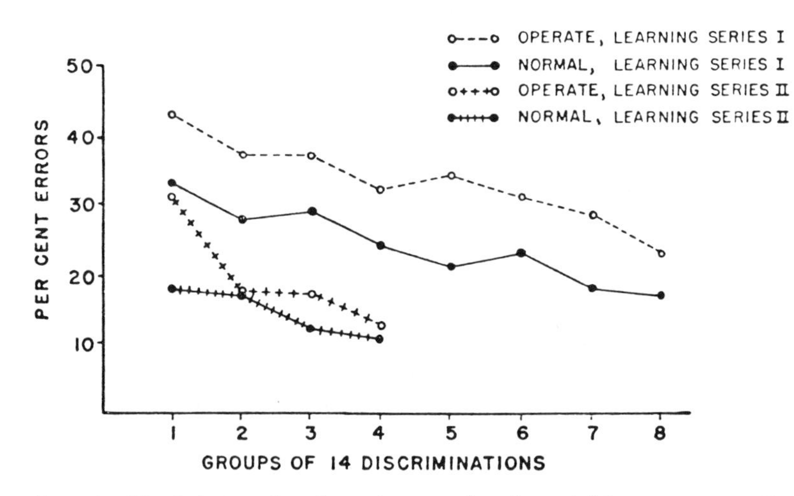

FIG. 15. Discrimination learning set curves based on total error responses: normal and operated monkeys.

Our observations on the formation of social-emotional learning sets have been entirely qualitative and informal, but there would appear to be no reason why they could not be studied experimentally.

The emphasis throughout this paper has been on the rôle of the historical or experience variable in learning behavior—the forgotten variable in current learning theory and research. Hull's Neo-behaviorists have constantly emphasized the necessity for an historical approach to learning, yet they have not exploited it fully. Their experimental manipulation of the experience variable has been largely limited to the development of isolated habits and their generalization. Their failure to find the phenomenon of discontinuity in learning may stem from their study of individual as opposed to repetitive learning situations.

The field theorists, unlike the Neobehaviorists, have stressed insight and hypothesis in their description of learning. The impression these theorists give is that these phenomena are properties of the innate organization of the individual. If such phenomena appear independently of a gradual learning history, we have not found them in the primate order.

Psychologists working with human subjects have long believed in the phenomenon of learning sets and have even used sets as explanatory principles to account for perceptual selection and incidental learning. These psychologists have not, however, investigated the nature of these learning sets which their subjects bring to the experimental situation. The determining experiential variables of these learning sets lie buried in the subjects' pasts, but the development of such sets can be studied in the laboratory as long as the human race continues to reproduce its kind. Actually, detailed knowledge of the nature of the formation of learning sets could be of such importance to educational theory and practice as to justify prolonged and systematic investigation.

In the animal laboratory where the experiential factor can be easily controlled, we have carried out studies that outline the development and operation of specific learning sets. We believe that the construct of learning sets is of importance in the understanding of adaptive behavior. Since this is our faith, it is our hope that our limited data will be extended by those brave souls who study *real* men and *real* women.

[MS. received June 23, 1948]

Editor's Comments on Paper 26

26 GARDNER and GARDNER
Teaching Sign Language to a Chimpanzee

Few questions generate more general interest than that of whether nonhuman animals can use language. Until recently Descarte's seventeenth-century pronouncement that language is the major differentiator between man and beast was widely accepted. Although important parallels could be drawn between the learning processes involved in, say, the acquisition of bird song dialects and the acquisition of human language (for example, Marler, 1970), the gulf between the communicatory behavior of nonhuman animals and the semantic and syntactic richness of human language seemed unbridgeable.

Furthermore, several attempts to instill something like human language in chimpanzees had been singularly unsuccessful (Furness, 1916; Hayes and Hayes, 1951). Undeterred, R. Allen Gardner and Beatrice T. Gardner set out to train a young chimpanzee, Washoe, to use sign language. Reasoning that the manipulative capacities of a chimp seemed well suited to a gestural language, the Gardners decided to use a version of a sign language used by the deaf. Their success is reported in Paper 26. Although not as rich as latter accounts (for example, Gardner and Gardner, 1971 and 1975), this paper was such a departure from what had been done previously that is opened a new line of inquiry, spawning a number of similar projects.

Like many of the papers reprinted in this book, the Gardners' paper tells us a great deal about learning and intelligence, yet virtually nothing about the normal behavior of the wild chimpanzee. Washoe lived in a highly contrived environment, one very different from any ever occupied previously by a chimp. The experiment tells us not what chimpanzees typically do, but rather what chimps are capable of doing given sufficient environmental support. This may seem to be an odd strategy for studying animal behavior, but learning often only becomes observable under unusual conditions, for which the animal's unlearned behavior patterns are inappropriate.

About the same time the Gardners were beginning "Project Washoe," David Premack and his colleagues began studying the use of plastic symbols by a chimpanzee, Sarah. Again, the training procedures were highly contrived. The behavioral results (summarized in Premack, 1976) tell us a great deal about the nature of chimpanzee intelligence. Another project used a computer-controlled communications panel to train a chimpanzee to use an artificial language (Rumbaugh et al, 1973).

Considerable controversy remains as to the degree to which chimpanzees have succeeded in learning to use language. Serious questions remain open, including the extent to which the chimpanzees depend on cues provided by the experimenters (Terrace et al, 1979). People engaged in this difficult line of inquiry can take solace from noting that similar problems plague the study of language acquisition in children. The relationships between the languages achieved by the chimpanzees and the languages of humans are not yet sufficiently detailed to permit firm conclusions. Experiments such as these help us understand both the nature of chimpanzees and the nature of ourselves.

REFERENCES

Furness, W. H., 1916, Observations on the mentality of chimpanzees and orang-utans, *Am. Philos. Soc. Proc.* **55**:281–290.

Gardner, B. T., and R. A. Gardner, 1971, Two-way communication with an infant chimpanzee, in *Behavior of Non-Human Primates,* vol. 4, A. Schrier and F. Stollnitz, eds., Academic Press, New York.

Gardner, B. T., and R. A. Gardner, 1975, Evidence for sentence constituents in the early utterances of child and chimpanzee. *J. Exp. Psychol.: General* **104**:244–267.

Hayes, K. J., and C. Hayes, 1951, The intellectual development of a home-raised chimpanzee, *Am. Philos. Soc. Proc.* **95**:105–109.

Marler, P., 1970, A comparative approach to vocal learning: song development in white crowned sparrows, *J. Comp. Physiol. Psychol.* **71:** monogr. suppl.

Premack, D., 1976, *Intelligence in Ape and Man.,* Lawrence Erlbaum Associates, Hillsdale, N. J..

Rumbaugh, D. M., T. V. Gill, and E. von Glaserfeld, 1973, Reading and sentence completion by a chimpanzee, *Science* **182**:731–733.

Terrace, H. S., L. A. Petitto, R. J. Sanders, and T. G. Bever, 1979, Can an ape create a sentence? *Science* **206**:891–902.

26

Reprinted from *Science* **165**:664–672 (1969)

Teaching Sign Language to a Chimpanzee

A standardized system of gestures provides a means of two-way communication with a chimpanzee.

R. Allen Gardner and Beatrice T. Gardner

The extent to which another species might be able to use human language is a classical problem in comparative psychology. One approach to this problem is to consider the nature of language, the processes of learning, the neural mechanisms of learning and of language, and the genetic basis of these mechanisms, and then, while recognizing certain gaps in what is known about these factors, to attempt to arrive at an answer by dint of careful scholarship (*1*). An alternative approach is to try to teach a form of human language to an animal. We chose the latter alternative and, in June 1966, began training an infant female chimpanzee, named Washoe, to use the gestural language of the deaf. Within the first 22 months of training it became evident that we had been correct in at least one major aspect of method, the use of a gestural language. Additional aspects of method have evolved in the course of the project. These and some implications of our early results can now be described in a way that may be useful in other studies of communicative behavior. Accordingly, in this article we discuss the considerations which led us to use the chimpanzee as a subject and American Sign Language (the language used by the deaf in North America) as a medium of communication; describe the general methods of training as they were initially conceived and as they developed in the course of the project; and summarize those results that could be reported with some degree of confidence by the end of the first phase of the project.

The authors are, respectively, (i) professor of psychology and (ii) research associate and lecturer in psychology at the University of Nevada, Reno 89507.

Preliminary Considerations

The chimpanzee as a subject. Some discussion of the chimpanzee as an experimental subject is in order because this species is relatively uncommon in the psychological laboratory. Whether or not the chimpanzee is the most intelligent animal after man can be disputed; the gorilla, the orangutan, and even the dolphin have their loyal partisans in this debate. Nevertheless, it is generally conceded that chimpanzees are highly intelligent, and that members of this species might be intelligent enough for our purposes. Of equal or greater importance is their sociability and their capacity for forming strong attachments to human beings. We want to emphasize this trait of sociability; it seems highly likely that it is essential for the development of language in human beings, and it was a primary consideration in our choice of a chimpanzee as a subject.

Affectionate as chimpanzees are, they are still wild animals, and this is a serious disadvantage. Most psychologists are accustomed to working with animals that have been chosen, and sometimes bred, for docility and adaptability to laboratory procedures. The difficulties presented by the wild nature of an experimental animal must not be underestimated. Chimpanzees are also very strong animals; a full-grown specimen is likely to weigh more than 120 pounds (55 kilograms) and is estimated to be from three to five times as strong as a man, pound-for-pound. Coupled with the wildness, this great strength presents serious difficulties for a procedure that requires interaction at close quarters with a free-living animal. We have always had to reckon with the likelihood that at some point Washoe's physical maturity will make this procedure prohibitively dangerous.

A more serious disadvantage is that human speech sounds are unsuitable as a medium of communication for the chimpanzee. The vocal apparatus of the chimpanzee is very different from that of man (*2*). More important, the vocal behavior of the chimpanzee is very different from that of man. Chimpanzees do make many different sounds, but generally vocalization occurs in situations of high excitement and tends to be specific to the exciting situations. Undisturbed, chimpanzees are usually silent. Thus, it is unlikely that a chimpanzee could be trained to make refined use of its vocalizations. Moreover, the intensive work of Hayes and Hayes (*3*) with the chimpanzee Viki indicates that a vocal language is not appropriate for this species. The Hayeses used modern, sophisticated, psychological methods and seem to have spared no effort to teach Viki to make speech sounds. Yet in 6 years Viki learned only four sounds that approximated English words (*4*).

Use of the hands, however, is a prominent feature of chimpanzee behavior; manipulatory mechanical problems are their forte. More to the point, even caged, laboratory chimpanzees develop begging and similar gestures spontaneously (*5*), while individuals that have had extensive contact with human beings have displayed an even wider variety of communicative gestures (*6*). In our choice of sign language we were influenced more by the behavioral evidence that this medium of communication was appropriate to the species than by anatomical evidence of structural similarity between the hands of chimpanzees and of men. The Hayeses point out that human tools and mechanical devices are constructed to fit the human hand, yet chimpanzees have little difficulty in using these devices with great skill. Nevertheless, they seem unable to adapt their vocalizations to approximate human speech.

Psychologists who work extensively with the instrumental conditioning of animals become sensitive to the need to use responses that are suited to the species they wish to study. Lever-pressing in rats is not an arbitrary response invented by Skinner to confound the mentalists; it is a type of response commonly made by rats when they are first placed in a Skinner box. The exquisite control of instrumental behavior by schedules of reward is achieved only if the original responses are well chosen. We chose a language based on gestures because we reasoned that gestures for the chimpanzee should be analogous to bar-pressing for rats, key-pecking for pigeons, and babbling for humans.

American Sign Language. Two systems of manual communication are used by the deaf. One system is the manual alphabet, or finger spelling, in which configurations of the hand correspond to letters of the alphabet. In this system the words of a spoken language, such as English, can be spelled out manually. The other system, sign language, consists of a set of manual configurations and gestures that correspond to particular words or concepts. Unlike finger spelling, which is the direct encoding of a spoken language, sign languages have their own rules of usage. Word-for-sign translation between a spoken language and a sign language yields results that are similar to those of word-for-word translation between two spoken languages: the translation is often passable, though awkward, but it can also be ambiguous or quite nonsensical. Also, there are national and regional variations in sign languages that are comparable to those of spoken languages.

We chose for this project the American Sign Language (ASL), which, with certain regional variations, is used by the deaf in North America. This particular sign language has recently been the subject of formal analysis (*7*). The ASL can be compared to pictograph writing in which some symbols are quite arbitrary and some are quite representational or iconic, but all are arbitrary to some degree. For example, in ASL the sign for "always" is made by holding the hand in a fist, index finger extended (the pointing hand), while rotating the arm at the elbow. This is clearly an arbitrary representation of the concept "always." The sign for "flower," however, is highly iconic; it is made by holding the fingers of one hand extended, all five fingertips touching (the tapered hand), and touching the fingertips first to one nostril then to the other, as if sniffing a flower. While this is an iconic sign for "flower," it is only one of a number of conventions by which the concept "flower" could be iconically represented; it is thus arbitrary to some degree. Undoubtedly, many of the signs of ASL that seem quite arbitrary today once had an iconic origin that was lost through years of stylized usage. Thus, the signs of ASL are neither uniformly arbitrary nor uniformly iconic; rather the degree of abstraction varies from sign to sign over a wide range. This would seem to be a useful property of ASL for our research.

The literate deaf typically use a combination of ASL and finger spelling; for purposes of this project we have avoided the use of finger spelling as much as possible. A great range of expression is possible within the limits of ASL. We soon found that a good way to practice signing among ourselves was to render familiar songs and poetry into signs; as far as we can judge, there is no message that cannot be rendered faithfully (apart from the usual problems of translation from one language to another). Technical terms and proper names are a problem when first introduced, but within any community of signers it is easy to agree on a convention for any commonly used term. For example, among ourselves we do not finger-spell the words *psychologist* and *psychology*, but render them as "think doctor" and "think science." Or, among users of ASL, "California" can be finger-spelled but is commonly rendered as "golden playland." (Incidentally, the sign for "gold" is made by plucking at the earlobe with thumb and forefinger, indicating an earring—another example of an iconic sign that is at the same time arbitrary and stylized.)

The fact that ASL is in current use by human beings is an additional advantage. The early linguistic environment of the deaf children of deaf parents is in some respects similar to the linguistic environment that we could provide for an experimental subject. This should permit some comparative evaluation of Washoe's eventual level of competence. For example, in discussing Washoe's early performance with deaf parents we have been told that many of her variants of standard signs are similar to the baby-talk variants commonly observed when human children sign.

Washoe. Having decided on a species and a medium of communication, our next concern was to obtain an experimental subject. It is altogether possible that there is some critical early age for the acquisition of this type of behavior. On the other hand, newborn chimpanzees tend to be quite helpless and vegetative. They are also considerably less hardy than older infants. Nevertheless, we reasoned that the dangers of starting too late were much greater than the dangers of starting too early, and we sought the youngest infant we could get. Newborn laboratory chimpanzees are very scarce, and we found that the youngest laboratory infant we could get would be about 2 years old at the time we planned to start the project. It seemed preferable to obtain a wild-caught infant. Wild-caught infants are usually at least 8 to 10 months old before they are available for research. This is because infants rarely reach the United States before they are 5 months old, and to this age must be added 1 or 2 months before final purchase and 2 or 3 months for quarantine and other medical services.

We named our chimpanzee Washoe for Washoe County, the home of the University of Nevada. Her exact age will never be known, but from her weight and dentition we estimated her age to be between 8 and 14 months at the end of June 1966, when she first arrived at our laboratory. (Her dentition has continued to agree with this initial estimate, but her weight has increased rather more than would be expected.) This is very young for a chimpanzee. The best available information indicates that infants are completely dependent until the age of 2 years and semi-dependent until the age of 4; the first signs of sexual maturity (for example, menstruation, sexual swelling) begin to appear at about 8 years, and full adult growth is reached between the ages of 12 and 16 (*8*). As for the complete lifespan, captive specimens have survived for well over 40 years. Washoe was indeed very young when she arrived; she did not have her first canines or molars, her hand-eye coordination was rudimentary, she had only begun to crawl about, and she slept a great deal. Apart from making friends with her and adapting her to the daily routine, we could accomplish little during the first few months.

Laboratory conditions. At the outset we were quite sure that Washoe could learn to make various signs in order to obtain food, drink, and other things. For the project to be a success, we felt that something more must be developed. We wanted Washoe not only to ask for objects but to answer questions about them and also to ask us questions. We wanted to develop behavior that could be described as conversation. With this in mind, we attempted to provide Washoe with an environment that might be conducive to this sort of behavior. Confinement was to be minimal, about the same as that of human infants. Her human companions were to be friends and playmates as well as providers and protectors, and they were to introduce a great many games and activities that would be likely to result in maximum interaction with Washoe.

In practice, such an environment is readily achieved with a chimpanzee; bonds of warm affection have always been established between Washoe and her several human companions. We have enjoyed the interaction almost as much as Washoe has, within the limits of human endurance. A number of human companions have been enlisted to participate in the project and relieve each other at intervals, so that at least one person would be with Washoe during all her waking hours. At first we feared that such frequent changes would be disturbing, but Washoe seemed to adapt very well to this procedure. Apparently it is possible to provide an infant chimpanzee with affection on a shift basis.

All of Washoe's human companions have been required to master ASL and to use it extensively in her presence, in association with interesting activities and events and also in a general way, as one chatters at a human infant in the course of the day. The ASL has been used almost exclusively, although occasional finger spelling has been permitted. From time to time, of course, there are lapses into spoken English, as when medical personnel must examine Washoe. At one time, we considered an alternative procedure in which we would sign and speak English to Washoe simultaneously, thus giving her an additional source of informative cues. We rejected this procedure, reasoning that, if she should come to understand speech sooner or more easily than ASL, then she might not pay sufficient attention to our gestures. Another alternative, that of speaking English among ourselves and signing to Washoe, was also rejected. We reasoned that this would make it seem that big chimps talk and only little chimps sign, which might give signing an undesirable social status.

The environment we are describing is not a silent one. The human beings can vocalize in many ways, laughing and making sounds of pleasure and displeasure. Whistles and drums are sounded in a variety of imitation games, and hands are clapped for attention. The rule is that all meaningful sounds, whether vocalized or not, must be sounds that a chimpanzee can imitate.

Training Methods

Imitation. The imitativeness of apes is proverbial, and rightly so. Those who have worked closely with chimpanzees have frequently remarked on their readiness to engage in visually guided imitation. Consider the following typical comment of Yerkes (*9*): "Chim and Panzee would imitate many of my acts, but never have I heard them imitate a sound and rarely make a sound peculiarly their own in response to mine. As previously stated, their imitative tendency is as remarkable for its specialization and limitations as for its strength. It seems to be controlled chiefly by visual stimuli. Things which are seen tend to be imitated or reproduced. What is heard is not reproduced. Obviously an animal which lacks the tendency to reinstate auditory stimuli—in other words to imitate sounds—cannot reasonably be expected to talk. The human infant exhibits this tendency to a remarkable degree. So also does the parrot. If the imitative tendency of the parrot could be coupled with the quality of intelligence of the chimpanzee, the latter undoubtedly could speak."

In the course of their work with Viki, the Hayeses devised a game in which Viki would imitate various actions on hearing the command "Do this" (*10*). Once established, this was an effective means of training Viki to perform actions that could be visually guided. The same method should be admirably suited to training a chimpanzee to use sign language; accordingly we have directed much effort toward establishing a version of the "Do this" game with Washoe. Getting Washoe to imitate us was not difficult, for she did so quite spontaneously, but getting her to imitate on command has been another matter altogether. It was not until the 16th month of the project that we achieved any degree of control over Washoe's imitation of gestures. Eventually we got to a point where she would imitate a simple gesture, such as pulling at her ears, or a series of such gestures—first we make a gesture, then she imitates, then we make a second gesture, she imitates the second gesture, and so on—for the reward of being tickled. Up to this writing, however, imitation of this sort has not been an important method for introducing new signs into Washoe's vocabulary.

As a methd of prompting, we have been able to use imitation extensively to increase the frequency and refine the form of signs. Washoe sometimes fails to use a new sign in an appropriate situation, or uses another, incorrect sign. At such times we can make the correct sign to Washoe, repeating the performance until she makes the sign herself. (With more stable signs, more indirect forms of prompting can be used—for example, pointing at, or touching, Washoe's hand or a part of her body that should be involved in the sign; making the sign for "sign," which is equivalent to saying "Speak up"; or asking a question in signs, such as "What do you want?" or "What is it?") Again, with new signs, and often with old signs as well, Washoe can lapse into what we refer to as poor "diction." Of course, a great deal of slurring and a wide range of variants are permitted in ASL as in any spoken language. In any event, Washoe's diction has frequently been improved by the simple device of repeating, in exaggeratedly correct form, the sign she has just made, until she repeats it herself in more correct form. On the whole, she has responded quite well to prompting, but there are strict limits to its use with a wild animal—one that is probably quite spoiled, besides. Pressed too hard, Washoe can become completely diverted from her original object; she may ask for something entirely different, run away, go into a tantrum, or even bite her tutor.

Chimpanzees also imitate, after some delay, and this delayed imitation can be quite elaborate (*10*). The following is a typical example of Washoe's delayed imitation. From the beginning of the project she was bathed regularly and according to a standard routine. Also, from her 2nd month with us, she always had dolls to play with. One day, during the 10th month of the project, she bathed one of her dolls in the way we usually bathed her. She filled her little bathtub with water, dunked the doll in the tub, then took it out and dried it with a towel. She has repeated the entire performance, or parts of it, many times since, sometimes also soaping the doll.

This is a type of imitation that may be very important in the acquisition of language by human children, and many of our procedures with Washoe were devised to capitalize on it. Routine activities—feeding, dressing, bathing, and so on—have been highly ritualized, with appropriate signs figuring prominently in the rituals. Many games have been invented which can be accompanied by appropriate signs. Objects and activities have been named as often as possible, especially when Washoe seemed to be paying particular attention to them. New objects and new examples of familiar objects, including pictures, have been continually brought to her attention, together with the appropriate signs. She likes to ride in automobiles, and a ride in an automobile, including the preparations for a ride, provides a wealth of sights that can be accompanied by signs. A good destination for a ride is a home or the university nursery school, both well stocked with props for language lessons.

The general principle should be clear: Washoe has been exposed to a wide variety of activities and objects, together with their appropriate signs, in the hope that she would come to associate the signs with their referents and later make the signs herself. We have reason to believe that she has come to understand a large vocabulary of signs. This was expected, since a number of chimpanzees have acquired extensive understanding vocabularies of spoken words, and there is evidence that even dogs can acquire a sizable understanding vocabulary of spoken words (*11*). The understanding vocabulary that Washoe has acquired, however, consists of signs that a chimpanzee can imitate.

Some of Washoe's signs seem to have been originally acquired by delayed imitation. A good example is the sign for "toothbrush." A part of the daily routine has been to brush her teeth after every meal. When this routine was first introduced Washoe generally resisted it. She gradually came to submit with less and less fuss, and after many months she would even help or sometimes brush her teeth herself. Usually, having finished her meal, Washoe would try to leave her highchair; we would restrain her, signing "First, toothbrushing, then you can go." One day, in the 10th month of the project, Washoe was visiting the Gardner home and found her way into the bathroom. She climbed up on the counter, looked at our mug full of toothbrushes, and signed "toothbrush." At the time, we believed that Washoe understood this sign but we had not seen her use it. She had no reason to ask for the toothbrushes, because they were well within her reach, and it is most unlikely that she was asking to have her teeth brushed. This was our first observation, and one of the clearest examples, of behavior in which Washoe seemed to name an object or an event for no obvious motive other than communication.

Following this observation, the toothbrushing routine at mealtime was altered. First, imitative prompting was introduced. Then as the sign became more reliable, her rinsing-mug and toothbrush were displayed prominently until she made the sign. By the 14th month she was making the "toothbrush" sign at the end of meals with little or no prompting; in fact she has called for her toothbrush in a peremptory fashion when its appearance at the end of a meal was delayed. The "toothbrush" sign is not merely a response cued by the end of a meal; Washoe retained her ability to name toothbrushes when they were shown to her at other times.

The sign for "flower" may also have been acquired by delayed imitation. From her first summer with us, Washoe showed a great interest in flowers, and we took advantage of this by providing many flowers and pictures of flowers accompanied by the appropriate sign. Then one day in the 15th month she made the sign, spontaneously, while she and a companion were walking toward a flower garden. As in the case of "toothbrush," we believed that she understood the sign at this time, but we had made no attempt to elicit it from her except by making it ourselves in appropriate situations. Again, after the first observation, we proceeded to elicit this sign as often as possible by a variety of methods, most frequently by showing her a flower and giving it to her if she made the sign for it. Eventually the sign became very reliable and could be elicited by a variety of flowers and pictures of flowers.

It is difficult to decide which signs were acquired by the method of delayed imitation. The first appearance of these signs is likely to be sudden and unexpected; it is possible that some inadvertent movement of Washoe's has been interpreted as meaningful by one of her devoted companions. If the first observer were kept from reporting the observation and from making any direct attempts to elicit the sign again, then it might be possible to obtain independent verification. Quite understandably, we have been more interested in raising the frequency of new signs than in evaluating any particular method of training.

Babbling. Because the Hayeses were attempting to teach Viki to speak English, they were interested in babbling, and during the first year of their project they were encouraged by the number and variety of spontaneous vocalizations that Viki made. But, in time, Viki's spontaneous vocalizations decreased further and further to the point where the Hayeses felt that there was almost no vocal babbling from which to shape spoken language. In planning this project we expected a great deal of manual "babbling," but during the early months we observed very little behavior of this kind. In the course of the project, however, there has been a great increase in manual babbling. We have been particularly encouraged by the increase in movements that involve touching parts of the head and body, since these are important components of many signs. Also, more and more frequently, when Washoe has been unable to get something that she wants, she has burst into a flurry of random flourishes and arm-waving.

We have encouraged Washoe's babbling by our responsiveness; clapping, smiling, and repeating the gesture much as you might repeat "goo goo" to a human infant. If the babbled gesture has resembled a sign in ASL, we have made the correct form of the sign and have attempted to engage in some appropriate activity. The sign for "funny" was probably acquired in this way. It first appeared as a spontaneous babble that lent itself readily to a simple imitation game—first Washoe signed "funny," then we did, then she did, and so on. We would laugh and smile during the interchanges that she initiated, and initiate the game ourselves when something funny happened. Eventually Washoe came to use the "funny" sign spontaneously in roughly appropriate situations.

Closely related to babbling are some gestures that seem to have appeared independently of any deliberate training on our part, and that resemble signs so closely that we could incorporate them into Washoe's repertoire with little or no modification. Almost from the first she had a begging gesture—an extension of her open hand, palm up, toward one of us. She made this gesture in situations in which she wanted aid and in situations in which we were holding some object that she wanted. The ASL signs for "give me" and "come" are very similar to this, except that they involve a prominent beckoning movement. Gradually Washoe came to incorporate a beckoning wrist movement into her use of this sign. In Table 1 we refer to this sign as "come-gimme." As Washoe has come to use it, the sign is not simply a modification of the original begging gesture. For example, very commonly she reaches forward with one hand (palm up) while she gestures with the other hand (palm down) held near her head. (The result resembles a classic fencing posture.)

Another sign of this type is the sign for "hurry," which, so far, Washoe has always made by shaking her open hand vigorously at the wrist. This first appeared as an impatient flourish following some request that she had made in signs; for example, after making the "open" sign before a door. The correct ASL for "hurry" is very close, and we began to use it often, ourselves, in appropriate contexts. We believe that Washoe has come to use this sign in a meaningful way, because she has frequently used it when she, herself, is in a hurry—for example, when rushing to her nursery chair.

Instrumental conditioning. It seems intuitively unreasonable that the acquisition of language by human beings could be strictly a matter of reiterated instrumental conditioning—that a child acquires language after the fashion of a rat that is conditioned, first, to press a lever for food in the presence of one stimulus, then to turn a wheel in the presence of another stimulus, and so on until a large repertoire of discriminated responses is acquired. Nevertheless, the so-called "trick vocabulary" of early childhood is probably acquired in this way, and this may be a critical stage in the acquisition of language by children. In any case, a minimal objective of this project was to teach Washoe as many signs as possible by whatever procedures we could enlist. Thus, we have not hesitated to use conventional procedures of instrumental conditioning.

Anyone who becomes familiar with

Table 1. Signs used reliably by chimpanzee Washoe within 22 months of the beginning of training. The signs are listed in the order of their original appearance in her repertoire (see text for the criterion of reliability and for the method of assigning the date of original appearance).

Signs	Description	Context
Come-gimme	Beckoning motion, with wrist or knuckles as pivot.	Sign made to persons or animals, also for objects out of reach. Often combined: "come tickle," "gimme sweet," etc.
More	Fingertips are brought together, usually overhead. (Correct ASL form: tips of the tapered hand touch repeatedly.)	When asking for continuation or repetition of activities such as swinging or tickling, for second helpings of food, etc. Also used to ask for repetition of some performance, such as a somersault.
Up	Arm extends upward, and index finger may also point up.	Wants a lift to reach objects such as grapes on vine, or leaves; or wants to be placed on someone's shoulders; or wants to leave potty-chair.
Sweet	Index or index and second fingers touch tip of wagging tongue. (Correct ASL form: index and second fingers extended side by side.)	For dessert; used spontaneously at end of meal. Also, when asking for candy.
Open	Flat hands are placed side by side, palms down, then drawn apart while rotated to palms up.	At door of house, room, car, refrigerator, or cupboard; on containers such as jars; and on faucets.
Tickle	The index finger of one hand is drawn across the back of the other hand. (Related to ASL "touch.")	For tickling or for chasing games.
Go	Opposite of "come-gimme."	While walking hand-in-hand or riding on someone's shoulders. Washoe usually indicates the direction desired.
Out	Curved hand grasps tapered hand; then tapered hand is withdrawn upward.	When passing through doorways; until recently, used for both "in" and "out." Also, when asking to be taken outdoors.
Hurry	Open hand is shaken at the wrist. (Correct ASL form: index and second fingers extended side by side.)	Often follows signs such as "come-gimme," "out," "open," and "go," particularly if there is a delay before Washoe is obeyed. Also, used while watching her meal being prepared.
Hear-listen	Index finger touches ear.	For loud or strange sounds: bells, car horns, sonic booms, etc. Also, for asking someone to hold a watch to her ear.
Toothbrush	Index finger is used as brush, to rub front teeth.	When Washoe has finished her meal, or at other times when shown a toothbrush.
Drink	Thumb is extended from fisted hand and touches mouth.	For water, formula, soda pop, etc. For soda pop, often combined with "sweet."
Hurt	Extended index fingers are jabbed toward each other. Can be used to indicate location of pain.	To indicate cuts and bruises on herself or on others. Can be elicited by red stains on a person's skin or by tears in clothing.
Sorry	Fisted hand clasps and unclasps at shoulder. (Correct ASL form: fisted hand is rubbed over heart with circular motion.)	After biting someone, or when someone has been hurt in another way (not necessarily by Washoe). When told to apologize for mischief.
Funny	Tip of index finger presses nose, and Washoe snorts. (Correct ASL form: index and second fingers used; no snort.)	When soliciting interaction play, and during games. Occasionally, when being pursued after mischief.
Please	Open hand is drawn across chest. (Correct ASL form: fingertips used, and circular motion.)	When asking for objects and activities. Frequently combined: "Please go," "Out, please," "Please drink."

young chimpanzees soon learns about their passion for being tickled. There is no doubt that tickling is the most effective reward that we have used with Washoe. In the early months, when we would pause in our tickling, Washoe would indicate that she wanted more tickling by taking our hands and placing them against her ribs or around her neck. The meaning of these gestures was unmistakable, but since we were not studying our human ability to interpret her chimpanzee gestures, we decided to shape an arbitrary response that she could use to ask for more tickling. We noted that, when being tickled, she tended to bring her arms together to cover the place being tickled. The result was a very crude approximation of the ASL sign for "more" (see Table 1). Thus, we would stop tickling and then pull Washoe's arms away from her body. When we released her arms and threatened to resume tickling, she tended to bring her hands together again. If she brought them back together, we would tickle her again. From time to time we would stop tickling and wait for her to put her hands together by herself. At first, any approximation to the "more" sign, however crude, was rewarded. Later, we required closer approximations and introduced imitative prompting. Soon, a very good version of the "more" sign could be obtained, but it was quite specific to the tickling situation.

In the 6th month of the project we were able to get "more" signs for a new game that consisted of pushing Washoe across the floor in a laundry basket. In this case we did not use the shaping procedure but, from the start, used imitative prompting to elicit the "more" sign. Soon after the "more" sign became spontaneous and reliable in the laundry-basket game, it began to appear as a request for more swinging (by the arms)—again, after first being elicited with imitative prompting. From this point on, Washoe transferred the "more" sign to all activities, including feeding. The transfer was usually spontaneous, occurring when there was some pause in a desired activity or when some object was removed. Often we ourselves were not sure that Washoe

Table 1. (continued)

Signs	Description	Context
Food-eat	Several fingers of one hand are placed in mouth. (Correct ASL form: fingertips of tapered hand touch mouth repeatedly.)	During meals and preparation of meals.
Flower	Tip of index finger touches one or both nostrils. (Correct ASL form: tips of tapered hand touch first one nostril, then the other.)	For flowers.
Cover-blanket	Draws one hand toward self over the back of the other.	At bedtime or naptime, and, on cold days, when Washoe wants to be taken out.
Dog	Repeated slapping on thigh.	For dogs and for barking.
You	Index finger points at a person's chest.	Indicates successive turns in games. Also used in response to questions such as "Who tickle?" "Who brush?"
Napkin-bib	Fingertips wipe the mouth region.	For bib, for washcloth, and for Kleenex.
In	Opposite of "out."	Wants to go indoors, or wants someone to join her indoors.
Brush	The fisted hand rubs the back of the open hand several times. (Adapted from ASL "polish.")	For hairbrush, and when asking for brushing.
Hat	Palm pats top of head.	For hats and caps.
I-me	Index finger points at, or touches, chest.	Indicates Washoe's turn, when she and a companion share food, drink, etc. Also used in phrases, such as "I drink," and in reply to questions such as "Who tickle?" (Washoe: "you"); "Who I tickle?" (Washoe: "Me.")
Shoes	The fisted hands are held side by side and strike down on shoes or floor. (Correct ASL form: the sides of the fisted hands strike against each other.)	For shoes and boots.
Smell	Palm is held before nose and moved slightly upward several times.	For scented objects: tobacco, perfume, sage, etc.
Pants	Palms of the flat hands are drawn up against the body toward waist.	For diapers, rubber pants, trousers.
Clothes	Fingertips brush down the chest.	For Washoe's jacket, nightgown, and shirts; also for our clothing.
Cat	Thumb and index finger grasp cheek hair near side of mouth and are drawn outward (representing cat's whiskers).	For cats.
Key	Palm of one hand is repeatedly touched with the index finger of the other. (Correct ASL form: crooked index finger is rotated against palm.)	Used for keys and locks and to ask us to unlock a door.
Baby	One forearm is placed in the crook of the other, as if cradling a baby.	For dolls, including animal dolls such as a toy horse and duck.
Clean	The open palm of one hand is passed over the open palm of the other.	Used when Washoe is washing, or being washed, or when a companion is washing hands or some other object. Also used for "soap."

wanted "more" until she signed to us.

The sign for "open" had a similar history. When Washoe wanted to get through a door, she tended to hold up both hands and pound on the door with her palms or her knuckles. This is the beginning position for the "open" sign (see Table 1). By waiting for her to place her hands on the door and then lift them, and also by imitative prompting, we were able to shape a good approximation of the "open" sign, and would reward this by opening the door. Originally she was trained to make this sign for three particular doors that she used every day. Washoe transferred this sign to all doors; then to containers such as the refrigerator, cupboards, drawers, briefcases, boxes, and jars; and eventually—an invention of Washoe's—she used it to ask us to turn on water faucets.

In the case of "more" and "open" we followed the conventional laboratory procedure of waiting for Washoe to make some response that could be shaped into the sign we wished her to acquire. We soon found that this was not necessary; Washoe could acquire signs that were first elicited by our holding her hands, forming them into the desired configuration, and then putting them through the desired movement. Since this procedure of guidance is usually much more practical than waiting for a spontaneous approximation to occur at a favorable moment, we have used it much more frequently.

Results

Vocabulary. In the early stages of the project we were able to keep fairly complete records of Washoe's daily signing behavior. But, as the amount of signing behavior and the number of signs to be monitored increased, our initial attempts to obtain exhaustive records became prohibitively cumbersome. During the 16th month we settled on the following procedure. When a new sign was introduced we waited until it had been reported by three different observers as having occurred in an appropriate context and spontaneously (that is, with no prompting other than a question such as "What is it?" or "What do you want?"). The sign was then added to a checklist in which its occurrence, form, context, and the kind of prompting required were recorded. Two such checklists were filled out each day, one for the first half of the day and one for the second half. For a criterion of acquisition we chose a reported frequency of at least one appropriate and spontaneous occurrence each day over a period of 15 consecutive days.

In Table 1 we have listed 30 signs that met this criterion by the end of the 22nd month of the project. In addition, we have listed four signs ("dog," "smell," "me," and "clean") that we judged to be stable, despite the fact that they had not met the stringent criterion before the end of the 22nd month. These additional signs had, nevertheless, been reported to occur appropriately and spontaneously on more than half of the days in a period of 30 consecutive days. An indication of the variety of signs that Washoe used in the course of a day is given by the following data: during the 22nd month of the study, 28 of the 34 signs listed were reported on at least 20 days, and the smallest number of different signs reported for a single day was 23, with a median of 29 (*12*).

The order in which these signs first appeared in Washoe's repertoire is also given in Table 1. We considered the first appearance to be the date on which three different observers reported appropriate and spontaneous occurrences. By this criterion, 4 new signs first appeared during the first 7 months, 9 new signs during the next 7 months, and 21 new signs during the next 7 months. We chose the 21st month rather than the 22nd month as the cutoff for this tabulation so that no signs would be included that do not appear in Table 1. Clearly, if Washoe's rate of acquisition continues to accelerate, we will have to assess her vocabulary on the basis of sampling procedures. We are now in the process of developing procedures that could be used to make periodic tests of Washoe's performance on samples of her repertoire. However, now that there is evidence that a chimpanzee can acquire a vocabulary of more than 30 signs, the exact number of signs in her current vocabulary is less significant than the order of magnitude—50, 100, 200 signs, or more—that might eventually be achieved.

Differentiation. In Table 1, column 1, we list English equivalents for each of Washoe's signs. It must be understood that this equivalence is only approximate, because equivalence between English and ASL, as between any two human languages, is only approximate, and because Washoe's usage does differ from that of standard ASL. To some extent her usage is indicated in the column labeled "Context" in Table 1, but the definition of any given sign must always depend upon her total vocabulary, and this has been continually changing. When she had very few signs for specific things, Washoe used the "more" sign for a wide class of requests. Our only restriction was that we discouraged the use of "more" for first requests. As she acquired signs for specific requests, her use of "more" declined until, at the time of this writing, she was using this sign mainly to ask for repetition of some action that she could not name, such as a somersault. Perhaps the best English equivalent would be "do it again." Still, it seemed preferable to list the English equivalent for the ASL sign rather than its current referent for Washoe, since further refinements in her usage may be achieved at a later date.

The differentiation of the signs for "flower" and "smell" provides a further illustration of usage depending upon size of vocabulary. As the "flower" sign became more frequent, we noted that it occurred in several inappropriate contexts that all seemed to include odors; for example, Washoe would make the "flower" sign when opening a tobacco pouch or when entering a kitchen filled with cooking odors. Taking our cue from this, we introduced the "smell" sign by passive shaping and imitative prompting. Gradually Washoe came to make the appropriate distinction between "flower" contexts and "smell" contexts in her signing, although "flower" (in the single-nostril form) (see Table 1) has continued to occur as a common error in "smell" contexts.

Transfer. In general, when introducing new signs we have used a very specific referent for the initial training—a particular door for "open," a particular hat for "hat." Early in the project we were concerned about the possibility that signs might become inseparable from their first referents. So far, however, there has been no problem of this kind: Washoe has always been able to transfer her signs spontaneously to new members of each class of referents. We have already described the transfer of "more" and "open." The sign for "flower" is a particularly good example of transfer, because flowers occur in so many varieties, indoors, outdoors, and in pictures, yet Washoe uses the same sign for all. It is fortunate that she has responded well to pictures of objects. In the case of "dog" and "cat" this has proved to be important because live dogs and cats can be too exciting, and we have had to use pictures to elicit most of the "dog" and "cat" signs. It is noteworthy that Washoe has transferred the "dog" sign to the sound of barking by an unseen dog.

The acquisition and transfer of the sign for "key" illustrates a further point. A great many cupboards and doors in Washoe's quarters have been kept secure by small padlocks that can all be opened by the same simple key. Because she was immature and awkward, Washoe had great difficulty in learning to use these keys and locks. Because we wanted her to improve her manual dexterity, we let her practice with these keys until she could open the locks quite easily (then we had to hide the keys). Washoe soon transferred this skill to all manner of locks and keys, including ignition keys. At about the same time, we taught her the sign for "key," using the original padlock keys as a referent. Washoe came to use this sign both to name keys that were presented to her and to ask for the keys to various locks when no key was in sight. She readily transferred the sign to all varieties of keys and locks.

Now, if an animal can transfer a skill learned with a certain key and lock to new types of key and lock, it should not be surprising that the same animal can learn to use an arbitrary response to name and ask for a certain key and then transfer that sign to new types of keys. Certainly, the relationship between the use of a key and the opening of locks is as arbitrary as the relationship between the sign for "key" and its many referents. Viewed in this way, the general phenomenon of transfer of training and the specifically linguistic phenomenon of labeling become very similar, and the problems that these phenomena pose for modern learning theory should require similar solutions. We do not mean to imply that the problem of labeling is less complex than has generally been supposed; rather, we are suggesting that the problem of transfer of training requires an equally sophisticated treatment.

Combinations. During the phase of the project covered by this article we made no deliberate attempts to elicit combinations or phrases, although we may have responded more readily to strings of two or more signs than to single signs. As far as we can judge, Washoe's early use of signs in strings was spontaneous. Almost as soon as she had eight or ten signs in her repertoire, she began to use them two and three at a time. As her repertoire increased, her tendency to produce strings of two or more signs also increased, to the point where this has become a common mode of signing for her. We, of course, usually signed to her in combinations, but if Washoe's use of combinations has been imitative, then it must be a generalized sort of imitation, since she has invented a number of combinations, such as "gimme tickle" (before we had ever asked her to tickle us), and "open food drink" (for the refrigerator—we have always called it the "cold box").

Four signs—"please," "come-gimme," "hurry," and "more"—used with one or more other signs, account for the largest share of Washoe's early combinations. In general, these four signs have functioned as emphasizers, as in "please open hurry" and "gimme drink please."

Until recently, five additional signs—"go," "out," "in," "open," and "hear-listen"—accounted for most of the remaining combinations. Typical examples of combinations using these four are, "go in" or "go out" (when at some distance from a door), "go sweet" (for being carried to a raspberry bush), "open flower" (to be let through the gate to a flower garden), "open key" (for a locked door), "listen eat" (at the sound of an alarm clock signaling mealtime), and "listen dog" (at the sound of barking by an unseen dog). All but the first and last of these six examples were inventions of Washoe's. Combinations of this type tend to amplify the meaning of the single signs used. Sometimes, however, the function of these five signs has been about the same as that of the emphasizers, as in "open out" (when standing in front of a door).

Toward the end of the period covered in this article we were able to introduce the pronouns "I-me" and "you," so that combinations that resemble short sentences have begun to appear.

Concluding Observations

From time to time we have been asked questions such as, "Do you think that Washoe has language?" or "At what point will you be able to say that Washoe has language?" We find it very difficult to respond to these questions because they are altogether foreign to the spirit of our research. They imply a distinction between one class of communicative behavior that can be called language and another class that cannot. This in turn implies a well-established theory that could provide the distinction. If our objectives had required such a theory, we would certainly not have been able to begin this project as early as we did.

In the first phase of the project we were able to verify the hypothesis that sign language is an appropriate medium of two-way communication for the chimpanzee. Washoe's intellectual immaturity, the continuing acceleration of her progress, the fact that her signs do not remain specific to their original referents but are transferred spontaneously to new referents, and the emergence of rudimentary combinations all suggest that significantly more can be accomplished by Washoe during the subsequent phases of this project. As we proceed, the problems of these subsequent phases will be chiefly concerned with the technical business of measurement. We are now developing a procedure for testing Washoe's ability to name objects. In this procedure, an object or a picture of an object is placed in a box with a window. An observer, who does not know what is in the box, asks Washoe what she sees through the window. At present, this method is limited to items that fit in the box; a more ingenious method will have to be devised for other items. In particular, the ability to combine and recombine signs must be tested. Here, a great deal depends upon reaching a stage at which Washoe produces an extended series of signs in answer to questions. Our hope is that Washoe can be brought to the point where she describes events and situations to an observer who has no other source of information.

At an earlier time we would have been more cautious about suggesting that a chimpanzee might be able to produce extended utterances to communicate information. We believe now that it is the writers—who would predict just what it is that no chimpanzee will ever do—who must proceed with caution. Washoe's accomplishments will probably be exceeded by another chimpanzee, because it is unlikely that the conditions of training have been optimal in this first attempt. Theories of language that depend upon the identification of aspects of language that are exclusively human must remain tentative until a considerably larger body of intensive research with other species becomes available.

Summary

We set ourselves the task of teaching an animal to use a form of human language. Highly intelligent and highly social, the chimpanzee is an obvious choice for such a study, yet it has not been possible to teach a member of this species more than a few spoken words. We reasoned that a spoken language, such as English, might be an inappropriate medium of communication for a chimpanzee. This led us to choose American Sign Language, the gestural system of communication used by the deaf in North America, for the project.

The youngest infant that we could obtain was a wild-born female, whom we named Washoe, and who was estimated to be between 8 and 14 months old when we began our program of training. The laboratory conditions, while not patterned after those of a human family (as in the studies of Kellogg and Kellogg and of Hayes and Hayes), involved a minimum of confinement and a maximum of social interaction with human companions. For all practical purposes, the only verbal communication was in ASL, and the chimpanzee was maximally exposed to the use of this language by human beings.

It was necessary to develop a rough-and-ready mixture of training methods. There was evidence that some of Washoe's early signs were acquired by delayed imitation of the signing behavior of her human companions, but very few if any, of her early signs were introduced by immediate imitation. Manual babbling was directly fostered and did increase in the course of the project. A number of signs were introduced by shaping and instrumental conditioning. A particularly effective and convenient method of shaping consisted of holding Washoe's hands, forming them into a configuration, and putting them through the movements of a sign.

We have listed more than 30 signs that Washoe acquired and could use spontaneously and appropriately by the end of the 22nd month of the project. The signs acquired earliest were simple demands. Most of the later signs have been names for objects, which Washoe has used both as demands and as answers to questions. Washoe readily used noun signs to name pictures of objects as well as actual objects and has frequently called the attention of her companions to pictures and objects by naming them. Once acquired, the signs have not remained specific to the original referents but have been transferred spontaneously to a wide class of appropriate referents. At this writing, Washoe's rate of acquisition of new signs is still accelerating.

From the time she had eight or ten signs in her repertoire, Washoe began to use them in strings of two or more. During the period covered by this article we made no deliberate effort to elicit combinations other than by our own habitual use of strings of signs. Some of the combined forms that Washoe has used may have been imitative, but many have been inventions of her own. Only a small proportion of the possible combinations have, in fact, been observed. This is because most of Washoe's combinations include one of a limited group of signs that act as combiners. Among the signs that Washoe has recently acquired are the pronouns "I-me" and "you." When these occur in combinations the result resembles a short sentence. In terms of the eventual level of communication that a chimpanzee might be able to attain, the most promising results have been spontaneous naming, spontaneous transfer to new referents, and spontaneous combinations and recombinations of signs.

References and Notes

1. See, for example, E. H. Lenneberg, *Biological Foundations of Language* (Wiley, New York, 1967).
2. A. L. Bryan, *Curr. Anthropol.* **4**, 297 (1963).
3. K. J. Hayes and C. Hayes, *Proc. Amer. Phil. Soc.* **95**, 105 (1951).
4. K. J. Hayes, personal communication. Dr. Hayes also informed us that Viki used a few additional sounds which, while not resembling English words, were used for specific requests.
5. R. M. Yerkes, *Chimpanzees* (Yale Univ. Press, New Haven, 1943).
6. K. J. Hayes and C. Hayes, in *The Non-Human Primates and Human Evolution*, J. A. Gavan, Ed. (Wayne Univ. Press, Detroit, 1955), p. 110; W. N. Kellogg and L. A. Kellogg, *The Ape and the Child* (Hafner, New York, 1967; originally published by McGraw-Hill, New York, 1933); W. N. Kellogg, *Science* **162**, 423 (1968).
7. W. C. Stokoe, D. Casterline, C. G. Croneberg, *A Dictionary of American Sign Language* (Gallaudet College Press, Washington, D.C., 1965); E. A. McCall, thesis, University of Iowa (1965).
8. J. Goodall, in *Primate Behavior*, I. DeVore, Ed. (Holt, Rinehart & Winston, New York, 1965), p. 425; A. J. Riopelle and C. M. Rogers, in *Behavior of Nonhuman Primates*, A. M. Schrier, H. F. Harlow, F. Stollnitz, Eds. (Academic Press, New York, 1965), p. 449.
9. R. M. Yerkes and B. W. Learned, *Chimpanzee Intelligence and Its Vocal Expression* (William & Wilkins, Baltimore, 1925), p. 53.
10. K. J. Hayes and C. Hayes, *J. Comp. Physiol. Psychol.* **45**, 450 (1952).
11. C. J. Warden and L. H. Warner, *Quart. Rev. Biol.* **3**, 1 (1928).
12. The development of Washoe's vocabulary of signs is being recorded on motion-picture film. At the time of this writing, 30 of the 34 signs listed in Table 1 are on film.
13. The research described in this article has been supported by National Institute of Mental Health grants MH-12154 and MH-34953 (Research Scientist Development Award to B. T. Gardner) and by National Science Foundation grant GB-7432. We acknowledge a great debt to the personnel of the Aeromedical Research Laboratory, Holloman Air Force Base, whose support and expert assistance effectively absorbed all of the many difficulties attendant upon the acquisition of a wild-caught chimpanzee. We are also grateful to Dr. Frances L. Fitz-Gerald of the Yerkes Regional Primate Research Center for detailed advice on the care of an infant chimpanzee. Drs. Emanual Berger of Reno, Nevada, and D. B. Olsen of the University of Nevada have served as medical consultants, and we are grateful to them for giving so generously of their time and medical skills. The faculty of the Sarah Hamilton Fleischmann School of Home Economics, University of Nevada, has generously allowed us to use the facilities of their experimental nursery school on weekends and holidays.

AUTHOR CITATION INDEX

SUBJECT INDEX

About the Editor

ROBERT W. HENDERSEN is Associate Professor of Psychology at the University of Illinois at Urbana-Champaign. A graduate of Reed College, he obtained the Ph.D. at the University of Pennsylvania in 1973. He has published a variety of research papers dealing with learning, acquired motivation, and animal memory.